1. Insecticides

1.1 Introduction

Insecticides are defined as chemicals which are used to control damage an annoyance from insects by poisoning then through oral ingestion of stomach poisons, by contact with the cuticle or by fumigant action through the air. Insects may also be controlled by chemicals such as attractants and repellents which adversely influence their behaviour, and by chemosterilants which prevent their reproduction.

The term "*insecticides*" has been now replaced by "*pesticides*" which include toxic chemicals, whether used against insect fungi, weed or rodents, etc. Agricultural disinfectants and animal health products are in many instances also included under the term "pesticides".

The use of insecticides has not only permitted the control of disease transmitted by insects but also has led to increased food production and better health.

1.2 Classification of Insecticides

On their mode of action

(a) Stomach or internal insecticides: Those insecticides which are eaten are as stomach insecticides. Those insecticides are generally applied against insects chewing mouthpart but under certain conditions they are also effective against insects with sponging, syphoning, lapping or mouthpart.

Stomach poisons are applied in the following manner:

(i)   The food of the insect is covered with a thick layer of poison so that the insect cannot feed without ingesting it.
(ii)   The poison is mixed with an attractant to from poison bait which the insect will seek out and feed upon.
(iii)   Finely divided power of insecticides is sprinkled over the runways of the insect and become lodged upon feet or antennae and is subsequently swallowed by the insect while it is cleaning their appendages with the mouthparts.
(iv)   The insecticide may be applied as a systemic poison which is absorbed and distributed throughout the tissues or plant or animal host so that insects feeding thereon are killed. By this means sucking insects many be controlled with stomach poisons.

(b) Contact or external insecticides: Those insecticides which kill the insect by means external contact are known as contact insecticides. These means penetrate the blood directly through the insect cuticle or by entrance through the spiracles of the respiratory system into the tracheae. Contact insecticides may be applied directly to the insect or as residue to plant surface, animal's habitations, or other places frequented by insects.

(c) Fumigant: Those insecticides which exert their action in the gaseous state are known as fumigants. Their application is generally limited to plants or products in tight enclosures or those which can be enclosed in gastight tents or wrappings or to soil. Fumigants are effective against nearly all insect, regardless of the type of mouthparts, since the gas enters the insect body through the spiracle during respiration.

(d) Attractants: Those are the insecticides which lure insect through olfactory stimulation. They may be food lures, sex lure or oviposition lures.

(e) Repellents: These are mildly poisonous or only offensive insecticides which make food or living condition unattractive for insect. They are used in a variety of ways such as poison barriers for which chinch bung and termites, repellent on plant or animal to prevent insect feeding and mothproofing agent to render woollen impervious to the feeding of clothes moths carpet beetles.

On their chemical nature and source

(a) Inorganic insecticides

 They generally act only as stomach poisons. Inorganic Insecticides dominant the market, the inorganic insecticides dominants the market, the inorganic insecticides still find useful application in several areas. Some important inorganic insecticides are described as follows:

(i) Lead arsenate: This compound, $PbHASO_4$ was first development in 1892 for the control of the gypsy moth, porthetria dispar, in forests of the eastern United States. It was widely used during the 1930_s and 1940_s for the control of the codding moth and many other chewing insects. It is still employed for cotton insect and codding moth control in apples as well as for other insect-control purposes.

It is manufactured from litharge (PbO) which is first dissolved in the calculated amount of acetic acid and nitric acid and then treated with a calculated quantity of arsenic acid when lead arsenate is precipitated out. The various reactions involved during the preparation of pbHSO$_4$ are follows:

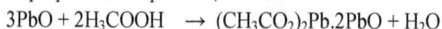

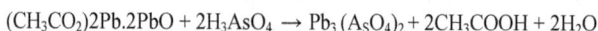

$$3PbO + 2H_3COOH \rightarrow (CH_3CO_2)_2Pb.2PbO + H_2O$$

$$(CH_3CO_2)2Pb.2PbO + 2H_3AsO_4 \rightarrow Pb_3(A_5O_4)_2 + 2CH_3COOH + 2H_2O$$

$$Pb_3(A_5O_4)_2 + 2HNO_3 \rightarrow 2PbHA_5O_4 + Pb(NO_3)_2$$

$$Pb\,(NO_3)_2 + H_3AsO_4 \rightarrow PbHAsO_4 + 2HNO_3$$

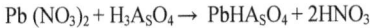

The commercial lead arsenic is a very fluffy pink powder. In water, especially water containing strong alkali, it undergoes the following reaction, which product plant burn:

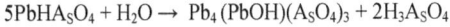

$$5PbHAsO_4 + H_2O \rightarrow Pb_4\,(PbOH)(AsO_4)_3 + 2H_3AsO_4$$

The addition of lime of zinc sulphate to the water as a precipitant for the arsenic acid serves as a "safener".

(ii) Calcium arsenates: Commercial calcium arsenates are mixture of $Ca_3\,(AsO_4)$ and $CaHASO_4$ with an excess of lime and calcium carbonate. They tend to decompose, as shown under lead arsenate above. A safe from of basic calcium arsenate is $[Ca_3\,(AsO_4)]\,Ca\,(OH)_2$

Calcium arsenate is cheaper than lead arsenate but suffers from the following disadvantages:

(i)   It is less effective then lead arsenate because it does not adhere well to leaves.

(ii)   It is hydrolysed more rapidly than lead arsenate.

(iii) It may product arsenate injury.

(iv)   It is used as dust wherever lead arsenate is used as spray.

Calcium arsenate is still employed of certain cotton insects and coddling moth controls in apples as well as for other insect control purposes.

(iii) Paris green: It is copper acetoarsenate $(CH_3COO)_2\,Cu.3Cu\,(ASO_2)_3$. It was used for the first time in 1867 against the Colorado month, potato beetles and surface feeding larvae. It is generally used as a dust powder.

(iv)Fluorides: They are salts of hydrofluoric acid, HF, fluoaluminic acid, H_2SiF_6 and fluoaluminic acid, H_3AIF_6. The insecticidal properties or those compounds are approximately related to the fluorine content, and the solubility in the digestive juices of the insect.

(v) Sulphur and its compound: Sulphur and its compound are among the oldest and most widely used pesticides. Sulphur is used as a wettable powder or dust, alone or in combination with other pesticides for foliar fungal control of mildews and other organisms.

Calcium polysulphide $(CaS_x, x= 4.5)$ is usually found in an about 30 per cent aqueous solution in many commercial preparations which are used for soil conditioning and fungal, mite and insect control. The ammonium and sodium variants are also used. Calcium polysulphide is obtained by boiling milk of lime with sulphur.

(vi)Miscellaneous compounds. Cuprous cyanides CuCN and zinc phosphide, Zn_3P_2 have been used as a stomach poisons for mosquito larvae and agricultural pests. Borax has been used to killed house fly maggots in manure or refuse. Boric acid has been as stomach poison for cockroaches. Sodium selenate, Na_2SeO_4 has been used as a systemic spider mites and aphids. Thaltium sulphate, Tl_2SO_4 is very effective poison bait for ants.

Organic insecticides

(1) Natural (plant) Insecticides

Some plant materials have been most widely used as insecticides. The roots steams, leaves or flower are first dried and then finely ground. These are used as such or their active constituents are extracted which are used alone or in combination with other toxicant and auxillary material. Sometimes the essential oil extracted from plants are used as attractants or repellent. The important natural insecticides are as follows:

(i) Nicotinoids: Nictotine from tobacco was one of the earliest insecticides was recommended for use in 1763 as a tea for the destruction of aphids. Nicotine is found in the leaves of *Nicotinoids tobacum* N. Rustica (2-14) % and is also found in *Duboisia hopwoodii* and *Aesclepias Syriaca.* These plants also contain nor-nicotine and anabasin which are also of insecticidal importance. Among these, nicotine sulphate is widely used because it is more stable and less volatile. It is also safer and more convenient to handle and the free alkaloid is rapidly liberated by the addition of soap, hydrated lime or ammonium hydroxide to the spray solution.

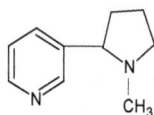
Nicotine
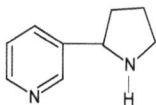
Nor-nicotine
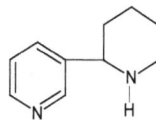
Anabasin

Nicotine sulphate is used as a contact insecticide for aphids attacking fruits, vegetable, and ornamentals, and as fumigants for greenhouse plant and poultry mites.

Mode of Action: Nicotine and anabasine affect the ganglia of the insect central nervous system, facilitating trans synaptic conduction at low concentration and blocking conduction at higher levels.

(ii) Pyrethroids: Pyrethroids is extracted from the ground Flowers of Chrysanthemum cinerariaefolium and C. Coccineum. It has been used in Europe and America since about 1800 for killing insect and worms. The active material is extracted from the dried and powdered flowers of the plant several times with kerosene oil or some other organic solvent.

The chemical investigation has revealed that the insecticidal activity of pyrethrum is associated with five ester, the pyrethrins I and II the cinerins I and II and Jasmolin II which are present in pyrethrum. The structure of pyrethrum is as follows:

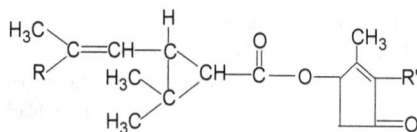

Pyrethrum

Ester   R   R'

Pyrethrin I	-CH$_3$	-CH$_2$CH=CH-CH=CH$_2$
Pyrethrin II	-COOCH$_3$	-CH$_2$CH=CH-CH=CH$_2$
Cinerin I	-CH$_3$	-CH$_2$CH=CH-CH$_3$
Cinerin II	-COOCH$_3$	-CH$_2$CH=CH-CH$_3$
Jasmolin II	-COOCH$_3$	-CH$_2$CH=CH-CH$_2$CH$_3$
Allethrin	-CH$_3$	-CH$_2$-CH=CH$_2$

Knowledge of the structure of the pyrethrins together with their high cost stimulated the synthesis of allethrin which is chapter and has high insecticidal activity. It is effective against insects resistant to DDT. Because of their very rapid knockdown, the pyrithroids have long been favoured for use in household fly cattle sprays at about 0.03-01%.

(iii) Rotenoids: Plants containing retenoids have been used as fish poisons by tropical natives for many centuries. It was not until, 1902 that the active principle, rotenoids, was isolated.

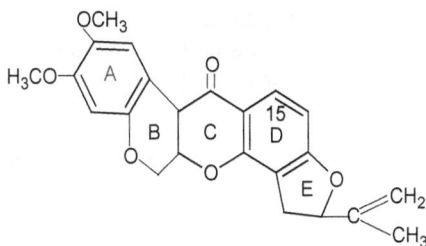

Rotenone

The other naturally occurring rotenoids are as follows:

(a) Elliptone: It has furan ring in place of E of rotenone.

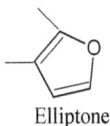

Elliptone

(b) Sumatrol: It is 15-hydroxyrotenone.

(c) Malaccol: It is 15-hydroxyelliptone.

(d) 1-α-Toxicarol: It has a hydroxyl group at carbon 15 and the following ring in place of ring E of rotenone.

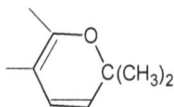

(e) *Deguelin:* It has hydrogen atom on carbon 15 in place of hydroxyl group of toxicarrol. Pure crystalline rotenone is prepared by extracting powdered rotenone containing roots with a solvent such as ether or carbon tetrachloride and concentrating the solution to produce. Insects poisoned with rotenone exhibit a steady decline in oxygen consumption and the insecticide has been shown to have a specific action in interfering with the electron transport involved in the oxidation of $NADH_2$ to NAD by cytochrome b. Poisoning therefore, inhibits the mitochondrial oxidation of Kreb's cycle intermediates which is catalysed by NAD.

(iv) *Sabadilia:* The seeds of sabadilia, family Liliaceae, have been in native louse powder for centuries. These have 2-4% veratrine and include about 13% cevadine 10% veratridine and lesser amount of cevadine, sinabate and sabadine. Cevadine is the angelic acid and ester of cevine of the alkaloids which are responsible for the insecticidal action. Sabadilla is generally used as a dust or wettable powder of the grounds seeds for the control of plant feeding Hemipter and with sugar as toxic bait for thrips.

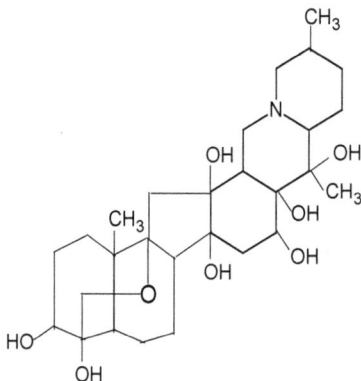

Cevine

(v) *Ryania:* The root and stem of the plant Ryania speciosa have been found to highly toxic to some insects, particularly caterpillars. The main constituent of this plant is the natural substance, ryanodine having the formula $C_{25}H_{35}NO_9$ and melting at 219-222 ^{0}C. This compound is effective both as a contact and a stomach poison.

(2) Synthetic organic insecticides

Synthetic organic insecticides very good have contact and stomach poison actions are sometime used as fumigants. However, the contact action predominates and possess broad spectrum of insecticidal property. Recently they are widely used in the field of house hold facilities, agriculture and many other fields.

(a) Nitrophenol

These were proposed for the control of harmful insects in the last century. From 1937 onward the nitrophenol compounds have been used also as selective contact herbicides. In order to study the insecticidal, fungicidal and herbicidal properties, a large number of alkyl cycloalkyl and arydinitrophenol compound have been synthesized but practical use have been made of only a few compound which are described below:

(i) P-Nitrophenol (m. p. 113.4 ^{0}C): It is produced by alkaline saponification of p-nitrochlorobenzene at an elevated temperature.

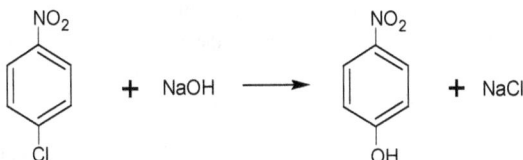

It is also obtained by direct nitration of phenol in a mixture with the o-isomer.

Use: It is used as an agent for preserving natural rubber and some other non-metallic material from destruction by micro or ganisms.

(ii) 2, 4-Dinitrophenol (m. p., 114-115 ^{0}C): It is prepared by alkaline hydrolysis of 2, 4-dinitro chlorobenzen or by oxidative nitration of benzene in the presence of mercury salts.

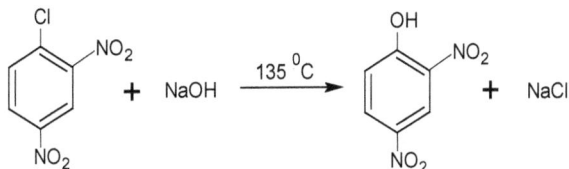

Use: The insecticides and herbicides properties of 2, 4-dinitrophenol are considerably weaker than those of the alkyldinitrophenols and, consequently it is scarcely used at all for the control of plant pests and weeds, but it employed in disinfectant compositions.

(iii) 2, 4-Dinitro-6-methylphenol (dinitro-o-cresol, DNOC, DINOC): It is a yellow crystalline substance, m. p. 86.4^{0}C vapour pressure 5.2 x 10^{-5} mm at 25^{0}C. It is manufactured by the direct nitration of o-cresol with a nitrating mixture at a low temperature. In some cases, the sulphonation of o-cresol is first carried out with 70-93% sulphuric acid.

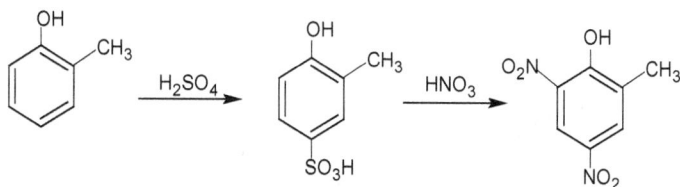

$$\text{o-cresol} \xrightarrow{H_2SO_4} \text{(SO}_3\text{H derivative)} \xrightarrow{HNO_3} \text{2,4-dinitro-6-methylphenol}$$

In small quantities, it is also produced by the oxidative nitration of toluene in the presence of mercuric nitrate. Dinitro-m-cresol is the main product in this reaction.

Use: In agriculture, it is used for controlling plant pests and diseases, and for treatment of fruit trees before opening of the bubs either in the form of oil spray in the form of solution of its salts. In the control of weeds it is generally used in the form of aqueous solution of the salts.

(iv) 2, 4-Dinitro-6-sec-butylphenol or 4, 6-Dinitro-o-secbutylphenol (Dinoseb or DNOSBP): It is a yellow crystalline substance, m. p. 38 ^{0}C. It is prepared by the following method:

(a) It is prepared form 2-sec-butyl-phenol. It is first treated with 80-90% sulphuric which to yield the sulpho mixture which is then added at 50°-100°C to 50% nitric acid.

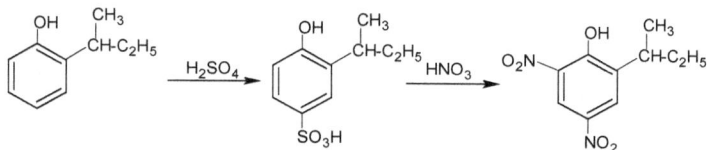

$$\text{2-sec-butylphenol} \xrightarrow{H_2SO_4} \text{(SO}_3\text{H derivative)} \xrightarrow{HNO_3} \text{2,4-dinitro-6-sec-butylphenol}$$

The 2-sec-butylphenol required for the above synthesis is obtained by the alkylation of phenol with butylene at 200-250 ^{0}C in the presence of aluminium phenolate.

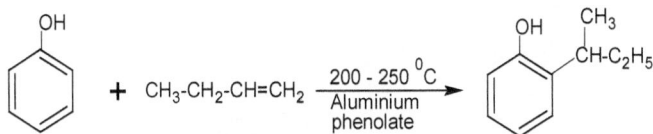

It is also possible to prepare 2-sec-butylphenol by alkylation of p-bromophenol with subsequent removal of bromine by reduction.

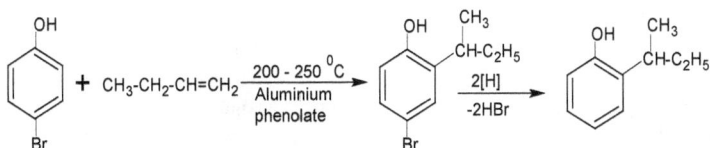

(b) It is also prepared by alkylation of p-phenolsulphonic acid with butylene or butyl alcohol and subsequent nitration of the 2-butylphenol-4-sulphonic acid without separation it from the reaction mixture.

Use: In order to control plants and weeds, dinoseb is generally used in the form of aqueous solution s of the phenolates of ammonia and organic amines. For the control of plant pests, 2, 4-dinitro-6-sec-butylphenyl acetate is used which somewhat less toxic than the phenol.

In order to control true powdery mildew on apple trees, 2, 4-dinitro-6-sec-butylphenyl methacrylate is used along with some other ester of 2, 4-dinitro-6-sec-butylphenol, 2, 4-dinitro-6-sec-amylphenol (b.p.149 ^{0}C) is similar to dinoseb in herbicidal effect.

(v) 2, D-Dinitro-6-sec-butylphenyl isopropyl carbonate or (dinobuton, acrex, Dessin, Dinofen): It is a pale yellow crystals, m. p. 61- 62 ^{0}C. It is prepared by the condensation of an alkali salt of dinoseb with isopropylchloformate.

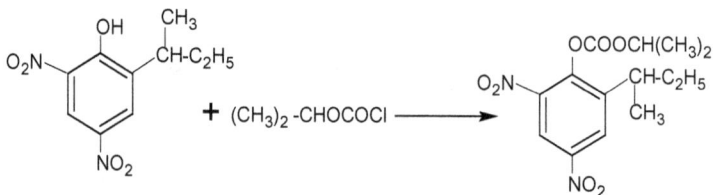

Use: It is a non-systemic acaricide and fungicide (active against powdery mildews).

(vi) 4, 6-dinitro-2-cyclohexylphenol or 4, 6-dinitro-o-cyclohexylphenol (Dinex or DNOCHP): It is a yellowish crystalline solid with m. p. 106 ^{0}C. It is solubility in water changes with the ph of the medium. It is prepared by the nitration of 2-cyclohexylphenol with a mixture of sulphuric acid nitric acids at 60-90 ^{0}C. 2-Cyclohexyphenol required for the above synthesis is prepared by the condensation of phenol with cyclohexene in the oresence of aluminium phenolate or other catalysts at 200-250 ^{0}C.

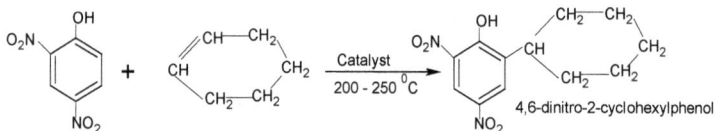

Use: It is used for the treatment of fruit and ornamental trees in the dormant state. However, it is poisonous for man and animal but somewhat less so than DNOC and dinoseb.

(vii) 4, 6-dinitro-2-caprylphenyl crotonate (Dinocap or Karathane): It is a dark-brown liquid (b. p. 138-140°C at 0.05mm). In order to prepare this, 2-sec-octylphenol is first nitrated and the dinitrophenol formed is further treated with crotonylchloride in the presences of HCl acceptors.

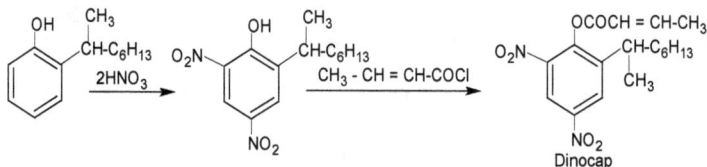

Use: It is used as an acaricidal and fungicide.

Mode of action of dinitrophenols: These chemicals have broad biological activity embracing the insecticides, fungicide, acaricide and herbicide fields. This broad activity results from their capacity to block oxidative phosphorylation, hence interfering with vital biochemical synthesis cycles found in animal life. As a result poisoned insects undergo pronounced increases in the rate of respiration and they die from metabolic exhaustion due to their in ability to utilise the energy provided by respiration and glycolysis for the conversion of o phosphate to high energy phosphate such as ATP. The 4, 6-diniotro-2-alkylphenols are so far the most active and activity depending upon the of the alkyl chain, reaching a maximum at six or seven carbon.

(b) Halogen Derivative of Aromatic Hydrocarbon

These were among the earliest synthetic organics employed or pesticides. DDT exhibited a broad spectrum of insecticidal activity and success stimulated the research that led to many of the other organic chlorine compounds. The simplicity of the chlorination of benzene led to the discovery of the usefulness of benzene hexachloride and lindane as insecticides. Few chlorinated organic are described as follows:

(i) DDT [1, 1, 1 trichloro-2, 2-bis (p-chlorophenylethane)]: It was first synthesized by Ziedler in 1834 but its insecticidal properties were only discovered in 1939 by Muller.

Preparation

(a) It is manufactured by condensing chlorobenzene (2 parts) with chloral (1 part) at about 30°C in the presence of condensing agents (10 parts) such as concentrated sulphuric acid, hydrogen fluoride, anhydrous ammonium chloride, etc. However, the most frequently used method is the condensation of chloral with chlorobenzene in the presence of concentrated sulphuric acid or weak oleum at a temperature not higher than 30°C because at higher temperature the amount of p-chlorobenzenesulphonic acid that is formed as a by product increase sharply. The temperature is kept at the some point by external cooling. The reaction takes 5-6 hours for completion. The crude product is removing from the top, washed with Na_2CO_3 solution, then distilled under vacuum to separated unreacted chlorobenzene. The product is pulverised and marketed as such. The product consists of rather more than 70% of the p, p-isomer together with a number of other by product including a trace of the o, o-isomer. For most application, these impurities have no deleterious effect and the product is used without purification.

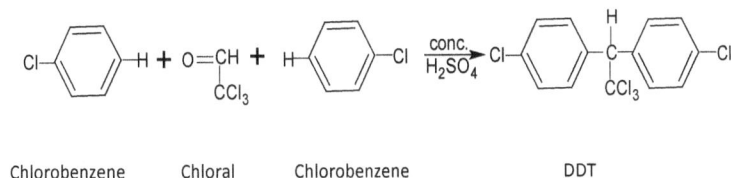

Chlorobenzene   Chloral   Chlorobenzene   DDT

The above synthesis of DDT involves the following mechanism:

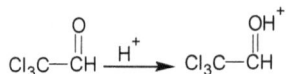

$$Cl_3C-\overset{\overset{\displaystyle O}{\|}}{C}H \xrightarrow{\ H^+\ } Cl_3C-\overset{\overset{\displaystyle OH^+}{\|}}{C}H$$

The chloral necessary for the production of DDT is obtained by chlorination if ethyl alcohol or acetaldehyde. The chlorination of acetaldehyde proceeds through an enol from and may be represented by the following general equation:

$$CH_3CHO + 3Cl_2 \longrightarrow CCl_3CHO + 3HCl$$

The mechanism of chlorination of ethyl alcohol is more complex:

$$C_2H_5OH + Cl_2 \longrightarrow [C_2H_5OCl] + HCl \longrightarrow CH_3CHO + HCl$$

$$CH_3CHO + Cl_2 + C_2H_5OH \longrightarrow CH_2ClCHClOC_2H_5 + H_2O$$

$$CH_3CHO + Cl_2 + 2C_2H_5OH \longrightarrow 2HC_2ClCH\ (OC_2H_5)_2 + HCl + H_2O$$

$$CH_2ClCH\ (OC_2H_5)_2 + Cl_2 \longrightarrow CHCl_2CH\ (OC_2H_5)_2 + HCl$$

$$CHCl_2CH\ (OC_2H_5)_2 + H_2O \longrightarrow CHCl_2CH\ (OH)\ OC_2H_5 + C_2H_5OH$$

$$CHCl_2CH\ (OH)\ OC_2H_5 + Cl_2 \longrightarrow CCl_3CH\ (OH)\ OC_2H_5 + HCl$$

$$CCl_3CH\ (OH)\ OC_2H_5 + H_2O \longrightarrow CCl_3CHO + C_2H_5OH$$

The product obtained by the chlorination of ethyl alcohol contains choral alchoholate and choral hydrate. This product on treatment with concentrated sulphuric acid yields choral.

$$CCl_3CH\ (OH)_2 + H_2SO_4 \longrightarrow CCl_3CHO + H_2SO_4\ H_2O$$
$$CCl_3CH\ (OH)\ OC_2H_5 + H_2SO_4 \longrightarrow CCl_3CHO + C_2H_5OSO_3H + H_2O$$

(b) DDT is also prepared by the condensation of chlorobenzen with pentachlorethane.

$$2C_6H_5Cl + CHCl_2CCl_2 \longrightarrow (ClC_6H_5)_2CHCCl_2 + 2HCl$$

DDT obtained by this method is strongly contaminated with by-products.
(c) An interesting method for the synthesis of DDT is the reaction of chlorobenzene with p-chlorophenyltrichloromethyl-carbinol.

$$C_6H_5Cl + ClC_6H_4CH\ (OH)\ CCl_3 \longrightarrow (ClC_6H_4)_2CHCCl_2 + H_2O$$

This reaction takes place readily in the presence of sulphuric acid or oleum. The p-chlorophenyl trichloromethyl-catbinol required for this synthesis is obtained by the reaction of chloroform with p-chlorobenzaldehyde.

$$ClC_6H_4CHO + CHCl_3 \longrightarrow ClC_6H_4CH\ (OH)\ CCl_3$$

Properties: it is white, crystalline, almost colourless solid of M.P.109°C. It is soluble in benzene kerosene, petrol, liquid Freon and ethanol but insoluble in ether. The pure p, p' isomer of DDT is thermally stable; However, its decomposition starts above 195°C in accordance to the following equation.

$$(ClC_6H_4)_2CHCCl_3 \longrightarrow (ClC_6H_4)_2C = CCl_2 + HCl$$

DDT undergoes decomposition in the presence of light in alcohol solution in accordance to the following reaction:

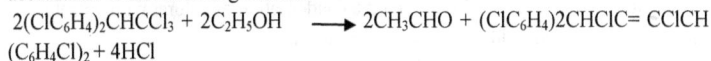

$$2(ClC_6H_4)_2CHCCl_3 + 2C_2H_5OH \longrightarrow 2CH_3CHO + (ClC_6H_4)2CHClC= CClCH$$
$$(C_6H_4Cl)_2 + 4HCl$$

In presence of air oxygen the above reaction proceeds further, thereby forming p, p'-dichloro benzophenone.

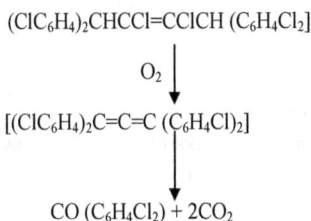

$$(ClC_6H_4)_2CHCCl=CClCH (C_6H_4Cl_2]$$

$$O_2 \downarrow$$

$$[(ClC_6H_4)_2C=C=C (C_6H_4Cl)_2]$$

$$\downarrow$$

$$CO (C_6H_4Cl_2) + 2CO_2$$

It is assumed that similar reaction also takes place on the leaves of plants.

Use: It has a wide use as a herbicide, fungicide and rodenticide in agriculture. It was found to be highly effective insecticide due to the following reasons:

(i) It is highly potent and kills a wide range of insects by both contact and ingestion.

(ii) It has little smell, and does not taint food.

(iii) It is stable and therefore persistent (This persistence is now considered to be a major disadvantage of DDT).Since it is made by a simple process from low cost raw materials it is cheap.

In spite of the above advantages, it has several grave shortcomings, viz, its storage in harmful concentration in animal tissues, its powder of transmission through cow's milk. DDT residues cannot strains also develop. Due to these adverse affect, many advanced countries like UK, USA, Norway, etc. have banned to use of DDT.

Some insects however possess an enzyme called DDT-dehydrochlorines which converts DDT in to a harmless product by eliminating HCl from chloral residue. But insects which are not destroyed by DDT are destroyed by BHC. The converse is also true.

Mode of action: Extensive investigations have shown that DDT has a highly insecticidal specific action which is largely due to the ease with which it is absorbed by the insect cuticle. Further DDT affects peripheral sensory organs in insect to product violent trains of afferent impulses which cause hyperactivity and then convulsions in the insects. The paralysis and death occur from metabolic exhaustion or from elaboration of a naturally occurring neurotoxin.

The insecticidal activity of DDT analogs is a function of the size, shape and polarity of molecule. There are large number of DDT analogs and homologs that have been studied a few, which are described below, have found practical use:

(a) Methoxychlor: Chemically, it is 1, 1, 1-trichloro-2, 2-bis (p-methoxyphenyl) ethane. It is obtained in good yields by the condensation of chloral with anisole in the presence of sulphuric acid other condensing agents.

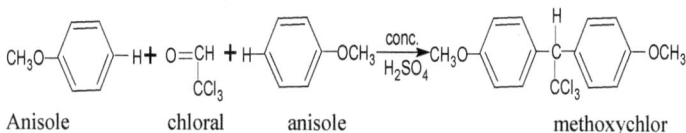

Anisole   chloral   anisole   methoxychlor

It is white consisting of about 88% of the p, p-isomer (m. p. 89 ^{0}C). It given a more rapid knock down of many insects than does DDT but it is lower chronic and acute toxicity. Therefore, it can be safely used on vegetables crops, cattles and against pests in the household. Unlike DDT it does not get stored in body and therefore it is favoured use on animal and animals forage.

(b) DDD: Chemically it is 1, 1-dichloro-2, 2-bis, (p-chlorohenyl) ethane. However, it is also known as TDE (tetrachlorodiphenyl ethane). It is prepared by the condensation of chlorobenzene with pure dichlorochetaldehyde.

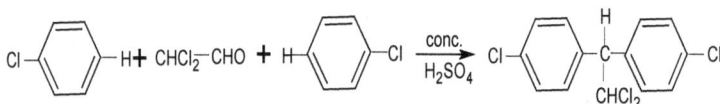

It consists if white crystals, m. p. 112 ^{0}C. In general DDD is less effective against most insects than DDT but it has the advantage of being one fifth to one tenth as acutely toxic to warm blooded animals and fish. DDD is superior to DDT for the control of such insects as mosquito larvae, tomato hornworms, and the red bonded leaf roller.

(c) Perthane: Chemically, it is 1, 1-dichloro 2, 2-bis (p-ethylphenyl) ethane. It is prepared by the condensation of dichloroacetaldehyde with ethylbenzen in the presence of sulphuric acid.

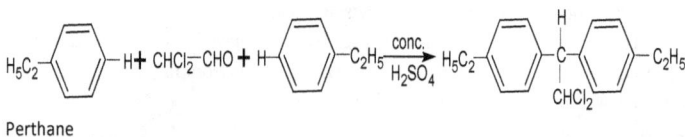

Perthane

It is a white crystalline substance, m. p. 56-57 ^{0}C. It is used in the form of emulsions and suspensions, and also on aerosols. Although it is one fifth to one tenth as active as DDT for different insects, it has been developed for application where low acute and chronic toxicity are desired.

(d) DFDT: Chemically, it is (1, 1, 1-trichloro-2, 2-bis (p-fluorophenyl) ethane. It is produced by the condensation of chloral with fluorobenzene in the presence of sulphuric acid.

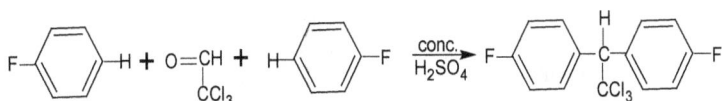

DFDT

Technical material is a colourless liquid containing about 10% of the o, p-isomer melts at 45 ^{0}C. Its use is similar to DDT. Its toxicity for insects is close to that of DDT. However, DEDT is less persistent than DDT, which in some cases is a great advantage. However, it is more expensive than DDT.

(c) Halogen Derivatives of Alicyclic Hydrocarbons

There are many halogen derivatives of the alicyclic hydrocarbons which have practical importance as valuable insectofungicides. Of these, benzene hexachlorid has been widely used. Other halogen derivatives of alicyclic hydrocarbons are represented by heptachlor, aldrin, isodrin, and dihydroheptachlor.

(i) Benzene hexachloride: benzene hexachlorid or more correctly 1, 2, 3, 4, 5, 6-hexachlorocyclo- hexane, was found to have strong insecticidal properties in about 1942, by ICI workers in U.K. and by Lebas in France. Its properties as an insecticide are broadly similar to those of DDT, and it has achieved substantial importance, though by no means to the some extent as DDT.

Benzene hexachloride is prepared by the addition of chlorine to chlorine to benzene under the influence of light without any catalyst. A mixture of isomer of benzene hexachloride is formed. The Y-isomer which has insecticidal properties only, makes up about 12% of this mixture. The Y-isomer is extracted with toluene, xylene, chlorohexane, or glacial acetic acid. Almost pure Y-isomer is obtained by repeated crystallisation. The 99 % pure γ-isomer of BHC is known as lindane.

The BHC is an almost white crystalline solid with a smell resembling phosgene. It causes tainting of foods with which it comes in contact. It is fairly toxic human being, because of its volatile nature. However, its action is short lived compared to DDT, it acts more rapidly than DDT is used control various harmful insects, plants pests and animal parasites.

$$C_6H_6 \ + \ 3Cl_2 \ \xrightarrow{\text{sunlight}} \ C_6H_6Cl_6$$

benzene   chlorine   A mixture of isomers of benzene hexachloride

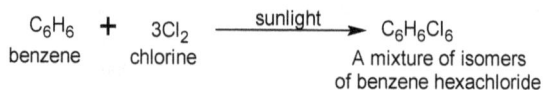

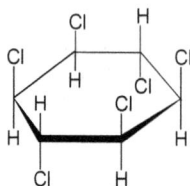

γ - Benzene hexachloride

It is primarily a contact poison but it may also kill the insects by vapour effect. In all countries it is predominantly lindane that is being manufactured because it convenient to use, has no odour, and leaves less residue in food lurage products.

Mode of action: Its mode if action is unknown. However, its specific toxicity is indication of the fact that like, DDT, it many interact with the pores of the lipoprotein structure of the insect nerve thereby causing distortion and consequent excitation of ionic transmission.

(ii) Polychloroterpenes: These were first used in U.S.S.R. for control of some parasites of animals. These have been produced by the chlorination of the naturally occurring terpenes. Some of these are described as follows:

(a) Oxaphene: It is the most well known terpene chlorination produced. It is prepared by the chlorination of the bycyclic terpene, camphene, to contain 67-69% chlorine, and has the empirical formula $C_{10}H_{10}C_{13}$. Actually, it is a complex mixture of poly chlorocamphenes and comphenes of different structure in which there may be present product of chlorination of tricyclene and related compound.

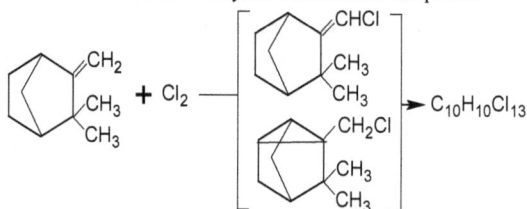

Toxaphene is a yellowish, semicrystalline gum, m. p. 65-90 ^{0}C. It is almost insoluble in water but dissolve well in many organic solvents. It is unstable in the presence of alkali, upon prolonged exposure to sunlight, and at temperature above 155 ^{0}C, liberation hydrogen chloride and losing some of its insecticide; it also gives satisfactory result in combating rodents.

(b) Strobane: Its composition is similar to that of toxaphene. It is prepared by the chlorination of a mixture of camphene and pinene to contain 66% chlorine. It is straw coloured liquid. It is almost insoluble in but highly soluble in organic solvent. It is

broad spectrum insecticide which is especially useful for the control of insect pests of cotton, field crops and animal. However, it is approximately one and a half weaker than toxaphene in insecticidal activity.

(c) Polychloropinene or chlorothene: It is produced by photochemical or initiated chlorination of bornyl chloride to a chlorine content of 64-67%. It is colourless viscous oil. It is almost insoluble in water and highly soluble in organic solvent.

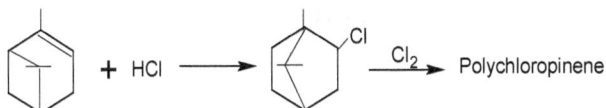

(iii) Polychlorocyclodienes: These are the derivatives of bi-tri and tetracyclic hydrocarbons. These compounds in most cases are produced by diene synthesis reactions using a hexachloro cyclopentadiene are described as follows:

(a) Chlordane: It is a prepared by the chlorination of chlordane at 50-80°C either in the presence of solvent or without one.

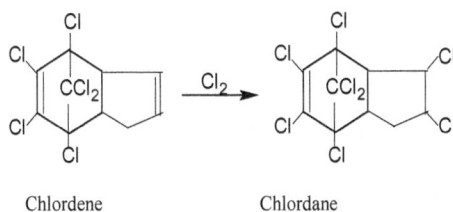

Chlordene   Chlordane

Technically, chlordane is an amber liquid b. p. 175 °C at 2 mm. It contains 60-75% of the cis and trans-isomer s of chlordane plus 40-25%of related compounds such as heptachlor, nonachlor and chlordane. Octachlor obtained as a result of substitutive chlorination of heptachlor many possible be present in technical grade chlordane. Structure formulae of nonachlor and octachlor are as follows:

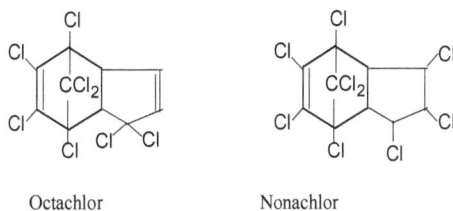

Octachlor   Nonachlor

Chlordane is used for the control of cockroaches, ants, termites and house hold pests, soli insects and certain pests of vegetable and field crops. It is gradually being displaced by the more effective heptachlor and other compound of diene synthesis.

17

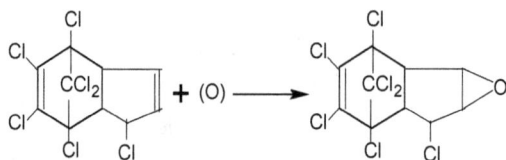

It is produced in good yields by the chlorination of chlordane with sulphuryl chloride. It is also produced by the action of thionyl chloride in hydroxychlordene which is formed by reacting chlordane with selenium dioxide.

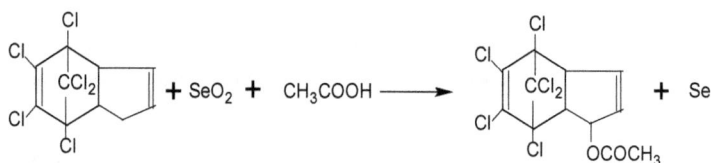

Chlordane

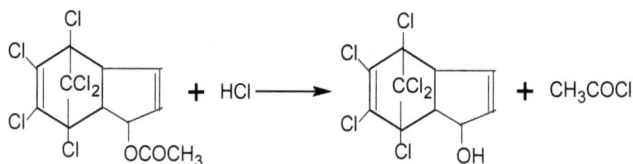

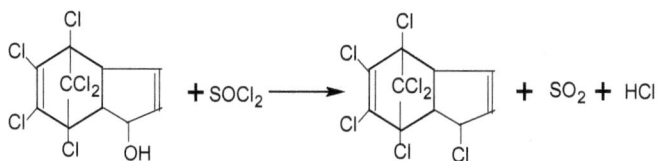

Hydroxychlordene   Heptachlor

The pure heptachlor is a white crystalline substance with weak camphor like odour, m. p. 95-96 ^{0}C. It is practically insoluble in water, moderately soluble in ethyl alcohol and highly soluble in derivatives of hydrocarbons. The technical grade heptachlor is a waxy mass, m. p. 46-74 ^{0}C.

Heptachlor is widely used for the control of parts of balfaea, corn, and for grosshopper control. It is also used as an insecticidal additive to seed disinfectants. This is not only protects the seeding from damage by insects but also stimulates the germination of seed. Heptachlor is readily oxidized to heptachlor epoxide which is somewhat more active. This reaction takes place readily under the influence of soil micro organisms in animals and probably insects. It is possible that the effect of heptachlor is based on the reaction of its epoxide with vitally important systems in animal and the insects.

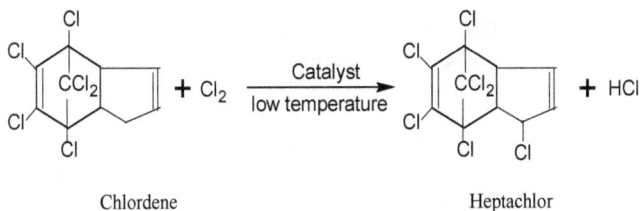

| Chlordene | | Heptachlor |

(c) Aldrin: It is produced by the reaction of hexachlorocyclopentadiene (I) with an excess of bicycle (2.2.1.) heptadiene-2, 5, (II) at 100 ^{0}C.

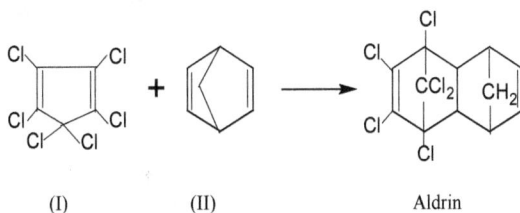

(I) (II) Aldrin

Biycyclo (2.2.1.) heptadiene-2, 5 required for the above synthesis is prepared by the reaction of cyclopentadinene with acetylene at 250-360 ^{0}C and 4-20 atmospheric pressure.

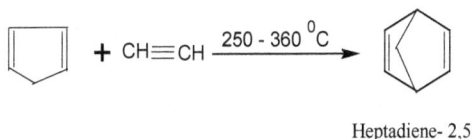

$$\square + CH\equiv CH \xrightarrow{250 - 360\ ^0C}$$

Heptadiene- 2,5

Pure aldrin is a white crystalline, m. p. 104-105 ^{0}C. However, technical preparation is a mass containing 92% aldrin, 12-13% analogs and about 5% various other compound. Technical preparation melts at 45-60 ^{0}C. Aldrin is broad spectrum insecticide used for the control of insect pests of fruits, vegetable cotton, and as a soil insecticide.

(d) Isodrin: It is obtained by the condensation of cyclopentadiene (II) with 1, 2, 3, 4, 7, 7-hexachloro bicycle (2.2.1) heptadiene 2,5 (I).

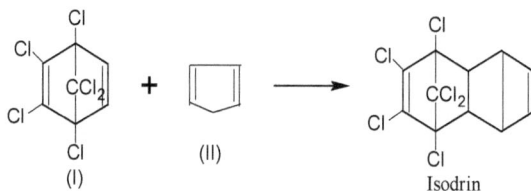

(I)   (II)   Isodrin

Isodrin is a white crystalline substance m. p. 240-242 ^{0}C. Although isodrin has a strong insecticidal effect, it has still not received much use in agriculture but it is employed for preparation of another insecticide endrin.

(e) Endrin: It is obtained by the oxidation of isodrin with hydrogen peroxide at low temperature. If the oxidation of isodrin is carried out at 100°C, the endrin rearranges to a ketone that does not possess any insecticidual effect.

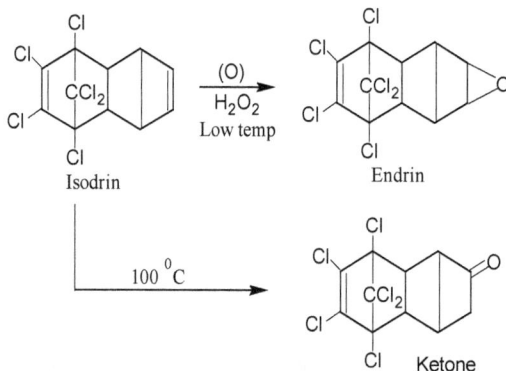

Endrin is especially for the control of lepidopterous larvae attacking cotton, field and vegetable crop. An advantageous property of endrin is its low persistence, since in the light it is relatively rapidly isomerised to the non-toxic ketone.

(f) Alodan: It is prepared by the condensation of hexachlorocyclopentadiene with cis-1, dichloro-butene-2 at 150-160 ^{0}C in an autoclave.

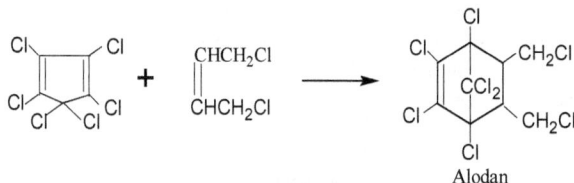

Alodan is a white crystalline substance m.p.104-105 ^{0}C it is insoluble in water but highly soluble in organic solvents. It is especially used for the control of ectoparasites of animal because of its low mammalian toxicity.

(g) Mirex and kepone: Mirex is obtained by the condensation of two molecules of hexachloro-cyclopentadiene.

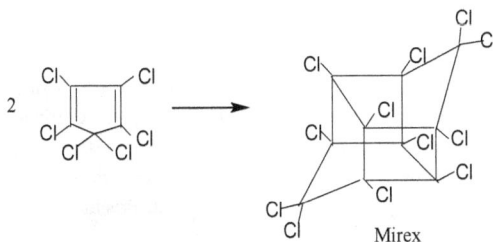

Mirex

Related to mirex is the compound kepon which is the 2-keto derivative of mirex. Kepone melts at 349 ^{0}C.

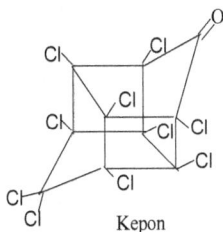

Kepon

Both mirex and kepone have largely stomach poison action and have been used to control ants, soil insects, and against chowing pests of ornamental.

(d) Organophosphours insecticides

The practical development of these compounds as insecticides is due to the original and extensive work of Schrader and co-workers beginning in 1937. This work attracted many other workers who synthesized more than 100,000 different organophosphours insecticides. Organophosphours insecticides are widely use as agents to plant and ectoparasites and in parts of domestic animals. The most important advantages of the organophosphours compounds as pesticides are as follows:

(a) They have high insecticidal and acricidal activity.

(b) They exhibit wide spectrum of action of plant pests.

(c) They possess low persistence.

(d) They breakdown to from products non toxic to man and animals.

(e) They have systemic action of a number of the compounds.

(f) Their low dosage is required.

(g) They act rapidly on plant pests.

(h) They have low chronic toxicity.

The main disadvantage of organophosphours insecticides is their relatively high toxicity to vertebrates, requiring suitable protective measures when using them.
General principles for organophosphors pesticide synthesis are as follows:
(i) The main reaction used in the Arbusove reaction which consists of formation of a phosphonate by reaction of a phosphite with an organic halide. However, if a halogen is alpha to a carbonyl group, a rearrangement called Parkow rearrangement, takes place to from a vinyl phosphate. This reaction forms the basis of synthesis of many organophosphate insecticides such as dichlorous, naled, mevinphos, dicrotophos, etc. This reaction is illustrated by the synthesis of naled.

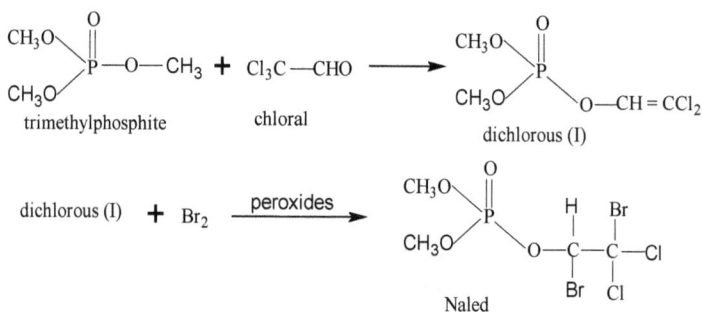

trimethylphosphite   chloral   dichlorous (I)

Naled

(ii) Another reaction used in the manufacture of organophosphorus insecticides is the addition of to an olefin or carbonyl or its displacement of a reactive halide. This reaction is illustrated by the manufacture of malathion.

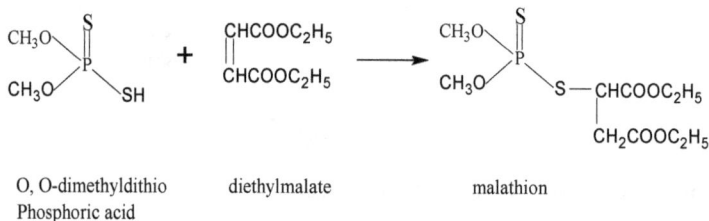

O, O-dimethyldithio   diethylmalate   malathion
Phosphoric acid

(iii) Another reaction used in the manufacture of organophosphorus insecticide is the displacement of a phosphorohalidate, usually a chloridate, by a nucleophilic reagent. A route to the synthesis of parathion is described as follows:

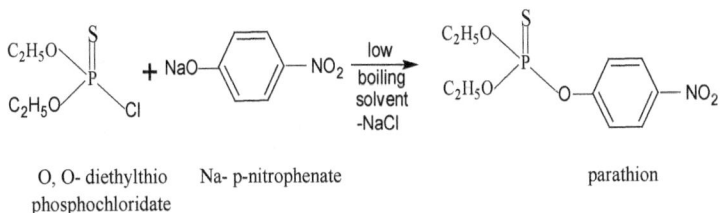

O, O- diethylthio   Na- p-nitrophenate   parathion
phosphochloridate

Let us now discuss the various organophosphorus pesticides one by one:

(i) TEPP (tetraethyl pyrophosphate): It was first synthesized by Clermont in 1854 but its insecticidal properties were discovered by Schrader in 1939.

Synthesis: There are numerous syntheses of TEPP which must be mentioned here because they are not only of historical signification but also of technical and scientific interest.

(a) Clermont's synthesis (1854). In 1854, Clermont, prompted by Wurts, synthesised TEPP by alkylating the silver salt of the pyrophosphorus acid ethyl iodide.

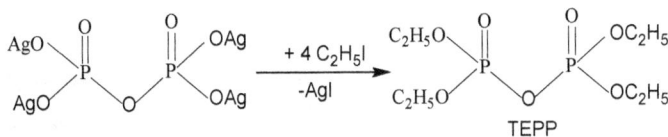

(b) Nylen's synthesis (1930), Nylen synthesized the TEPP in the following way:

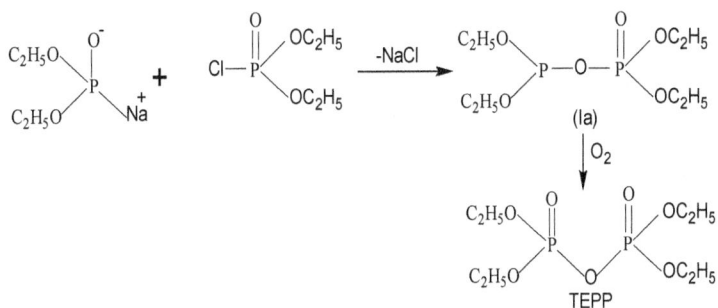

The oxidation can also be carried out with chlorine in place of oxygen.

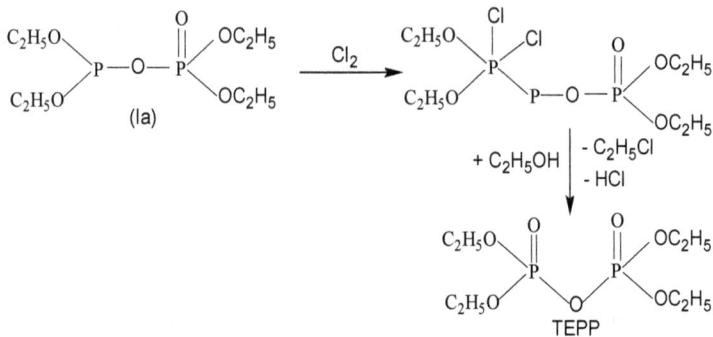

(c) Schrader's synthesis (1946): Schrader prepared TEPP by reaction phosphoryl chloride with triethyl phosphate in a mole ratio of 1:5.

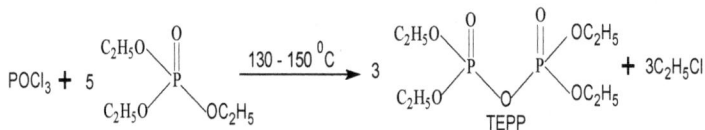

If the Schrader's process is carried out with a mol ratio of 1:5, hexaethyl tetraphosphate (HETP) is obtained as the main product.

(d) Woodstock's synthesis (1946): TEPP can be prepared by reaction phosphorus pentoxide with trietylphosphate in a mol ratio of 1:4.

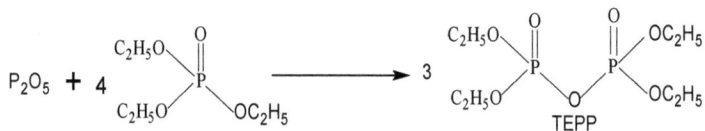

(e) Anand's synthesis (1969): This synthesis is as follows:

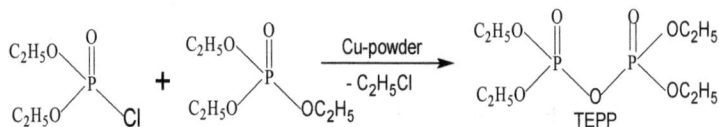

Properties: It is clear oily liquid m. p. 6.1 ^{0}C, b. p. 113 ^{0}C at 0.05 mm of mercury. It is miscible in water, ethanol, ethyl acetate carbon tetrachloride, benzene, xylene and methylated naphtaleness but not in kerosene or mineral oils.

Use: The principal uses of TEPP have been for the control of aphides and red spider mites on agriculture and ornamental crops and in greenhouse. It is extremely poisonous to mammals. Therefore, a respirator should be worn at all times for spraying or dusting with TEPP.

(ii) Bladan (Hexaethyl tetraphospate, HETP): Schrader (1943) react phosphoryl chloride with triethyl phosphate in a mole ratio 1:3. This reaction is known as the Schrader process. The reaction product was formulated as hexaethyl phosphate "HETP" this was introduced in 1943 under the name "Bladan" as the first contact insecticidal organophosphate Janning (1946) was the first to discover a synthesis of "HETP" from POCl$_3$ and alcohol which was later described also by Thurston (1864).

HETP

According to some other author, "HETP" is not a uniform product but rather a mixture of HETP, TEPP, triethyl phosphate and ethyl metaphosphate, the ratio of which depends upon the quantities of phosphoryl chloride and triethyl phosphate used. In all product mixture, TEPP would appear to be the actual active constituent.

Woodstock (1946) synthesized HETP by reacting phosphorus pentoxide with triethylphosphate in a mole ratio of 1:2.

HETP posses insecticidal and acaricidal activity, recognized by kiikenthal in 1938 due to its low stability towered hydrolysis, it is now a day's seldom used. Occasionally, however, a rapid degradation is desirable for such preparation may be applied until shortly before harvest. The oral LD$_{30}$ were 1.12 mg/kg.

(iii) Bladafun (O, O, O, O-tetraethyl pyrophosphorodithionate): This compound also known as sulphotepp. This is also obtained by the addition of sulphur to O, O, O, O'-tetraethyl pyrophosphite (Schrade, 1947).

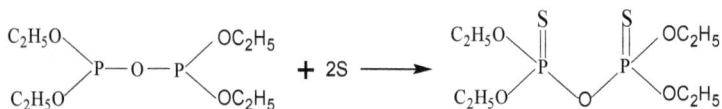

$$\underset{C_2H_5O}{\overset{C_2H_5O}{>}}P-O-P\underset{OC_2H_5}{\overset{OC_2H_5}{<}} \quad + \ 2S \ \longrightarrow \quad \underset{C_2H_5O}{\overset{C_2H_5O}{>}}\overset{S}{\overset{\|}{P}}-O-\overset{S}{\overset{\|}{P}}\underset{OC_2H_5}{\overset{OC_2H_5}{<}}$$

Bladafun

Bladafun is prepared on an industrial scale by the partial hydrolysis of o, o-diethyl phosphorothio chloridate

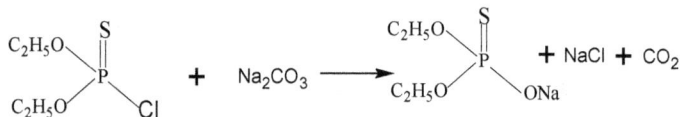

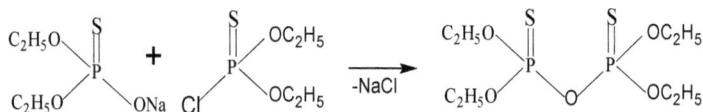

$$\underset{C_2H_5O}{\overset{C_2H_5O}{>}}\overset{S}{\overset{\|}{P}}-Cl \quad + \ Na_2CO_3 \ \longrightarrow \quad \underset{C_2H_5O}{\overset{C_2H_5O}{>}}\overset{S}{\overset{\|}{P}}-ONa \quad + \ NaCl \ + \ CO_2$$

$$\underset{C_2H_5O}{\overset{C_2H_5O}{>}}\overset{S}{\overset{\|}{P}}-ONa \ + \ Cl\overset{S}{\overset{\|}{P}}\underset{OC_2H_5}{\overset{OC_2H_5}{<}} \quad \xrightarrow{-NaCl} \quad \underset{C_2H_5O}{\overset{C_2H_5O}{>}}\overset{S}{\overset{\|}{P}}-O-\overset{S}{\overset{\|}{P}}\underset{OC_2H_5}{\overset{OC_2H_5}{<}}$$

Bladafun

This compound has insecticidal, acricidal activity and an oral LD_{50} of 5 mg/kg for the rat. On account of its high vapour pressure thermal stability, Bladafun is used in greenhouses as a fumigant. Because of the thio group, the compound is more stable towered hydrolysis the TEPP.

(iv) Schradan (OMPA; octamethyl pyrophosphereamidate): Schrader in 1941 synthesized OMPA in the following manner.

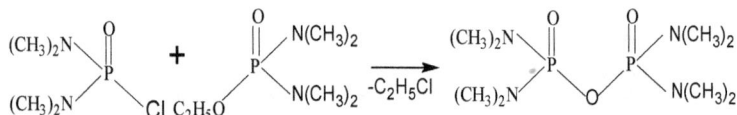

$$\underset{(CH_3)_2N}{\overset{(CH_3)_2N}{>}}\overset{O}{\overset{\|}{P}}-Cl \ + \ C_2H_5O\overset{O}{\overset{\|}{P}}\underset{N(CH_3)_2}{\overset{N(CH_3)_2}{<}} \quad \xrightarrow{-C_2H_5Cl} \quad \underset{(CH_3)_2N}{\overset{(CH_3)_2N}{>}}\overset{O}{\overset{\|}{P}}-O-\overset{O}{\overset{\|}{P}}\underset{N(CH_3)_2}{\overset{N(CH_3)_2}{<}}$$

OMPA

On a large scale, it is manufactured by the partial hydrolysis of tetramethyl phosphorodiamido chloridate.

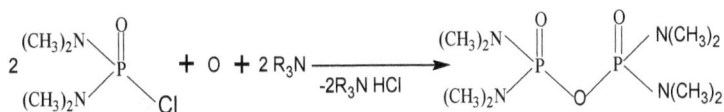

OMPA

OMPA was introduced by pest control Ltd, under the name "Pestox III". It is well known for its systematic insecticidal properties. The insecticide is of historical significance in so far as it was the first systemic phosphoric ester insecticide to be recognized.

(v) Parathion ({O, O-diethyl} O-p-nitrophenyl-phosphorothionate): It was discovered by Shcader in 1944 and has become the most widely used of all the organophosphorous insecticide. Two methods generally used for the preparation of parathion (I) (i) German method and (ii) American method. The two methods differ only in the preparation of O, O-diethylphosphorothiochloridate (II). The last step which involves the condensation of (II) with p-nitrophenol in the presence of an alkali is the same in both the cases.

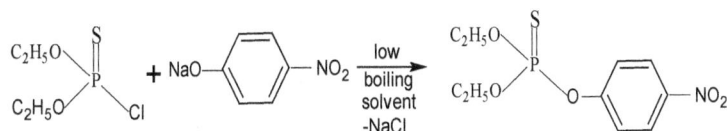

This process is carried out in chlorobenzene or in aqueous medium in the presence of emulsifiers.It is obtained in good yields also when O,O-diethylphosphorothichloridate (II) reacts with p-nitrophenol in the presence of potassium carbonate (HCl acceptor) in acetone or other solvents (the accepter is tertiary amine).

$$C_2H_5O \overset{S}{\underset{C_2H_5O}{\underset{(II)}{\overset{\|}{P}}}} Cl \;+\; HO\!-\!\!\bigcirc\!\!-\!NO_2 \;+\; K_2CO_3 \longrightarrow$$

$$C_2H_5O \overset{S}{\underset{C_2H_5O}{\overset{\|}{P}}} \!-\!O\!-\!\!\bigcirc\!\!-\!NO_2 \;+\; KHCO_3 \;+\; HCl$$
(I)

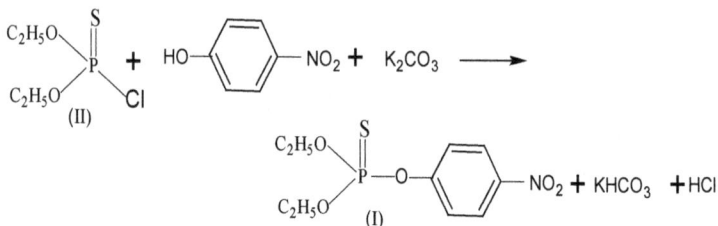

(a) German process: This makes use of thiophsphoryl chloride (PSCl$_3$) for the preparation of (II). However PSCl$_2$ were previously prepared by heating phosphourus trichloride with sulphur in a sealed tube at 130^{0}C. This method could not be applied on a large scale because of the very large size of the autoclave required for the synthesis. However, PSCl$_3$ is now prepared on an industrial scale by passing vapours of phosphorus trichloride over liquid sulphur of low viscosity at 140 ^{0}C. The yield of this method can be further increased by using catalysis such as charcoal, zinc chloride, etc.

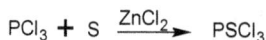

$$PCl_3 \;+\; S \xrightarrow{\;ZnCl_2\;} PSCl_3$$

Thiophsphoryl chloride is then made to react with ethyl alcohol when the compound(II) is obtained provided the hydrogen chloride produced in the reaction is continuously removed from the reaction mixture.

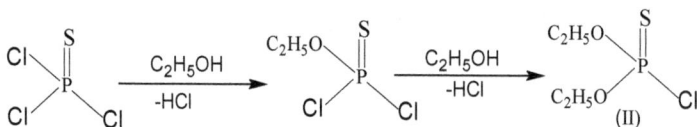

The compound (II) is then made to react with sodium salt of nitrophenol when (I) is formed.

(b) American method: In this method, the compound (II) is obtained by reacting P$_4$S$_{10}$ with adequate quantities of ethyl alcohol, followed by reaction with chlorine.

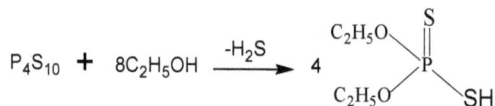

$$P_4S_{10} \;+\; 8C_2H_5OH \xrightarrow{\;-H_2S\;} 4\; C_2H_5O \overset{S}{\underset{C_2H_5O}{\overset{\|}{P}}} SH$$

$$2 \quad \overset{C_2H_5O}{\underset{C_2H_5O}{>}}\overset{S}{\underset{}{\overset{\|}{P}}}\text{—SH} \quad + \quad 3Cl_2 \quad \longrightarrow \quad 2 \quad \overset{C_2H_5O}{\underset{C_2H_5O}{>}}\overset{S}{\underset{(II)}{\overset{\|}{P}}}\text{—Cl} \quad + \quad S_2Cl_2 \quad + \quad 2HCl$$

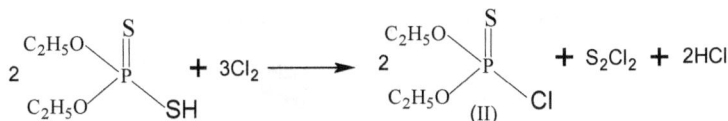

The use of phosphorus pentachloride as the chlorinating agent yields thiophosphoryl chloride as the by-product.

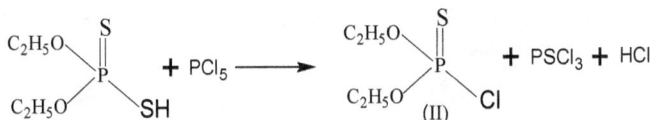

$$\overset{C_2H_5O}{\underset{C_2H_5O}{>}}\overset{S}{\underset{}{\overset{\|}{P}}}\text{—SH} \quad + \quad PCl_5 \quad \longrightarrow \quad \overset{C_2H_5O}{\underset{C_2H_5O}{>}}\overset{S}{\underset{(II)}{\overset{\|}{P}}}\text{—Cl} \quad + \quad PSCl_3 \quad + \quad HCl$$

Phosphorylation of sodium salt of p-nitrophenol with intermediate (II), parathion (I)is obtained.

Properties and uses: Technical material is a dark-brown liquid which has an unpleasant garlic odour. Pure compound is a clear oily liquid m. p. 6.1^{0}C and b. p. 113.1^{0}C at 0.05 mm of Hg. It's solubility in water is 24 mg/1. However, it is highly soluble in most organic solvents with the exception of paraffinic hydrocarbons.

Parathion is an excellent insecticide and acts both as a contact and a stomach poison. It has proved effective against a wider variety of insect than any other insecticide and is generally applied at concentrations of 0.01-0.1%. A limiting factor in the use of parathion has been its high toxicity to mammals and much effort has been developed to finding substitutes. Several closely related compounds have proved especially useful.

(vi) Methyl parathion (o, o-dimethyl o-p-nitrophenyl-phosphorothionate): It is synthesized by the condensation of (o, o-dimethyl o-p-nitrophenyl-phosphorothionate) with p-nitrophenol in the presence of an alkali:

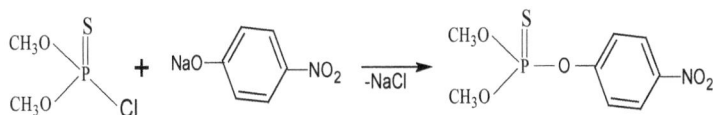

$$\overset{CH_3O}{\underset{CH_3O}{>}}\overset{S}{\underset{}{\overset{\|}{P}}}\text{—Cl} \quad + \quad NaO\text{—}\langle\!\!\!\bigcirc\!\!\!\rangle\text{—}NO_2 \quad \xrightarrow{-NaCl} \quad \overset{CH_3O}{\underset{CH_3O}{>}}\overset{S}{\underset{}{\overset{\|}{P}}}\text{—O—}\langle\!\!\!\bigcirc\!\!\!\rangle\text{—}NO_2$$

Methyl parathion

It is white solid, m. p. 36.^{0}C it is more effective than parathion against aphids and beetles. It acts as a stomach as well as contact poison. However, because of the lesser permeability of the skin of the mammals to the dimethyl esters as a compared to the diethyl esters, methylparathion is much less toxic to mammals than parathion.

(vii) Analogues of Parathion: Two chlorinated analogues of methyl parathion have greatly reduced acute mammalian toxicities and are useful as household insecticides and for certain agriculture pests.

(a) Dicapton (o-2-chloro-4-ntrophenyl (o, o-dimethyl o-p-nitrophenyl-phosphorothionate): It is white crystalline powder, m. p. 52-53 ^{0}C highly soluble in aromatic hydrocarbons, halogenated aromatic as soil as aliphatic hydrocarbons, etc. Dicapton is synthesized by the reaction of (o, o-dimethyl o-p-nitrophenyl-phosphorothionate) 2-chloro-4-nitrophenol in methyl ethyl ketone in the presence of anhydrous potassium carbonate.

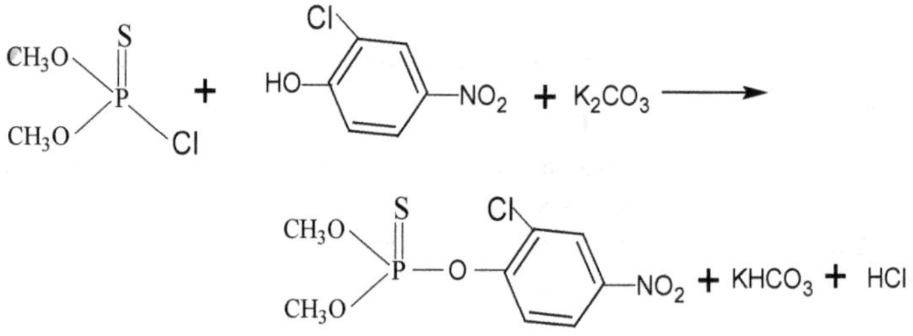

Dicapton

Dicapton is a white solid 9 m. p. 53C. It is used mainly in the form of emulsions or suspensions to control files.

(b) Chlorthion (o-3-chloro-4-nitrophenyl o, o – dimethylphosphorothionate): It is synthesized by the reaction of o, o-dimethylphosphorochloridate with 3-chloro-4-nitrophenol in methyl ethyl ketone in the presence of anhydrous potassium carbonate.

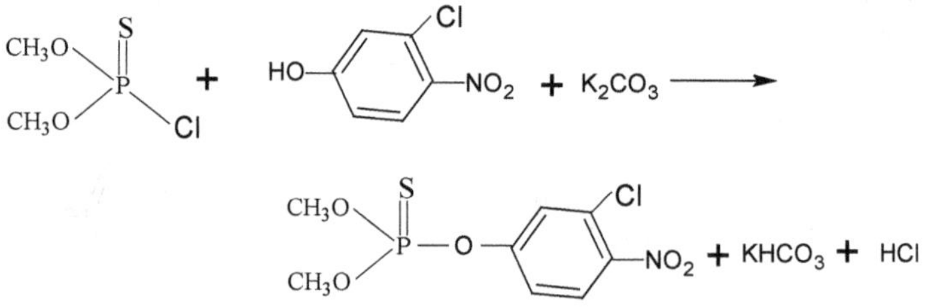

Chlorthion

It is yellow crystalline substance, m. p. 21°C. it is highly soluble in aromatic hydrocarbons and their halogen derivatives. It is somewhat less active than methyl parathion.

(viii) Abate (4-bis-(o, o-dimethyl thionophosphoryl-oxy) diphenyl sulphide: it is prepared by reacting bis (p-hydroxyphenyl-sulphide (I) with o, o-dimethyl phosphorothio chloridate (II) in the presence of aqueous sodium hydroxide.

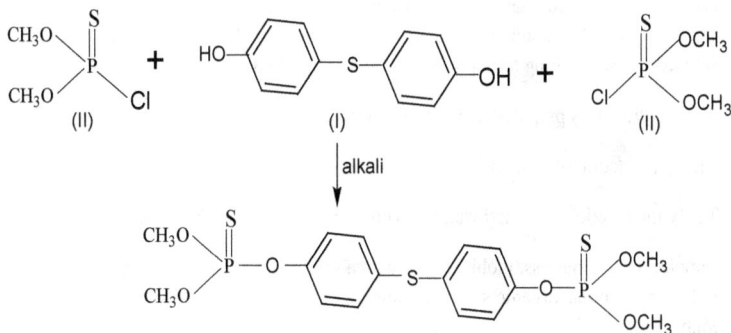

Abate

The bis-(p-hydroxyphenyl) sulphide is prepared by the reaction of phenol with sulphur dichloride.

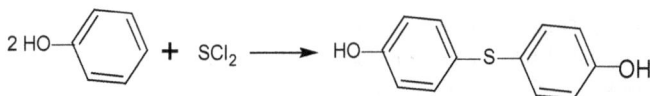

Technical grade is a brown viscose liquid insoluble in water but soluble in carbon tetrachloride, toluene etc. It is used to control mosquitoes and agricultural pests.

(e) Systematic insecticides for plants

These are the insecticides which when applied to seeds, roots or leaves of plants are absorbed and translocated to the various plants part in amounts lethal to insects feeding thereon. This method of plant protection has following advantages over the other conventional methods.

(a) These minimize the inequalities of spray coverage.

(b) This increases the length of a residual control by protection of the spray residue from attrition by weathering.

(c) This has less damaging effects on beneficial predatory and pollinating insects.

Various systematic insecticides for plants are as follows.

(i) Schradan: It is octamethyl pryophosphoramide. It is prepared by the reaction of tetramethyldiamido chlorophosphate with water in the presence of tertiary amines. It is also prepared by the reaction of ethyl tetramethyldiamidophosphate with tetramethylamidochloro phosphate at $40\,^0C$.

Technical grade Schradan is a brown liquid b. P. $154\,^0C$ at 2 mm of Hg. It is water miscible and also soluble in most organic solvent. When absorbed into plants Schradan provides long term protection from aphids and mites.

(ii) Dimefox: It is prepared by the following methods:

 (a) By the action of salts of HF on bis (dimethylamino) phosphoryl chloride.

(b) By the reaction of dimethylamine with dimethyl amidodiflurophosphate.

Dimefox is a colourless, mobile liquid with a weak odour. It is highly soluble in water and in many polar organic solvent. Dimefox is a powerful systematic insecticide for controlling sucking pests of hops and vectors of a virus disease of cacao trees. Because of its volatility, Dimefox has a short residual activity and is most suited for foliage application.

(iii) Demeton or Systox : It is a mixture of two parts of O,O-diethyl O-2(ethylthio) ethyl phosphorothionate(I) and one part of O, O – diethyl S-2 (ethylthio)ethylphosphorothionate(II).

In industry Demeton is produced by the reaction of dithylchlorothiophosphate with 2-hydroxydiethyl sulphide in the presence of an HCl acceptor. The compound (I) is partially isomerised directly to (II) in the process of preparation, and higher the reaction temperature more isomerisation occurs.

Technical grade Demeton is a yellowish liquid soluble in organic solvents. Demeton provides a long lasting systematic insecticide rapidly absorbed by roots, steam and foliage.

(iv) Dimethoate: [O, O-dimethyl S-(N-methyl carbamoyl) methyl phosphorodithionate]: It is produced by the reaction of salts of dimethyldithiophosphoric acid with N-methylchloroacetamide. This method is carried out in medium consisting of water and some organic solvent.

It is white crystalline solid, m. p. $51\,^0C$. It is soluble in organic solvents and soluble in water. Because of its water solubility it has proved especially useful as a persistent systematic for fruit larvae and for and for side dressing of soil about plants.

(v) Mevinphos or Phosdrin: It is dimethyl-2-carbomethoxy-1-methylvinyl phosphate. It exists in the form of cis and trans isomers, of which the trans isomer is the more stable. It is obtained by the reaction of sodium enolate of methyl acetoacetate with dimethylchlorophosphate.

Technically material is colourless liquid b. p. 106 -107.5 ^{0}C at 1 mm and is composed of about two parts of the trans and one part of cis-isomer. It is useful for treatment of edible produce close to harvest since it is rapidly dissipated by the volatilation and enzymatic decomposition in the plant.

(f) Derivatives of carbamic acid

Many of the more recently developed insecticides possess the carbamate structure. The carbamates are used in grain production and are involved with the inhibition of mitosis in the offending weed species. The insecticides having the dithiocarbamate group are standard fungicides of a wide variety of fungal plant diseases. The mixed oxygen-sulphur carbamates possessing the following general structure contribute important examples of herbicidal chemicals.

(i) Baygon (2-isopropoxy phenyl N-methyl carbamate): It is prepared by the reaction of methyl carbamoyl chloride or methyl isocynate with 2-isopropoxyphenol.

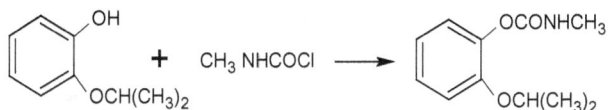

It is white solid, m. p. 91^{0}C. It is soluble in polar solvents and slightly soluble in water. It possess broad spectrum of insecticidal property and used to control of agricultural pests, household insects and insects that would menace public health.

(ii) Mesurol (4-methylmercapto-3, 5-dimethylphenyl N-methyl carbamate): It is prepared by the reaction of 4-methyl mercapto-3, 5-dimethylphenol with methyl isocynate in the presence of triethylamine. The reaction is carried out in any organic solvent like benzene.

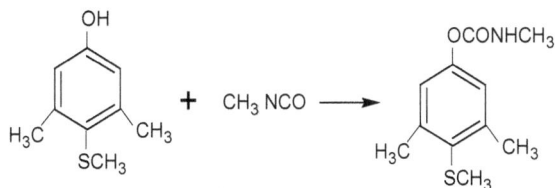

It is crystalline substance particularly odourless m. p. 121.5 ^{0}C. It is highly soluble in most organic solvents but almost insoluble in water. It is a broad spectrum insecticide for the control of pests of fruits and vegetables.

(iii) Temik (2-methyl-2-methyl thiopropionaldoxime O-N-methyl carbamate): It is prepared by the reaction of the corresponding oxime with methyl isocynate.

It is white crystalline substance m. p. 100 ^{0}C. In spite of its high toxicity, it has been poisoned as a systematic nematicide for introduction into the soil in the form of special granules.

(iv) Romate (3,4-dichloro benzyl N-methyl carbamate): It is prepared by reacting 3,4-dichloro benzyl alcohol with methyl isocynate in benzene at room temperature in the presence of dibutylene acetate as a catalyst.

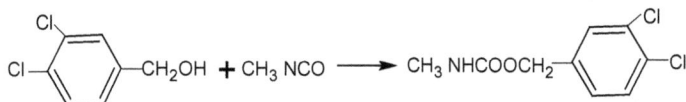

$$Cl-C_6H_3(Cl)-CH_2OH \ + \ CH_3\,NCO \longrightarrow CH_3\,NHCOOCH_2-C_6H_3(Cl)(Cl)$$

It is crystalline substance m. p. 53-54 ^{0}C, b. p. 139 ^{0}C at 0.5 mm of Hg. It is almost insoluble in water but highly soluble in most organic solvents. It is a herbicide for the control of weeds in planting of cotton, potatoes, tobacco and some other crops.

(v) Eptam or EPTC (S-ethyl N, N-di-n-propyl thiocarbamate): It is liquid with unpleasant odour b. p. 127 ^{0}C at 20 mm of mercury. It is highly soluble in most organic solvents. It is widely used as a pre-emergence herbicide to control weeds and plantings of alpha-alpha, beans, beets, carots, cabbage, flax, potatoes and many other crops.

(vi) Di-allate or Avadex (S-2,3-dichloroallyl N, N-diisopropylthiocarbamate): It is synthesized by the reaction of alkali salts of diisoproyl thiocarbamic acid with 1, 2, 3-trichloro propylene.

It is liquid b. p. 149-150 ^{0}C at 9 mm of Hg. It is miscible in all proportion with most aromatic hydrocarbons, ketones and halogen derivatives of hydrocarbons. It acts as a selective monocotyledonous weeds and some dicotyledonous. It is used for the control of wild oats in such crops as flax, barley, corn, peas, lentils, sugar, beets, beans etc.

(vii) Zineb [Zinc ethylene bis (dithiocarbamate)]: It is precipitated from aqueous solutions of diammonium or disodium ethylene bis (dithiocarbamate) by zinc sulphate. It is white crystalline substance. It is one of the most important fungicides used in agriculture.

(g) Fumigants

These are chemical substances which at a given temperature and pressure must exist in the gaseous state in sufficient concentration to be lethal to the insect pest. Compounds boiling at about room temperature such as hydrogen cyanide, methyl bromide and ethylene oxide are the most useful general fumigants.

The various fumigants often exhibit considerable specificity towards insect pests. The fumigants may be used individually or in combination. For example carbon tetrachloride is often used in combination with carbon disulphide to decreases flammability.

Citrus and decitrus fruit trees have been have been fumigated for the control of scale insects for many years by halogen cyanide introduced under relatively gaslight tents. Buildings have been fumigated by methyl bromide for the control of termites or powder post beetles.

In our villages ethylene bromide is used for controlling stored food pests. Although ethylene bromide is a liquid at ordinary temperature, it has a sufficiently high vapour pressure to display fumigant action. It is contained in glass ampoules of various capacities enveloped within an absorbent material like cotton wool. When the ampoule is crushed and introduced into the grain mass held in an air tight bin, the absorbed chemical slowly vaporizes and diffuses destroying the insect population.

The other fumigants for stored grain, ethylene dichloride (EDC) which is not toxic to insects compared to ethylene dibromide.

(h) Repellents

There are certain substances which are only mildly toxic or which may not be active poisons but they prevent damage to plants or animals by making the food or living conditions of the insects unattractive or offensive to them. Such substances are called repellents.

Repellents are generally employed when it is not possible to use an insecticide and may afford a greater or lesser degree of protection to manufactured products, growing plants or the bodies of animals and human being.

Repellents include mosquito repellent creams (based on dimethyl phthalate and or pyrethrum),mosquito coils(made from pyrethrum marc), mothballs or flakes(consisting naphthalene or p-dichlorobenzene) tablets for protecting stored grain(containing mercury) and smoke generators(formulated with an active ingredient such as lindane an oxidant and a combustible material). A very good mosquito repellent is N, N- diethyl-m-toluamide, it can be applied in alcoholic solution directly on the skin and gives pleasant lotion feeling.

The protection against various species of mosquitoes and flies as well as other biting arthropods is greatly extended by the use of mixtures such as the following.

1. Dimethylphthlate (3 parts) + Indalone (1 part) + 2- ethyl-1, 3-hexanediol (3 parts).

2. Dimethylphthlate (4 parts) + 2- ethyl-1, 3-hexanediol (3 parts) + Dimethyl carbamate(3 parts).

3. Dimethylphthlate (7 parts) + Magnesium stearate (3 parts).

4. Dimethylphthlate (25 parts) + White wax (19 parts) + Peanut oil (56 parts).

2. Odoriferous compounds

2.1 Introduction

The term perfume is originated from the Latin word *perfumare* means to fill with smoke. A perfume may be defined as a mixture of pleasantly odorous substances incorporated in a suitable solvent. Earlier, all the substances used in perfumery were of natural origin. After these synthetic compounds having the same odour as natural were discovered and recently scientists synthesized a variety of compounds having a wide range of fragrances. Chemically a perfume is composed of three ingredients *viz.* vehicle, fixative and odoriferous substances.

2.2 Vehicle or solvent

The solvent is required to keep the odoriferous substance in solution. A good solvent used in perfumery must be volatile, inert, non-irritant to human skin and almost odourless. The most commonly used solvent is the highly refined ethyl alcohol mixed with more or less water according to the solubility of the oil employed. The solvent (alcohol) with its volatile nature carries the odoriferous substances to the cells of the nose. The slight natural odour of the alcohol is deodorized by adding a small amount of gun benzoin or other resinous fixatives to it and allowing keeping for a week or two week.

2.3 Fixative

Generally, a perfume has two or more odoriferous substances of different volatility. In such cases the more volatile fragment gives in odor first, then the less volatile and so on and so forth; the result is that such perfumes will give a series of impressions (odors) rather than the desired fragrance. To overcome this difficulty, a fixative is added. Fixatives are the substances of lower volatility than the perfume oils. Their main function is to equalize the rate of evaporation of the various odors constituents of the perfume by retarding or increasing their volatilities. All the fixatives are classified into four groups, *viz.* animal fixatives, resinous fixatives, essential oil fixatives and synthetic fixatives. The important members of the various groups are given below.

(a) Animal fixatives

Sr. No.	Name of the oil	Source	Remark	Components
1.	Castor or castoreum	Excudate of the perineal gland of beaver.	Under in largest quantity.	Benzyl alcohol, Acetophenone, *l*-borneol and

				castorin.
2.	Civet	Secretion of the perineal gland of the civet cats.	Due to the presence of skatole the crude civet has disagreeable odor, which disappear on dilution and aging.	Civetone (acyclic ketone).
3.	Musk	Secretion of the perineal gland of the male musk deer.	Most useful animal fixative.	Muscone (acyclic ketone).
4.	Ambergris	Secretion of certain whales.	Although it is the best known animal fixative, yet least used. It is used as tincture.	Ambrein (a triterpe noid) in 80-85% and ambergris oil (active components) in 12-15%.
5.	Muskzibata	Glands of the musk rate.	Latest animal fixative.	Macrocyclic ketones.

(b) Resinous fixatives

Resinous fixatives are the normal or pathological excudates of certain plants. They are more important historically rather than commercially. The various important resinous fixatives are given below;

(i) Hard resin, e.g. benzoin.

(ii) Softer resins, e.g. myrrh and labdanum.

(iii) Moderately soft resins, e.g. *perubalsam, tolubalsam* etc.

(iv) Oleoresins, e.g. terpenoids.

(v) Extracts from resins, e.g. *ambrein.*

(c) Essential oil fixatives

Certain essential oils are used for their fixatives as well as odoriferous properties. The important examples are *clary sage, vetiver, patchouli, orris* and *sandalwood.* The boiling points of these oils are in the range of 285-290 ^{0}C, a higher range than that of normal essential oils.

(d) Synthetic fixatives

Synthetic fixatives may be grouped under two headings according to their applications.

(i) Those which are used only as fixatives to replace some of the animal fixatives. They are high boiling odorless esters. The important examples are glyceryl diacetate (b. p. 259 ^{0}C), ethyl phthalate (b. p. 295 ^{0}C) and benzyl benzoate (b. p. 323^{0}C).

(ii) Those which function as fixatives as well as odorous substances, i.e. in addition to their fixative properties they also impart their fragrances to the perfume in which they are used. Some of the important examples are; Amyl benzoate, phenethyl phenylacetate, cinnamic alcohol esters, acetophenone, musk ketone, musk ambrette, benzophenone, vanillin, coumarin, heliotropin, indole skatole, hydroxycitronellal.

It must be noted that any of the fixatives of any of the four main groups may or may not contribute to the odor of the finished product, but if they do so, they must blend with and complement the main odor.

2.4 Odorous substances

Most of the odorous substances used in perfumery may be grouped under either of the following headings.

(i) Essential oils

(ii) Isolates

(iii) Synthetics and semisynthetics

(i) Essential oils: Essential oils are volatile, odoriferous substances obtained from the flowers, fruits, leaves and roots of many plants. They are usually a mixture of different types of compounds, *viz.* esters alcohols, aldehydes, ketones, acids, phenols, lactone and hydrocarbons and are optically active. Some of the important method for their extraction is;

(a) Expression method: In one of the important expression methods, which is adopted in California, the plant material is crushed and the juice is screened to remove the coarse particles. The oil is separated from the extract by centrifuge of high speed; but by this method only half of the oil is extracted and the rest half of the oil remains in the residue which is utilized for the isolation of inferior quality of oil by distillation. The various oil which are extracted by this method are: citrus oil, lemon oil, grass oil etc.

(b) Steam distillation: This is the most commonly used method and based upon the steam volatile property of the essential oils. The material is taken in suitable state of division and steam distilled. The essential oil are separated from the distillate with the help of the purified volatile solvents, *viz.* light petroleum which is later on removed by evaporation. But this method suffers from some disadvantages:

(a) It can't be employed for the extraction of heat sensitive oils.

(b) Ester which are constituents of some natural oils and are responsible for the fragrance may be hydrolyzed by steam to give non or less fragrant products.

(c) Extraction by means of volatile solvents: This method is also commonly used in perfumery industry; especially when the steam distillation method is fails. The plant material is directly treated with volatile solvents (ether or petroleum ether) at room temperature. The filtrate is evaporated to remove the solvent leaving behind the essential oils.

(d) Effleurage process: This method is widely used in Southern France. It consists in exposing the petals of the flowers on the glass plates, convert with fat (mixture of lard and tallow) for several days. The old petals are replaced by the fresh and after some weeks the fat becomes rich in essential oils of the flowers. Now the petals are removed and the fat is extracted with ethanol; the extract is decanted leaving behind fat which is dried and reused. The alcohol is evaporated from the extract at 0 ^{0}C in vacuum to give essential oils. Now a day's fat is replaced by activated coconut charcoal, which has some advantages on fat as an adsorbent: (i) it provides more surface to the oils for adsorption. (ii) Chemically it is more stable than fat and hence the odour of the oil is maintained. (iii) The oils obtained by this method are free from glyceride of fats which are extracted along with essential oils in effleurage process.

The important essential oils along with their chief constituents and method of isolation are given below.

Sr. No.	Name of an oil	Method of isolation	Chief constituents
1.	Bitter almond	Steam distillation	Benzaldehyde, HCN
2.	Bay	Steam distillation	Eugenol
3.	Bergamot	Expression	Linalyl acetate, linalool
4.	Caraway	Steam distillation	Carvone, d-limonine
5.	Cassia	Steam distillation	Cinnamic aldehyde
6.	Cedarwood	Steam distillation	Cadrene, cedral
7.	Cinnamon	Steam distillation	Cinnamic aldehyde, eugenol
8.	Citronella	Steam distillation	Geraniol, Citronellol
9.	Clove	Steam distillation	Eugenol
10.	Coriander	Steam distillation	Linalool, pinene
11.	Eucalyptus	Steam distillation	Cineole
12.	Geranium	Steam distillation	Geraniol esters, citronellol
13.	Jasmine	Effleurage	Benzyl acetate, linalool and esters
14.	Lavender	Distillation	Linalool
15.	Lemon	Expression	d-Limonene, citral
16.	Sweet orange	Expression	d-Limonene
17.	Peppermint	Steam distillation	Menthol and its esters

18.	Rose	Effleurage	Geraniol and citronellol
19.	Sandalwood	Steam distillation	Santalol
20.	wintergreen	Steam distillation	Methyl salicylate

Isolates: Isolates are pure chemical compounds obtained either from an essential oil or other natural perfume material. The important example of isolates is eugenol from the clove oil, pinene from turpentine, and ethanol from anise oil and linalool from *linaloa oil*.

Synthetics and semisynthetics: Some odors substances are completely synthetic where as others are prepared from an isolate or other natural starting materials. The later are known as semisynthetics, some important examples of semisynthetics odorous substances are vanillin, prepared from eugenol from clove oil; ionone prepared from citral from lemon grass oil and terpineols from turpentine and pine oil. Only the most important synthetic and semisynthetics are discussed below.

2.4.1. Esters

(i) **Anthranilates:** Walbaum in 1899 reported the occurrence of methyl anthranilates in oil of neroli. Although anthranilic acid esters possess powerful and pleasant odors, they are not used in perfume compositions because the amino group in the molecule tends to react with various aldehydes to form highly colored Schiff bases and thus large amount of anthranilic acid esters cause decolouration. The problem of decolouration is reduced by the use of N, -methylmethyl anthranilate. But since the N-methyl derivatives possess very little odour, they are not of much value in perfumery.

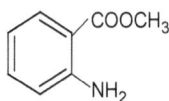

Methyl anthranilate   N-methyl methyl anthranilate

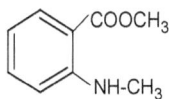

Anthranilic acid is manufactured from phthalimide in the following way:

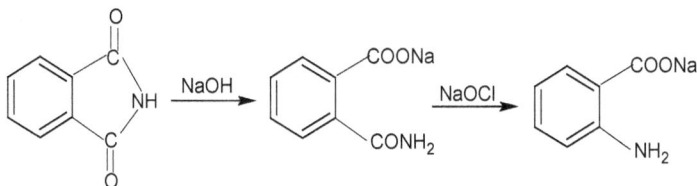

Methyl anthranilate is used in soaps and in floral blending, *viz* .orange flower, gardenia and jasmine.

(ii) Ester of benzyl alcohol

Although benzyl alcohol itself is of limited value in perfumery, its ester s is highly fragrant substances and is used in many types of compositions. Benzyl acetate has a fresh floral stable odor and is used in very large quantities. Benzyl propionate and isobutyrate are oftenly used to improve the quality of jasmine fragrances. Benzyl benzoate is used as a solvent and fixative. Benzyl cinnamate and salicylate are stable fixative fragrances.

(iii) Ester of cinnamic acid

Although cinnamic acid itself is not an important perfumery material, its various esters, *viz.* methyl, ethyl, benzyl and cinnamyl have pleasant long lasting and stable odors. Cinnamic acid is synthesized by the well known Perkin or Knoevenagel reaction.

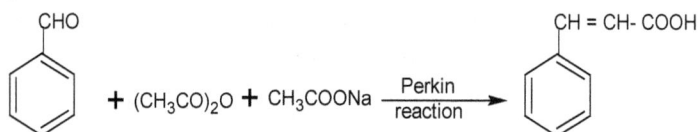

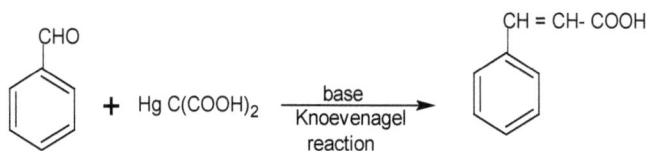

(iv) Salicylates

Among the salicylates, amyl salicylate is the most important esters. It is a low priced material and possesses a long lasting trifle type odor. It is used in many types' fragrances. Benzyl salicylate is used as a solvent and fixative for artificial (nitro) musk. Methyl salicylate (oil of wintergreen) is used as a flavoring ingredient and as a soap perfume. Salicylic acid is produced by Kolbe's synthesis.

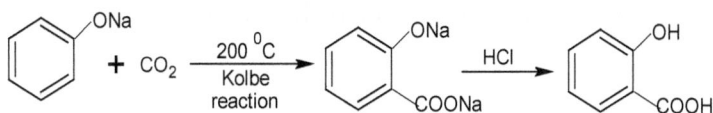

2.4.2. Alcohols

(i) Aliphatic alcohol

Aliphatic primary alcohol having seven to twelve carbon atoms possess pleasant odor and have been used in the perfumery since a long time. Among the unsaturated aliphatic alcohols used in perfumery the most important are 2-hexene-1-ol, 3-hexene-1-ol, 1-octene-3-ol and 2,6-nonadiene-1-ol.

(ii) Phenyl ethyl alcohol

It is one of the most widely used of the aromatic perfumery compounds. Because of its mild, pleasant and persistent Rose odor. It is frequently used in rose and many other types of perfume compositions. Its stability towards alkali makes it especially suitable for imparting perfume to various types of soap and cosmetics. It occurs in the volatile oils of rose, orange flowers, and others. Although there are a large number of methods are given below; out of these two, one involves Grignard reaction and other Friedel-Craft reaction.

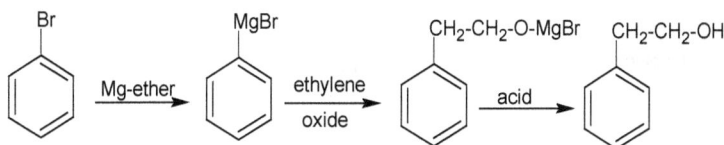

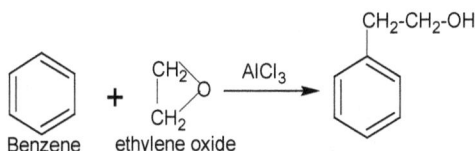

Several esters of phenyl ethyl alcohol are also used in perfume.

(iii) Citonellol (rhodinol): Citrenellol occurs in rose, germanium and citronella oils. It possesses the rose like odor. Citronellol and especially its esters, *viz.* formate, acetate

isobutyrate are frequently used in soap, detergent and cosmetics. On commercial scale citronellol is prepared by the catalytic reduction of citranellal.

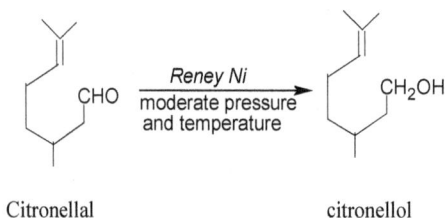

Citronellal   citronellol

(iv) Terpineols : Like the ionones and irones the terpenoids also exist in three isomeric forms, *viz.* α, β and γ, α-Terpeneols occurs frequently in natural sources, *viz.* yellow pine oil, oil of cajeput, niaouli etc.

Terpeneols have pleasant lilac-type stable odor. It is cheapest synthetic. Earlier, all terpeneols were used to be obtained from turpentine oil, which consists largely of α-pinene, but recently pine oil, terpeneols are obtained by following methods.

(a) One step process: The pine oil of α-pinene is heated with sulphuric acid and acetone for six hours at 35-40 ^{0}C and the product is purified by fractional distillation.

(b) Two step process: In this process the first step is the formation of a terpine hydrate from α-pinene and sulpuric acid in the presence of an emulsifying agent. The terpine hydrate is very easy to purify as compared to that of terpeneol. In the second step the purified terpine hydrate is dehydrated to terpineol by using oxygenated carboxyl acids.

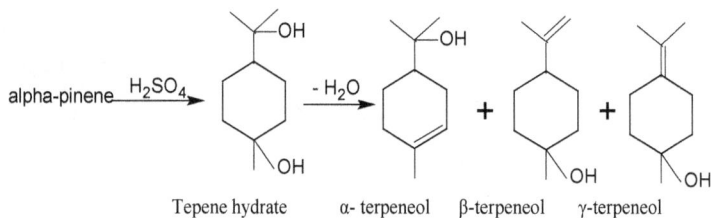

Tepene hydrate   α- terpeneol   β-terpeneol   γ-terpeneol

(v) Linalool: Linalool and its esters impart a type of fragrances which no other material can provide. Linalool has a soft sweetness odor. Linalool and its esters, *viz.* formate, acetate, propionate and isobutyrate are used in soap, detergent and cosmetics. Linalool and its esters are also found to be present in various natural and artificial flavorings in beverages, candies, ice creams and backed goods. Although linalools occur in the essential oils of several plants, *viz.* linae oil rose oil, orange oil, ylang oil and bergamot oil; the synthetic linalools are pure. Commercially the linalool is synthesized either from naturally occurring pinene or entirely synthetically through a series of complex reactions.

From β-pinene

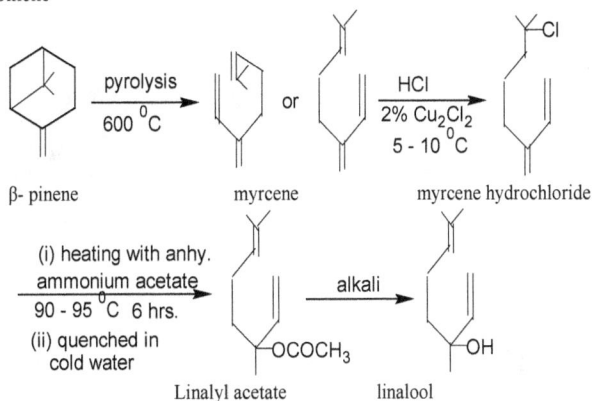

β- pinene   pyrolysis 600 °C   myrcene   or   HCl 2% Cu_2Cl_2 5 - 10 °C   myrcene hydrochloride

(i) heating with anhy. ammonium acetate 90 - 95 °C 6 hrs. (ii) quenched in cold water

Linalyl acetate   alkali   linalool

From acetone and acetylene

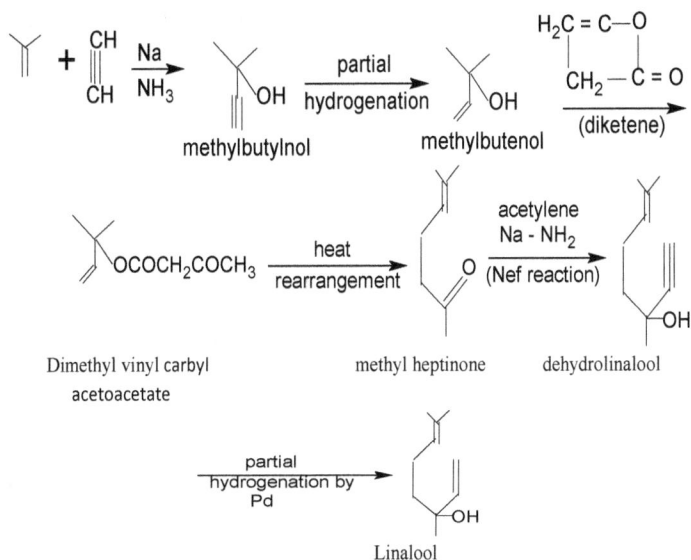

methylbutylnol   partial hydrogenation   methylbutenol $H_2C = C-O$ $CH_2-C = O$ (diketene)

Dimethyl vinyl carbyl acetoacetate   heat rearrangement   methyl heptinone   acetylene Na - NH_2 (Nef reaction)   dehydrolinalool

partial hydrogenation by Pd   Linalool

(vi) Geraniol and Nerol: These two isomeric primary alcohols occur much more widely in nature; the important natural sources for geraniol are rose oil and Turkish geranium oil, similarly the important sources for nerol are neroli oil and bergamot oil. Most of the geraniol and nerol today are prepared synthetically. Geraniol is one of the most widely used perfumery chemicals in soaps, detergent and cosmetics. It possess fresh rose like stable and non-discoloring fragrance in desired. Its lower esters are also used in perfumery. The commercial geraniol and nerol are prepared from β-pinene.

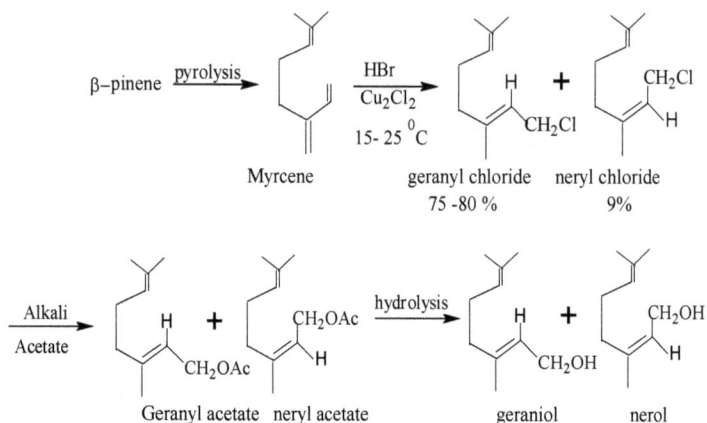

β–pinene $\xrightarrow{\text{pyrolysis}}$ Myrcene $\xrightarrow[\text{15- 25 }^0\text{C}]{\dfrac{\text{HBr}}{\text{Cu}_2\text{Cl}_2}}$ geranyl chloride 75 -80 % + neryl chloride 9%

$\xrightarrow[\text{Acetate}]{\text{Alkali}}$ Geranyl acetate + neryl acetate $\xrightarrow{\text{hydrolysis}}$ geraniol + nerol

2.4.3. Ketones

(i) Civetone: Civetone occurs in scent gland of African *civet cat* and is the cause of civet odour. It was isolated by Sack from the civet cat. Ruzika established its structure and suggested it to be a macrocyclic ketone. The civetone can synthesized by a number of methods, the most important of which is acyloin synthesis.

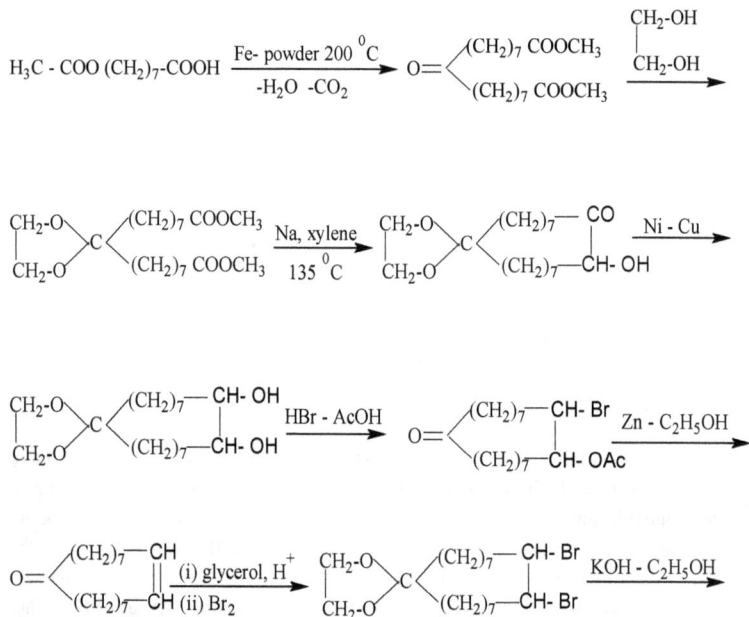

$H_3C - COO\,(CH_2)_7-COOH \xrightarrow[-H_2O\ -CO_2]{\text{Fe- powder 200 }^0\text{C}} O{=}C\begin{cases}(CH_2)_7\ COOCH_3 \\ (CH_2)_7\ COOCH_3\end{cases} \xrightarrow{\overset{CH_2-OH}{\underset{CH_2-OH}{|}}}$

$\begin{matrix}CH_2-O \\ CH_2-O\end{matrix}{>}C\begin{cases}(CH_2)_7\ COOCH_3 \\ (CH_2)_7\ COOCH_3\end{cases} \xrightarrow[135\ ^0\text{C}]{\text{Na, xylene}} \begin{matrix}CH_2-O \\ CH_2-O\end{matrix}{>}C\begin{cases}(CH_2)_7 - CO \\ (CH_2)_7 - CH-\ OH\end{cases} \xrightarrow{\text{Ni - Cu}}$

$\begin{matrix}CH_2-O \\ CH_2-O\end{matrix}{>}C\begin{cases}(CH_2)_7 - CH-\ OH \\ (CH_2)_7 - CH-\ OH\end{cases} \xrightarrow{\text{HBr - AcOH}} O{=}C\begin{cases}(CH_2)_7 - CH-\ Br \\ (CH_2)_7 - CH-\ OAc\end{cases} \xrightarrow{\text{Zn - C}_2\text{H}_5\text{OH}}$

$O{=}C\begin{cases}(CH_2)_7 - CH \\ (CH_2)_7 - CH\end{cases} \xrightarrow[\text{(ii) Br}_2]{\text{(i) glycerol, H}^+} \begin{matrix}CH_2-O \\ CH_2-O\end{matrix}{>}C\begin{cases}(CH_2)_7 - CH-\ Br \\ (CH_2)_7 - CH-\ Br\end{cases} \xrightarrow{\text{KOH - C}_2\text{H}_5\text{OH}}$

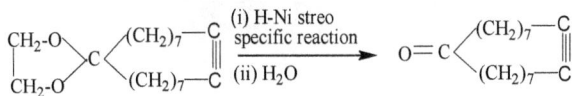

Cis- civetone

Civetone has been used in perfumery, but recently it is largely replaced by replaced by cyclopetadecanone which is used under the name of *exaltone* in perfumery. Exaltone is synthesized from thapsic acid chloride.

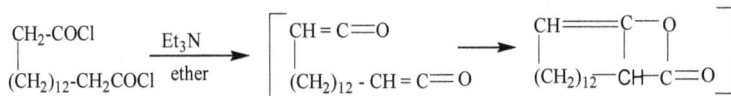

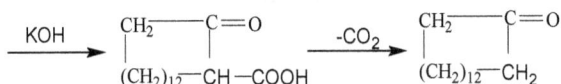

Exaltone

(ii) Muscone: Muscone is the active principle of Tibetian musk. Like the civetone it is also macrocyclic ketones. It is synthesized as follows.

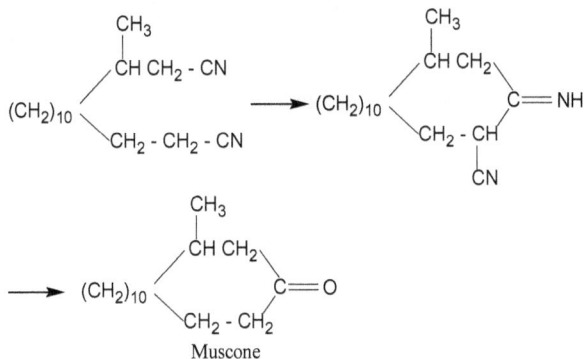

Muscone

Muscone is largely used in perfumery.

(iii) Artificial (nitro) Musk : Various synthetic compounds are found to possess the musk like odour and their use in perfumery is so frequent that nearly all the perfumes have some amount of nitro musk. The important nitro musks are described below.

(a) Musk xylene and musk ketone: Both of these artificial musks are synthesized from m-xylene.

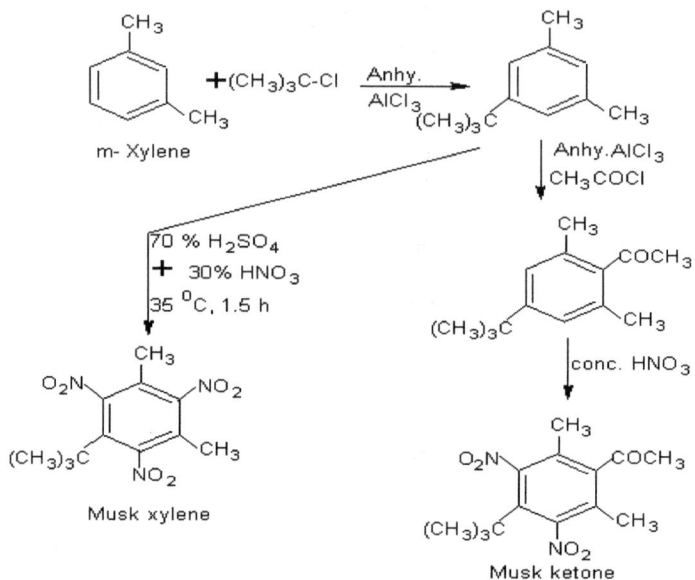

Musk xylene is a powerful natural musk like odour and is extensively used; while the musk ketone is the closest in odour to the natural musk. Musk xylene is an economical sweetener for floral soap and detergent perfumes. It accounts for nearly half of the nitro musks used in the perfumery.

(b) Musk ambrette: It is synthesized from m-cresol.

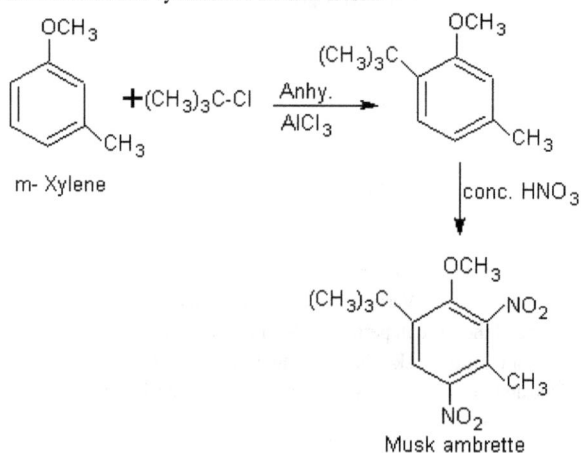

Among all the nitro musks, musk ambrette is the strongest in odour and is widely used in soaps. The most serious disadvantages of musk ambrette is its discolouring characteristics in the presence of heat or light.

(c) Moskene: It is prepared from p-cymene.

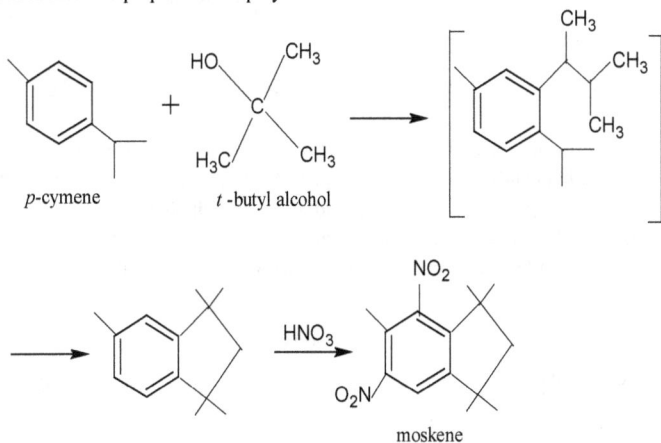

p-cymene	t-butyl alcohol

moskene

Since the moskene has musk ambrette like odour and discolourises moderately, it is extensively used to obtain the musk ambtrette odour.

Other important artificial musks are musk tibetine and phantolid.

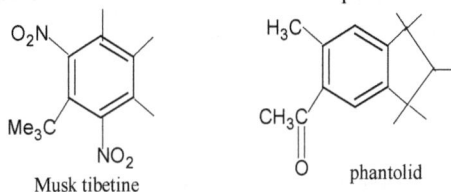

Musk tibetine   phantolid

(iv) Coumarin: Coumarin occurs widely in nature (nearly in 66 plants) but the chief natural source of coumarin is Tonka beans. However, commercially coumarin is the synthetic product. The synthetic coumarin may be obtained in a number of ways; commonly used method involves the Perkin reaction.

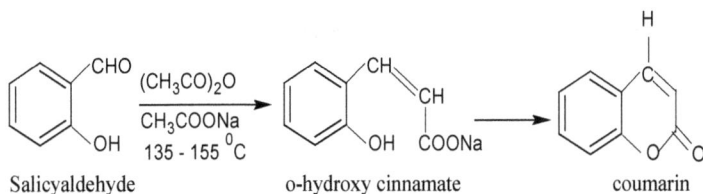

Salicyaldehyde   o-hydroxy cinnamate   coumarin

Coumarin is also prepared commercially by heating phenol with malic or fumaric acid in presence of sulphuric acid or zinc chloride.

$$\text{C}_6\text{H}_5\text{OH} + \underset{\text{CH - COOH}}{\overset{\text{CH - COOH}}{|}} \xrightarrow{\text{H}_2\text{SO}_4} \text{coumarin} + \text{CO} + 2\text{H}_2\text{O}$$

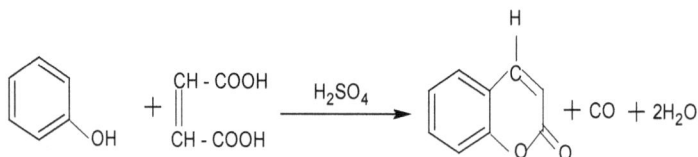

The addition of catalytic amount of pyridine is reported to increase the yield of coumarin.

Its application to impart sweet hay like odour is almost universally accepted. Coumarin is frequently used as a masking agent for disagrreable odour in industrial products, *viz.*, soaps, detergents, plastics and rubber household materials.

(v) Ionones: There are two important isomers of ionones; α- and β- both of which occur naturally in *boromia megastigma oil* but usually they are obtained as follows:

(a) From citral

$$\text{citral (CH - CHO)} + \text{CH}_3\text{CO CH}_3 \xrightarrow{\text{Ba(OH)}_2} \phi\text{-ionone (CH - CH = CH CO CH}_3)$$

$$\xrightarrow[\text{+ 2H}_2\text{O}]{\text{H}_2\text{SO}_4\text{-glycerol}} \left[\begin{array}{c} \overset{\text{OH}}{|} \\ \text{CH}_2 \text{ - CH = CH CO CH}_3 \\ | \\ \underset{\text{CH}_3}{\overset{\text{OH}}{|}} \end{array} \right] \xrightarrow{\text{- 2H}_2\text{O}}$$

β –ionone (CH = CH CO CH₃)   +   α –ionone (CH = CH CO CH₃)

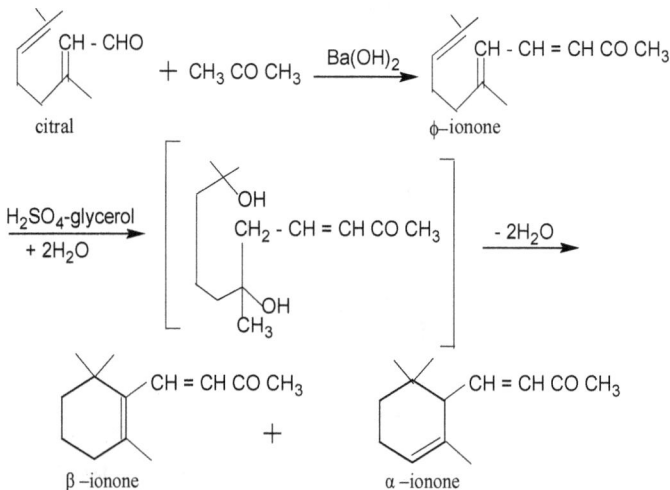

(b) From dehydrolinalool: Commercially dehydro linalool is converted into ionone by the following two methods.

(i) By the pyrolysis of dehydralinalool acetoacetate (Teisseire, 1957).

dehydrolinalool + $CH_3\,CO\,CH_2COOC_2H_5$ → dehydrolinalyl acetoacetate

$\xrightarrow{\text{pyrolysis}}$

$\xrightarrow{H_2SO_4\text{ - glycerol}}$

β –ionone   +   α –ionone

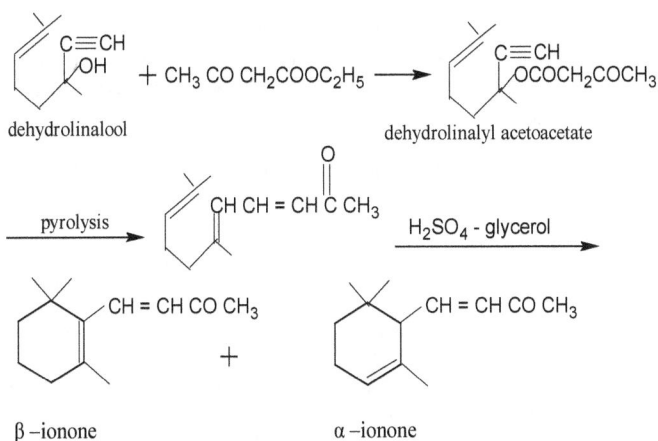

(ii) Dehydrolinalool when treated with isopropenyl ethyl ether in an acidic medium gives allenic ketones which yields pseudoionone on treatment with traces of alkali (Saucy and Marbet, 1960).

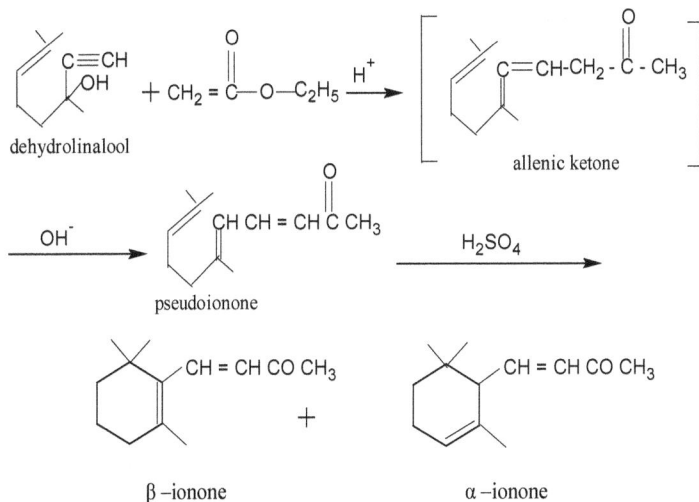

dehydrolinalool + $CH_2 = C-O-C_2H_5$ (with O) $\xrightarrow{H^+}$ [allenic ketone]

$\xrightarrow{OH^-}$ pseudoionone

$\xrightarrow{H_2SO_4}$

β –ionone   +   α –ionone

Although all commercial samples of ionones are mixtures of ionones containing a preponderance amount of one of the isomers, they can be separated from each other by steam distilling their bisulphate solutions when the β-ionone bisulphate product remains in solution and the α-ionone derivatives distilled over. The bisulphate derivatives can be easily converted into ionones by treatment with sodium carbonate solution. Moreover, the yield of either of the ionones can be increased by changing the nature of cyclising agent of pseudo ionone, e. g. when pseudo ionone is treated with sulphuric acid β-ionone predominates in mixture but if sulphuric acid is replaced by phosphoric acid α-ionone is the main product.

Recently in 1962 Theimer have isolated another ionone γ-ionone from the mixture of α-ionone and β-ionone.

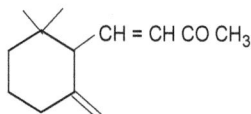

γ- ionone

Ionones possess the odour of cedar wood but the dilute alcoholic solution of ionones has the odour of violets. A large amount of ionones is consumed annually in the perfume, soap and cosmetics industries. Ionones from the important constituent of almost all the practically used perfumes and indeed it is difficult today to find a formula of perfume in which appropriate amount of ionones are not present.

(vi) Methyl ionones

These are prepared by the condensation of citral with methyl ethyl ketone (CH_3 CO CH_2CH_3) followed by cyclization. Now since the condensation of citral with acetone forms the three isomeric ionones. The condensation of citral with methyl ethyl ketone will form six isomeric methyl ionones; three each by the condensation of the aldehyde group of citral with either the methyl or methylene group of methyl ethyl ketone.

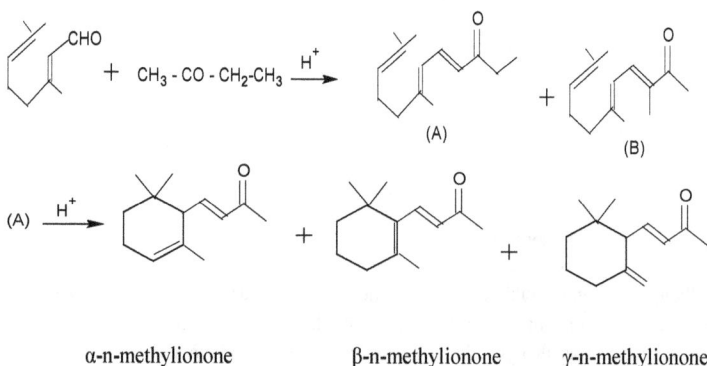

α-n-methylionone   β-n-methylionone   γ-n-methylionone

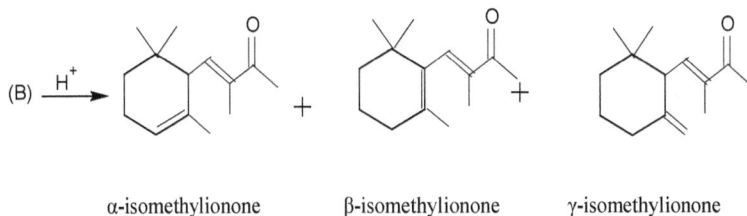

(B)

α-isomethylionone   β-isomethylionone   γ-isomethylionone

Among the six possible isomers the α-iso-methylionone is by far the most important. It has the finest odour quality and is the most of orris and violet like odour of all the methyl ionones and is sold under the various trade names. The odours of all the ionones and methyl ionones are given in the following table.

Name of the ionones	Odour
α- Ionone	Powerful floral odour, which becomes violet like on dilution. Its odour is preferred to that of β-ionone.
β- Ionone	Cedar wood, violets odour on dilution. Its odour is fruiter than the α- Ionone.
γ- Ionone	Similar to that of α- Ionone
α-n-methylionone	Dull, fruity, wood and orris root. Its odour is the weakest among all the methyl ionones.
α-iso-methylionone	Its odour is two or three times more than that of α-n-methylionone. It is the most important among all the methyl ionones.
β-n-methylionone	Its odour is lithery and raspberry like, but relatively weak in intensity.
α-iso-methylionone	Its odour is woody and vetivery like. It is generally mixed with other ionone.
γ-Ionone	Similar to that of α-isomer.

(vii) Irones: Like ionones, irones also exist in three isomeric forms, *viz.,* α, β and γ-irones, the last isomer predominates in the mixture which occurs in the oil of orris root *(Iris florentina).* Irones also possess the odour of violet and hence used in perfumery. Ionones are synthesized in the following manner.

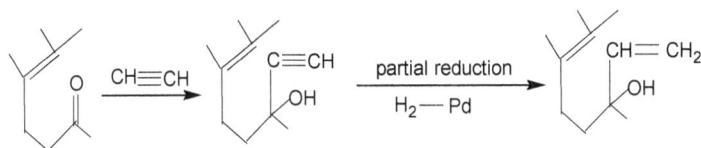

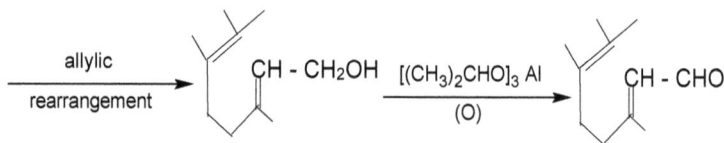

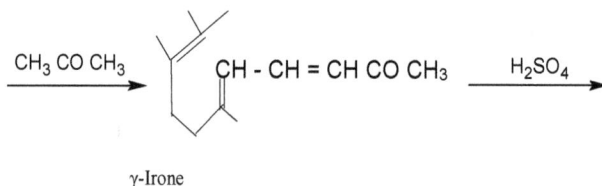

γ-Irone

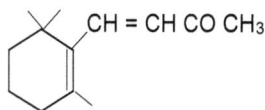 +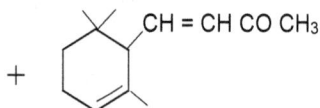

β-Irone   α-Irone

2. 4. 4 Aldehydes

(i) Aliphatic aldehyde: Fatty aldehyde is used nearly in all the perfumes. Among the important saturated aliphatic aldehydes the following are more important.

Name	Formula
Octyl aldehyde	$CH_3(CH_2)_6CHO$
Nonyl aldehyde	$CH_3(CH_2)_7CHO$
Decyl aldehyde	$CH_3(CH_2)_8CHO$
Undecynyl aldehyde	$CH_2 = CH(CH_2)_8CHO$
Undecylic aldehyde	$CH_3(CH_2)_9CHO$
Lauric aldehyde	$CH_3(CH_2)_{10}CHO$
Methylnonyl acetaldehyde	$CH_3(CH_2)_8 CH(CH_3)CHO$

(ii) Phenyl acetaldehyde: Phenyl acetaldehyde and its homologus are very essential in perfumery for the hyacinth and jonquil fragrance. Because of the instability of the parent compounds, sometimes its acetals are used. It is synthesized by the following methods.

(i)

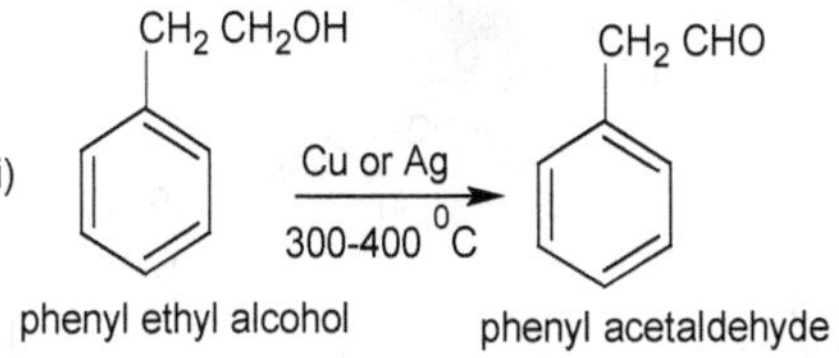

$CH_2 CH_2OH$

$\xrightarrow[300\text{-}400\ ^0C]{Cu\ or\ Ag}$

$CH_2 CHO$

phenyl ethyl alcohol   phenyl acetaldehyde

(ii)

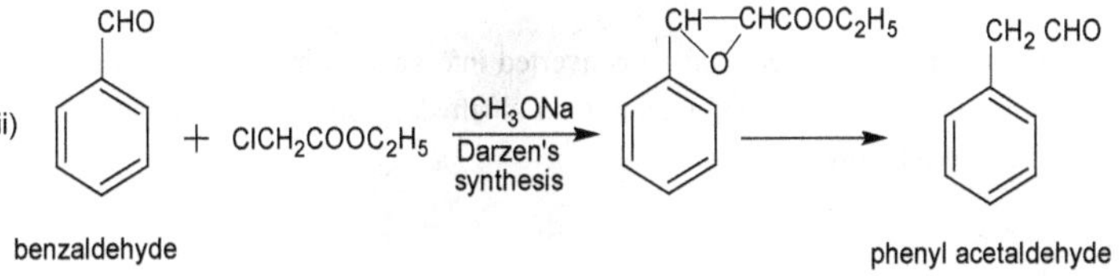

CHO

$+ ClCH_2COOC_2H_5$ $\xrightarrow[\substack{Darzen's \\ synthesis}]{CH_3ONa}$

$CH\!-\!CHCOOC_2H_5$ (with O bridge)

$\longrightarrow$

$CH_2 CHO$

benzaldehyde   phenyl acetaldehyde

Hydratropic aldehyde: The higher homologue of phenyl acetaldehyde is also synthesized with the help of Darzen's glycidic ester synthesis.

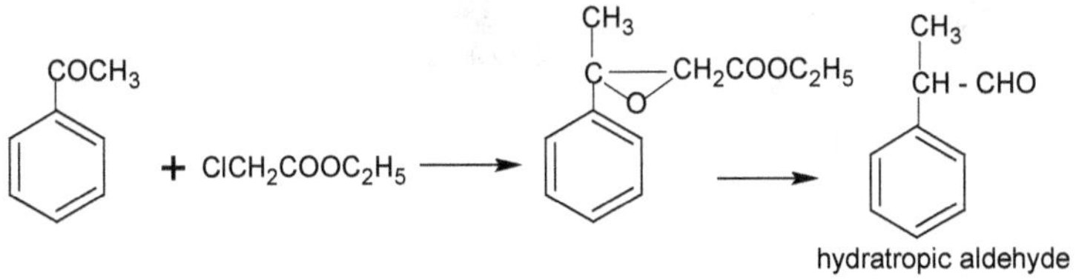

$COCH_3$

$+ ClCH_2COOC_2H_5 \longrightarrow$

$\underset{C}{\overset{CH_3}{|}}\!\!-\!\!CH_2COOC_2H_5$ (with O bridge)

$\longrightarrow$

$\underset{CH\text{-}CHO}{\overset{CH_3}{|}}$

hydratropic aldehyde

(iii) Vanillin: Vanillin is one of the most widely used flavours. It is used in a chocolate, candy, bakery products, ice creams and perfumery. Due to its good flavour it is also used as a valuable deodourant to mask the unpleasant odour of many manufactured goods, *viz.* wearing apparel, rubber goods, paper products and plastics. It must be noted that due to its serious discolouring property it is not used in soaps and most of the cosmetics. In addition to its use in the perfume and flavour industry, it serves as a starting material for the manufacture of pharmaceuticals. It is synthesized by the variety of methods.

(a) From eugenol: Eugenol occurs in the oil of cloves to the extent of 80-90 %. For converting eugenol into vanillin it is first isomerised to isoeugenol by meance of alkali which is then oxidized to vanillin, using nitrobenzene as the oxidizing agent.

CH$_2$CH=CH$_2$ CH=CH-CH$_3$

eugenol   KOH →   isoeugenol (2 moles)

OCH$_3$ OH

2 [benzene ring]–NO$_2$, (O) → 2 [CHO / OCH$_3$ / OH] Vanillin   + 2 CH$_3$ CHO + azobenzene

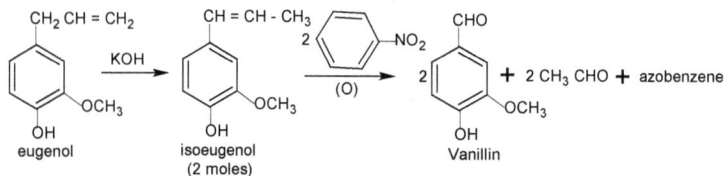

(b) From guaiacol: Guaiacol can be converted into vanillin by various methods, *viz.* Reimer-Tiemann synthesis and Gattermann aldehyde synthesis, but more promising method is given below.

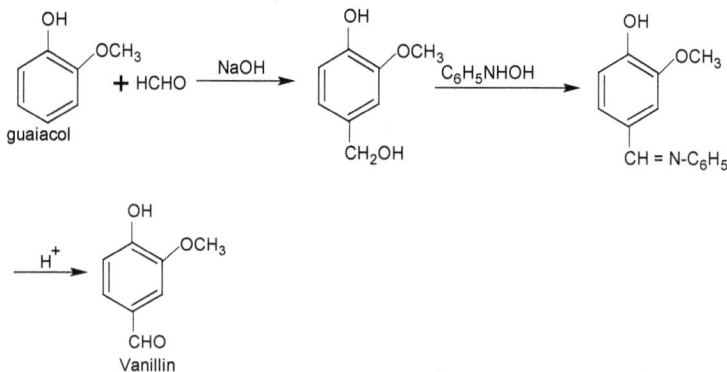

OH / OCH$_3$ guaiacol   + HCHO —NaOH→ OH / OCH$_3$ / CH$_2$OH —C$_6$H$_5$NHOH→ OH / OCH$_3$ / CH=N-C$_6$H$_5$

—H$^+$→ OH / OCH$_3$ / CHO   Vanillin

(c) From lignin: This is the most important method for manufacturing vanillin. In this process lignin is isolated the from sulphite waste liquor obtained in the paper industry during the manufacturing of paper pulp. Actually in the manufacture of paper pulp the chips of wood are digested with a solution of calcium bisulphate and sulphurous acid during which the lignin (binding material of cellulose fibers) is dissolved as calcium lignosulphonate. The solution obtained which possess more than 5% of calcium lignosulphonate and known as sulphite waste liquor.

The calcium lignosulphonate obtained in the above is treated with sodium salt, *viz.* NaOH, Na$_2$CO$_3$, Na$_2$SO$_3$ or Na$_2$SO$_4$ when sodium lignin sulphonic acid is obtained along with the formation of inorganic calcium salt. The calcium salt is removed and additional amount of NaOH is added to sodium lignin sulphonic acid solution and the solution is cooked under pressure for 1-2 hours at 130-200 lbs. During the cooking nearly 2 to 3 % of the lignin is converted vanillin which is present as sodium vanillate. The sodium vanillilate extracted by the meance of n-butyl alcohol, the solvent is removed by distillation. The sodium vanillate is now acidified by meance of SO$_2$ when vanillin is converted to the soluble vanillin bisulphite addition product. Now the free vanillin can be obtained by the acidification with sulphuric acid followed by expulsion of SO_2. The vanillin thus produced can be purified by recrystalisation, distillation or as a sodium bisulphite addition product.

(iv) Anisaldehyde or anisic aldehyde: In small amounts it occurs in anise, fennel, cassia, acasia and goldlacks oil. Its odour is resembles with that of Coumarin particularly on dilution. It can be prepared by the various methods, out of which first is the most important .

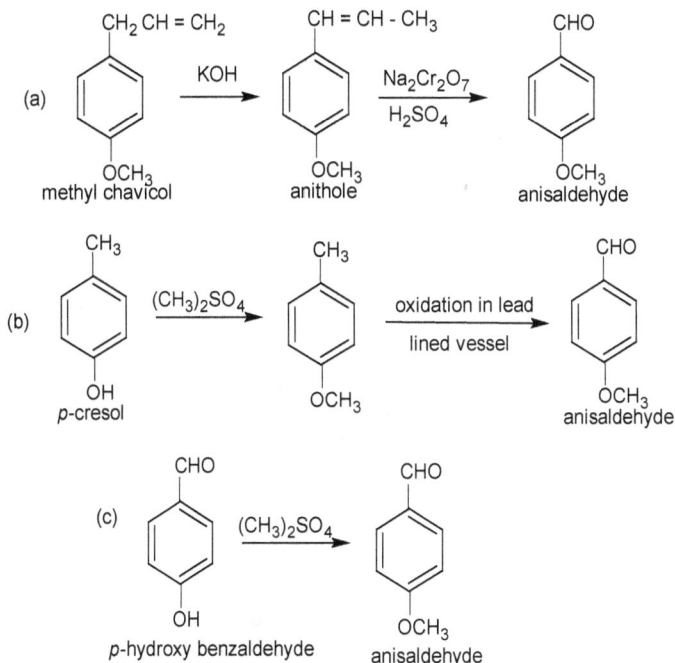

(a)

$CH_2 CH = CH_2$ / OCH_3
methyl chavicol

$\xrightarrow{KOH}$

$CH = CH - CH_3$ / OCH_3
anithole

$\xrightarrow[H_2SO_4]{Na_2Cr_2O_7}$

CHO / OCH_3
anisaldehyde

(b)

CH_3 / OH
p-cresol

$\xrightarrow{(CH_3)_2SO_4}$

CH_3 / OCH_3

$\xrightarrow[\text{lined vessel}]{\text{oxidation in lead}}$

CHO / OCH_3
anisaldehyde

(c)

CHO / OH
p-hydroxy benzaldehyde

$\xrightarrow{(CH_3)_2SO_4}$

CHO / OCH_3
anisaldehyde

The largest amount of anisaldehyde is used in soaps and detergents.

(v) Cinnamic aldehyde: Cinnamic aldehyde is occurs in the essential oils of cinnamon family. It has cinnamon odour. It is manufactured in large quantities for perfume and flvour industry by the condensation of benzaldehyde and acetaldehyde in the presence of alkali (Aldol condensation).

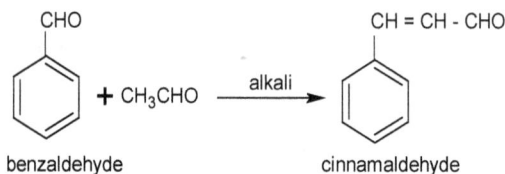

benzaldehyde + CH₃CHO —alkali→ cinnamaldehyde

Cinnamaldehyde is also constitutes the raw material for other perfumery compound, e. g. its aldehyde group on reduction gives cinnamyl alcohol possessing a pleasant and long lasting spicy odour, on hydrogenation its double bond reduced and phenyl propyl aldehyde is obtained, on complete hydrogenation of the side chain phenyl propyl alcohol is obtained, which itself and along its esters is used in perfume compositions.

(vi) Piperonal or heliotropin: Heliotropin occurs in the popular garden variety of heliotrope. It has a persistent odour without any discolourisation effect. In addition, it is used in the perfume and flavour industry. It is used in electroplating of metals. Moreover its various derivatives have been used as fungicides and insecticides. It is prepared from safrole.

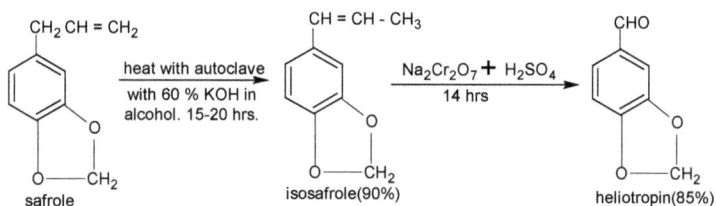

(vii) Cyclamen aldehyde: It is a valuable odorous compound having an odour reminiscent of cyclamen, lily of the valley, lilac etc. It is used in soaps and detergents. It does not occur in nature and is solely a synthetic product. Commercially cyclamen aldehyde is prepared by the following methods.

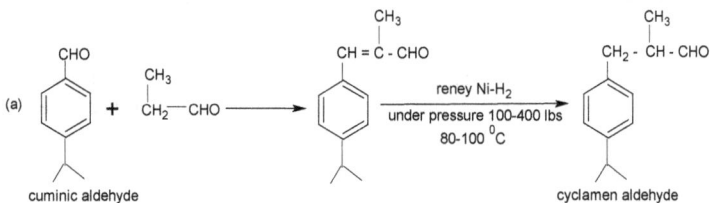

(b)

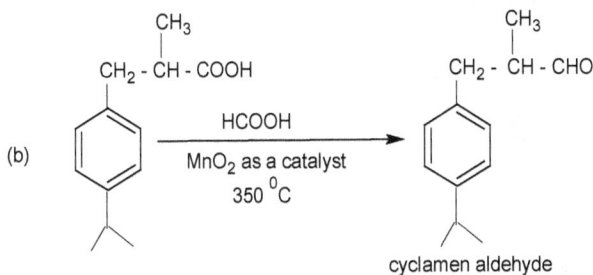

CH$_3$
|
CH$_2$ - CH - COOH

HCOOH
MnO$_2$ as a catalyst
350 ^{0}C
→
CH$_3$
|
CH$_2$ - CH - CHO

cyclamen aldehyde

(viii) Hydroxy citronellal: However, in general it has been observed that no single compound has the identical complex odour of a flower but hydroxycitronellal gives exactly the odour of lilly of the valley *(convallaria majalis)* flowers, although it does not occur in this flower or any other plant and thus it is completely synthetic product. Hydroxycitronellal is one of the appreciated synthetics discovered by the chemist. It has become indispensable components of most of the perfume formulation. It is widely used in soaps and cosmetics fragrances. Sometimes its dimethyl acetal is used in perfumery because of unstability of the parent compound (having a –CHO group). Commercially it is manufactured from citronellal or β-pinene.

(i)

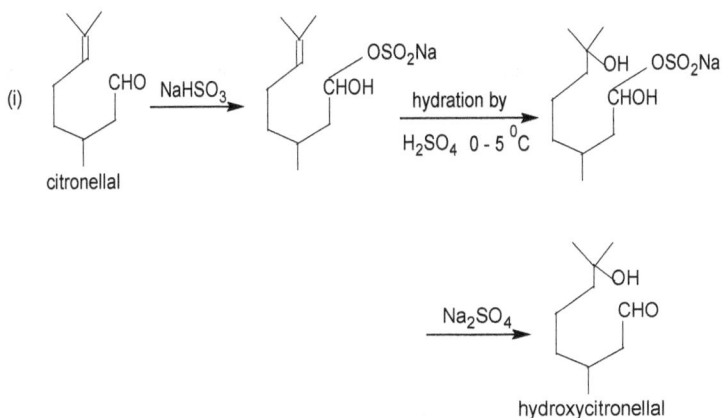

citronellal → CHO, NaHSO$_3$ → OSO$_2$Na, CHOH

hydration by H$_2$SO$_4$ 0 - 5 ^{0}C → OH, OSO$_2$Na, CHOH

Na$_2$SO$_4$ → OH, CHO

hydroxycitronellal

(ii)

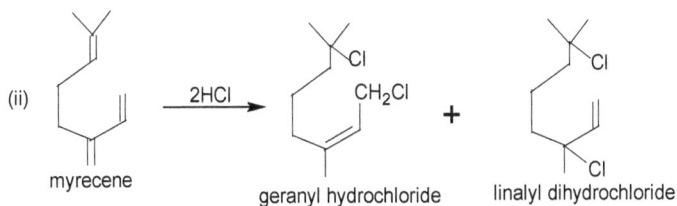

myrecene → 2HCl → Cl, CH$_2$Cl geranyl hydrochloride + Cl, Cl linalyl dihydrochloride

hydrolysis

CH$_2$OH HCOOH

hydroxy geraniol

OH

OH

OH

hydroxy linalool

H$_2$

isomerisation

CH$_2$OH HCOOH

OH

CHO

OH

hydroxy citronellol

hydroxy citronellal

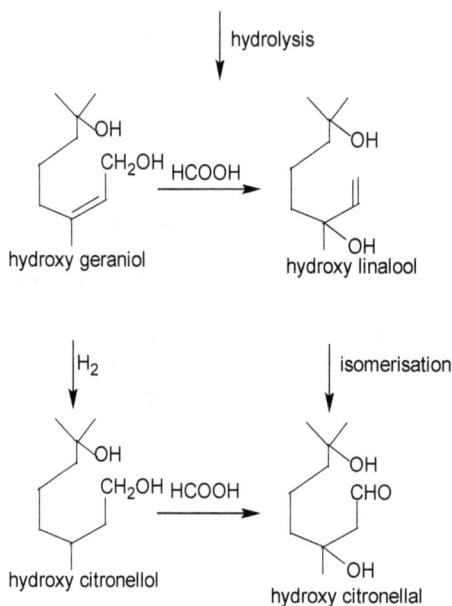

2.4.5 Diphenyl compounds: Among the various important diphenyl compounds in perfumery the most important are diphenyl oxide, diphenyl methane and benzophenone.

(i) Diphenyl oxide or ether: It has a powerful harsh geranium odour. Due to its low price and high stability, it is largely used in much industrial perfume formulation. Several methods have been developed for the synthesis of diphenyl oxide, and the most widely used involves the heating of alkali phenolate with chloro or bromo-benzene at elevated temperature.

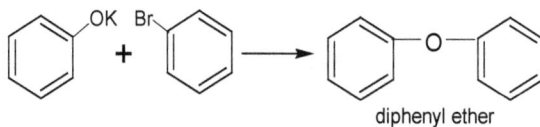

OK Br

+

O

diphenyl ether

Because of harsh odour of diphenyl oxide, some related compounds, *viz.* diphenyl methyl ether, dibenzyl ether and di-o-crecyl ether are sometimes used to obtained softer geranium fragrance.

(ii) Diphenyl methane: It is also a stable compound and possesses a soft fragrance, so it is also used in various perfume composition. Commercially diphenyl methane is obtained from benzene and methylene chloride by meance of Friedel-Craft reaction.

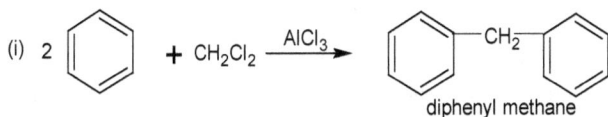

diphenyl methane

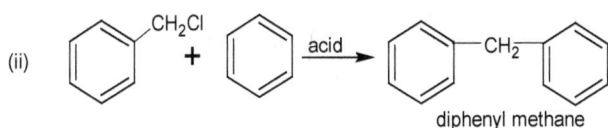

diphenyl methane

(iii) Benzophenone: It has a softer rose like fragrance and hence it is applied to give a long lasting sweet odour to several types of perfumes. Moreover, as it lowers the rate of evaporation of perfume constituents, it can also be used as a fixative is. Benzophenone synthesized by the following methods.

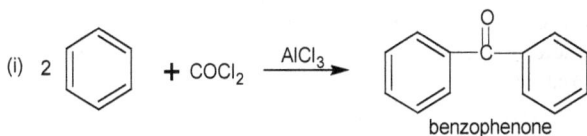

benzophenone

benzophenone dichloride   benzophenone

3. Explosives

3.1 Introduction

Explosives are those substances which, when subjected to a mechanical or thermal decomposition, undergo extremely rapid, self propagating decomposition with the gases. The gases occupy a much larger volume and volume further increases because of the high temperature produced in the reaction. Due to which extremely high pressure is exerted on the surroundings giving a destructive effect.

It appears to us that explosives are generally used for the destructive purposes. But it is not true. Actually, explosives are mainly used for the creation of dams, tunnels, roads in mountain regions, minig of coal, metals and non metals, etc.

Explosives material may be gaseous (e.g., a mixture of oxygen and acetylene), liquid (e.g., nitroglycerine) or solid (e.g., TNT). An explosive material may consist of a single substance (e.g., TNT) or a mixture of two or more substances, none of which is itself need to be an explosive (e.g., a mixture of sulphur, charcoal and nitrate). Explosives differ from widely in their sensitivity and power.

The majorities of explosives contain oxygen, present in nitro, nitrate perchlorate groups, etc, and develop their energy by a process of combustion, producing oxides of carbon, water, and nitrogen on explosion. However, oxygen is not essential; for example, the decomposition energy of lead azide results from the rupture of weak linkages between nitrogen atoms which subsequently recombine to form more stable arrangements.

3.2 Classification

Explosives classification is based on the types of reaction that produce explosions, namely, mechanical, chemical and nuclear. However, in the present article we are only concerned with the organic chemical explosives which comprise two main types:
(a) Low or deflagranting explosives.
(b) High or detonating explosives.
The latter are further classified as primary and secondary detonating explosives.
We will now discuss the various types of organic chemical explosive.

3.2.1. Low or deflagranting Explosives.

These are also known as propellants. Low explosives do not undergo explosive suddenly but they only burn. The chemical reactions take place comparatively slowly. The burning proceeds from the surface inwards in layers. Low explosives evolve large amount of gases and due to which they help explosion. The relatively low rates of pressure development and peak pressure of the explosives, in addition to rendering them helpful in guns rockets as propellents, give them a desirable "blasting action" for coal mining and other commercial blasting operations.

The smokeless powder propellants, which represent the only other "low" explosives in extensive use today, are indispensable in the military field, but owing to their high cost they have little commercial value. The smokeless powders are the basis of the most modern artillery, small arms and rocket ammunition. Three general types are available which are as follows:

(i) Single base powders: In these, nitrocotton is the basic ingredient.

(ii) Double base powders: In these, nitrocotton and nitroglycerine are the main basic ingredients.

(iii) Triple base powders: In these, nitroguanidine, nitroglycerine and nitrocellulose are the main basic ingredients.

3.2.2. High or Detonating Explosives.

These explosives are characterized by very high rates of reactions and pressure. These explosives are further subdivided into the following two type:

(a) Primary high explosives.

These explosives are also known as initiating explosives.

These explosives are very sensitive and may be exploded by application of flame, spark, impact and other primary heat sources of appropriate magnitudes. These are used in somewhat smaller quantities to initiate the explosion of large quantities of less sensitive secondary explosives.

Primary explosives are very dangerous to handle. Primary explosives are lead azide, mercury fulminate, tertracene, dinol, lead styphnate, etc.

(b) Secondary high explosives

These explosives are quite insensitive to mechanical shock and flame. However, these explode with great violence when they are initiated by initiating explosives. Secondary explosives develop detonation pressure from a minimum of 2500 to a maximum of about 350,000 atmospheres. The detonation temperatures of secondary explosives may range from as low as about 1500^0C to 5500^0C or higher depending on the nature of the explosive.

Important examples of secondary explosives are TNT, cyclonite, tetryl, picric acid, dinitrotoluene, ethylene dinitramine, etc.

(i) Nitro Explosives

(a) Nitrobenzene: It is prepared by the nitration of benzene with conc. nitric and sulphuric acid mixture at a temperature below 60^0C.

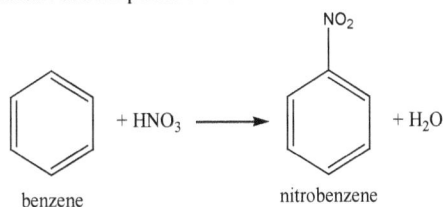

benzene   + HNO$_3$ $\longrightarrow$ nitrobenzene   + H$_2$O

Nitrobenzene is the main constituent of American Explosive *Rack-a Rack* (nitrobenzene-21%,potassium chlorate). Nitrobenzene is also the main constituent of an American smokeless powder called *indurate* (nitrobenzene 50%, gun powder 40%). When mixed with gun cotton, nitrobenzene has been used in the manufacture of propellants.

(b) m-Dinitrobenzene: When 1 part of nitrobenzene is heated with a mixture of equal amounts of conc. sulphuric acid and conc. nitric acid, the main product is m-dinitrobenzene but o-isomer is also obtained (7%). This is removed during crystallization.

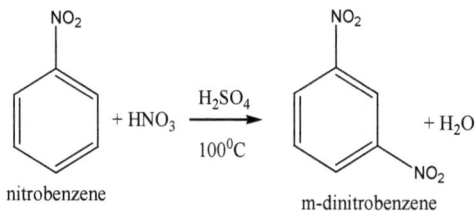

nitrobenzene   m-dinitrobenzene

When mixed with chlorates, it has been widely used in blasting explosives. It is also used in commercial explosives like *belite No.1*(ammonium nitrate 82-85%, dinitrobenzene 18-15%), *tonite No.3* (barium nitrate 68%, gun cotton 19%, dinitrobenzene 13%). It is also used in making Colt's pistol powder and dynamite.

(c) s-Trinitrobenzene (1, 3, 5-trinitrobenzene): It may be prepared by the nitration of *m*-dinitrobenzene with mixed acid consisting of fuming nitric acid and fuming sulphuric acid. This reaction is very slow and generally takes five days to complete.

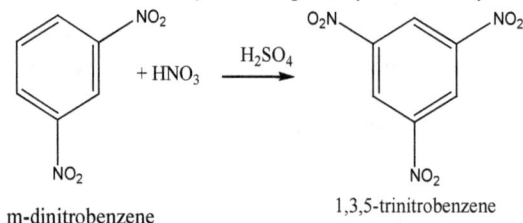

m-dinitrobenzene   1,3,5-trinitrobenzene

However, a better method of preparing 1, 3, 5-trinitrobenzene is to oxidize 2,4,6-trinitrotoluene and to decarboxylate the trinitrobenzoic acid so produced by heating it in acetic acid solution.

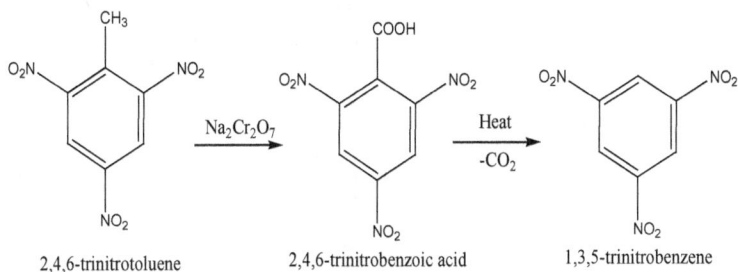

$$\text{2,4,6-trinitrotoluene} \xrightarrow{Na_2Cr_2O_7} \text{2,4,6-trinitrobenzoic acid} \xrightarrow[-CO_2]{Heat} \text{1,3,5-trinitrobenzene}$$

| 2,4,6-trinitrotoluene | 2,4,6-trinitrobenzoic acid | 1,3,5-trinitrobenzene |

As trinitrobenzene is more powerful than TNT and can be expoded more easily than dinitrobenzene. This has proposed as a substituted for picric acid. However, its high melting point and difficult of preparation mitigate against its use.

(d) Trinitrotoluene (TNT): It is the most important high explosive. It is manufactured by three stage nitration of toluene with mixed acids.

Raw materials: The essential raw materials for the manufacture of TNT are toluene, nitric acid, and sulphuric acid. Toluene is obtained as a byproduct of the distillation of coal. Sulphuric acid is manufactured either by Contact process or Chamber process.

Process: About 5000 kg. of mixture of acids is taken in a nitrator and 275 kg of toluene is slowly added into it. The temperature is allowed to rise about 140 ^{0}C. the batch is held in the nitrator for 20-30 minutes so as to achieve the required specific gravity of acids. At this stage, mononitrotoluene is obtained (first stage). Now the temperature is raised to 180 ^{0}C, when a mixture of dinitrotoluene is obtained (second stage). Finally, the temperature is raised at determined rate 230 ^{0}C, where it is held for 30 minutes. At this stage, a mixture is removed by pumps while waste acid is removed by air pressure.

The mixture of trinitrotoluene is in liquid state. It is washed with ammoniated solution of sodium sulphite. Ammonia neutralizes residual acids while sulphite removes the unsymmetrical trinitrotoluene as water soluble sodium dinitrosulphonates. Then 2, 4, 6-trinitrotoluene is crystallize by rapid addition of cold water and filtered. Now, 2,4,6-trinitrotoluene is first washed with hot water, then dried at 90 ^{0}C and finally converted into flakes which are packed into paper-lined boxes.

Chemical reaction: The three steps of nitration of toluene may be represented as follows:

CH₃

toluene

| First stage | HNO₃(28%) H₂SO₄(56%) H₂O(6%) |

o-nitrotoluene 60% + p-nitrotoluene 36.5% + m-nitrotoluene 45%

| Second stage | HNO₃(32%) H₂SO₄(61%) H₂O(7%) |

2:5-dinitrotoluene 0.9% | 2:3-dinitrotoluene 1.1% | 3:4-dinitrotoluene 2.5% | 2:6-dinitrotoluene 20.7% | 2:4-dinitrotoluene 74.8%

| Third stage | HNO₃(49%) H₂SO₄(49%) |

2:3:6- 0.3% | 2:3:4- 1.3% | 2:4:5- 2.9% | 2:4:6- 95.5%

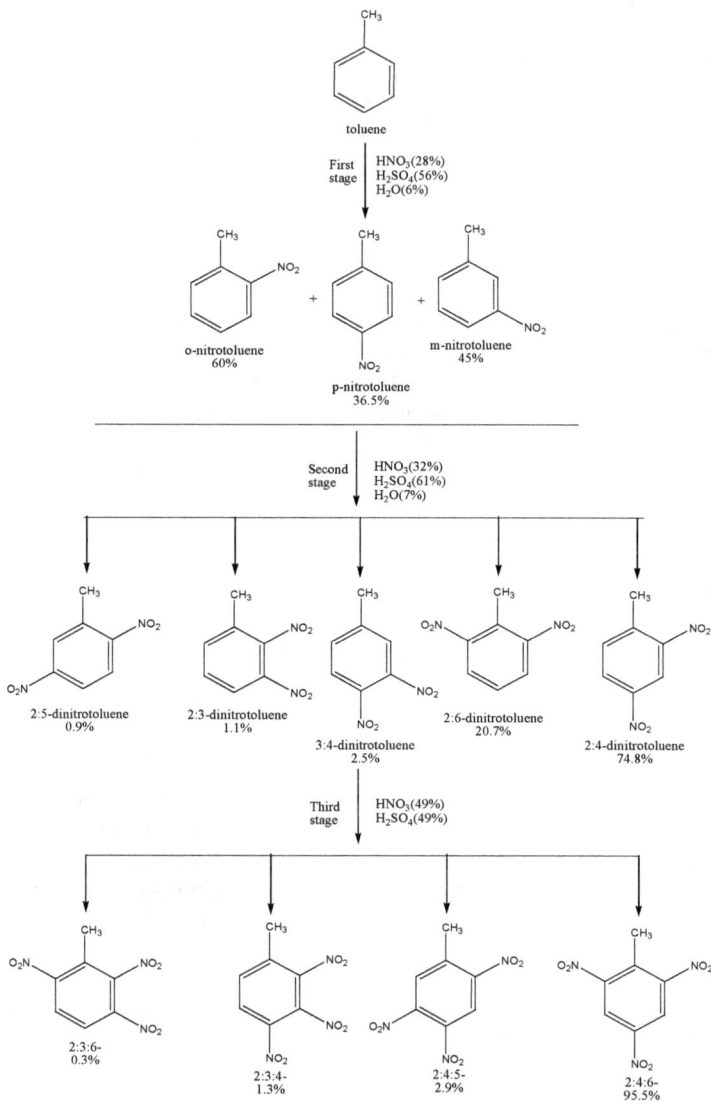

Uses: TNT is a yellow crystalline solid with melting point 80.4°C and its velocity of detonation (VOD) is 6700 m/sec at 1.58 densities. In spite of other new and most powerful explosives such as cynolnite and pentril it is still the most important military high explosive due to the following reasons:

(i) It is not sensitive and hence it is safe to handle during manufacture and storage.

(ii) It does not absorb moisture.

(iii) It does not react with metals to form unstable compounds.

(iv) It has high brisance value (shattering action).

Its two disadvantages, i.e. reactivity with alkali and toxicity towards humans have been overcome by selecting proper technical methods of its preparation.

TNT is mainly used in the manufacture of high explosives for shell and air borne demolition bombs. It undergoes decomposition in the following manner.

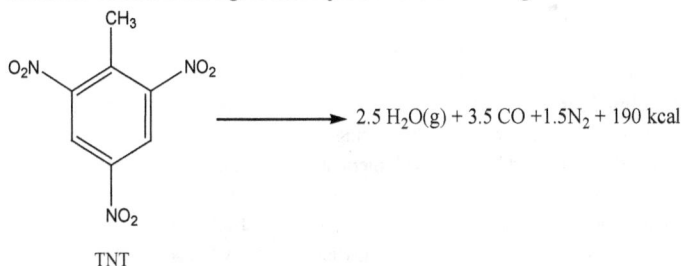

$$2.5\ H_2O(g) + 3.5\ CO + 1.5 N_2 + 190\ kcal$$

TNT

When mixed with ammonium nitrate, it forms the explosive amatol.

(i) Picric acid (2,4,6-Trinitrophenal): It was first prepared by Woulff in 1771 but it was not used until 1885 as a military explosive in shell.

Preparation: It is prepared as follows:

(a) From chlorobenzene: Picric acid is manufactured from chlorobenzene as follow:

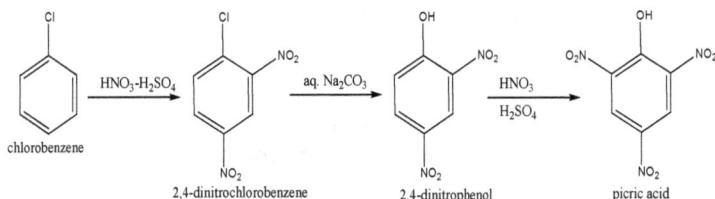

(b) From phenol: Picric acid is manufactured from phenol as follow:

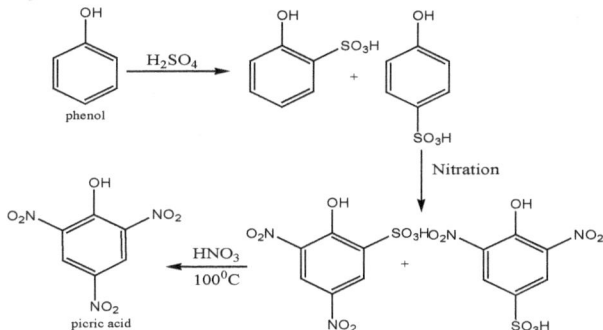

The direct nitration of phenol is not carried out because of violent reaction and consequent low yields. Therefore, phenol is first sulphonated by treating it with conc. H_2SO_4 at 100^0C. Then, phenol sulphonic acid is added to mixture of HNO_3 and H_2SO_4 in a nitrator and temperature is maintained at 110^0C by cooling coils and a agitation. During nitration the red colour of phenol sulphonic acid is separated by filtration after cooling. The separated picric acid is repeatedly washed with cold water to remove acid contents.

Use: Picric acid is yellow crystalline solid having m. p. 121^0C. Its velocity of detonation (VOD) is 7600 m/sec at 1.69 densities. Picric acid was the first high explosive used for melt loading but it was gradually replaced by TNT because high temperature required for its melt loading (120 ^0C) and tetryl (booster) gets desensitized on prolonged contact with picric acid and hence tetryl booster cannot be used with picric acid.

(ii) Ammonium picrate of Explosive D: It was prepared by Marchand in 1841. It was used as an explosive in admixture with potassium nitrate by Brugere in 1869.

It is manufacture by neutralizing an aqueous hot solution of picric acid by aqueous ammonia.

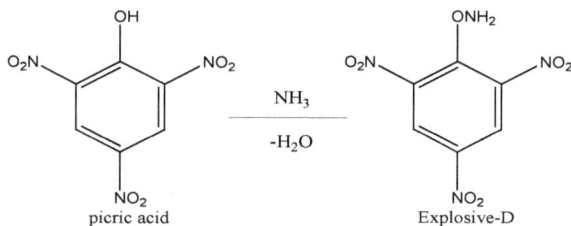

picric acid   Explosive-D

Ammonium picrate is one of the constituents of *picratol.* Ammonium picrate is insensitive to shock. Therefore, it is used in *armor-pierincing shells.*

(iii) Emmen's acid: It is prepared by dissolving picric acid in fuming nitric acid to saturation then heating. Emmen's acid alone or mixed with chlorates or nitrates can be used as an explosive. Emmen's acid mixed with chlorates or nitrates is more powerful explosive than picric acid.

(iv) s-Trinitroanisole: This is prepared by nitration of dinitroanisole. The latter compound in turn is prepared from 2,4-dinitro 1-chlorobenzene.

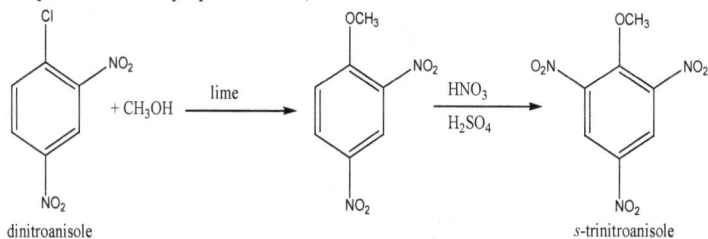

dinitroanisole   s-trinitroanisole *s*-trinitroanisole is a quite stable explosive, but is having slightly less power than picric acid. It is used with other high explosives or ammonium nitrate as filling for shells and bombs.

(ii) Nitric Esters:

(a) Ethyleneglycol dinitrate (EGDN): It was first made by Henry in 1870. It is prepared by the nitration of glycol.

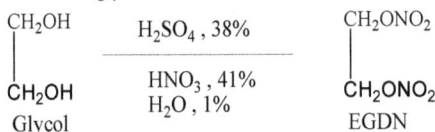

$$
\begin{array}{c}
CH_2OH \\
| \\
CH_2OH \\
\text{Glycol}
\end{array}
\quad
\begin{array}{c}
H_2SO_4 , 38\% \\
\overline{} \\
HNO_3 , 41\% \\
H_2O , 1\%
\end{array}
\quad
\begin{array}{c}
CH_2ONO_2 \\
| \\
CH_2ONO_2 \\
\text{EGDN}
\end{array}
$$

It is used to lower the freezing point of nitroglycerine in some dynamites but is too volatile to be used in double-base cordites. It is less sensitive and more stable then nitroglycerine.

(b) Diethylene glycol dinitrate (DEGDN): It was first described by Rinkenbach in 1927. It is made by nitrating diethylene glycol.

$$
\begin{array}{c}
CH_2CH_2OH \\
O \\
CH_2CH_2OH \\
\text{diethylene glycol}
\end{array}
+ 2HNO_3 \quad\longrightarrow\quad
\begin{array}{c}
CH_2CH_2ONO_2 \\
O \\
CH_2CH_2ONO_2 \\
\text{DEGDN}
\end{array}
+ 2H_2O
$$

(c) Nitroglycerine (glyceryl trinitrate, NG): It was first prepared by Sobrero in 1846 and used as an explosive by Nobel in 1864. It is an important ingredient in commercial explosives, dynamites and propellents. Nitroglycerine is prepared by the nitration of glycerol with the nitrating mixture having 59.5% H_2SO_4, 40% HNO_3 and 0.5% water.

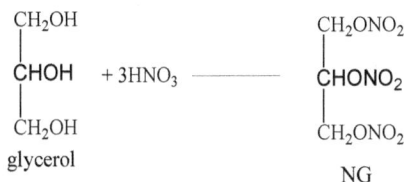

$$
\begin{array}{c}
CH_2OH \\
| \\
CHOH \\
| \\
CH_2OH \\
\text{glycerol}
\end{array}
+ 3HNO_3 \quad\longrightarrow\quad
\begin{array}{c}
CH_2ONO_2 \\
| \\
CHONO_2 \\
| \\
CH_2ONO_2 \\
\text{NG}
\end{array}
$$

Nitroglycerine is a colourless oily liquid. It is very sensitive to shock especially, when it contains air bubbles. The decomposition of NG may be represented as follow:

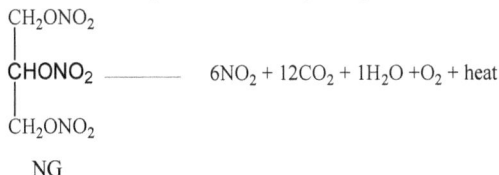

$$
\begin{array}{c}
CH_2ONO_2 \\
| \\
CHONO_2 \\
| \\
CH_2ONO_2 \\
\text{NG}
\end{array}
\quad\longrightarrow\quad 6NO_2 + 12CO_2 + 1H_2O + O_2 + \text{heat}
$$

An NG has high sensitivity, it is not used such but most of it is used in the manufacture dynamite and smokeless powder.

Dynamite: As pointed out earlier, NG is extremely sensitive to shock and therefore it is very hazardous to handle. Alfred Noble in 1867 discovered that NG could be absorbed by porous siliceous earth called kieselguhr. Solid explosive called dynamite is resulted. Now a days kieselguhr is not all used because the content of inert kieseguhr in dynamite is about 15% which decreases blasting capacity. Now a day's dynamite is prepared by absorbing nitroglycerine in saw dust or charcoal in the presence of sodium or ammonium nitrate (oxidizer). A small amount of antacid like ZnO or $CaCO_3$ may be added.

Dynamite is easy to handle and can accommodate up to 75% NG with the retension of the solid form. The demand of non-freezing dynamite for use in cold weather has caused the addition of certain other compounds which lower the freezing point of the NG. Ethylene glycol dinitrate is the most important example among this compound. Non freezing dynamites are equal in potentially to that of sample dynamite.

Gelatin dynamite, blasting gelatin or gelignite: A mixture of 1 part of nitrocellulose and 9 parts of NG forms a clear jelly like substance called gelatin dynamite. The name gelatin dynamite is given to the product because the mixture is jelly like.

Smokeless powder: It is colloidal cellulose nitrate containing about 1% diphenylamine (stabilizer) and a small amount of plasticizer (dibutyl phthalate). Smokeless powder contains about 13.2% nitrogen.

Cordite: It is prepared by mixing NG, cellulose and Vaseline. It is used as smokeless powder.

(iii) Nitrocellulose (cellulose nitrate, NC): It was first prepared by Peluze in 1836 by nitration of paper. In 1846 Schonpein and Botter, independently, disanguised NC from nitrostarch and described its explosive values.

Nitrocellulose may be prepared by a displacement procedure in which the cellulose is dipped into the nitrating acid in a shallow earthenware dish. Nitration proceeds at 17^0C to 21^0C for 1 to 2.30 hours without stirring, and then spent acid is slowly displaced by water and completely recovered. In the mechanical process the reactants are stirred for 20 to 25 minutes, then spent acid is separated by centrifuging and product is dipped into water.

$$[C_6H_7O_2(OH)_3]_3 + 3HNO_3 \longrightarrow [C_6H_7O_3(ONO_2)_2]_3 + 3H_2O$$

NC

Complete nitration of cellulose would correspond to a nitrogen content of 14.4% by weight, but it is customary to use products containing 12.2, 12.4 and 13% nitrogen in propellent compositions.

In the dry state, NC is a powerful and sensitive explosive but is insensitive to impact and friction when water-wet. **Gun cotton** is a cellulose nitrate having 13.2% of nitrogen.

(a) Nitrostarch: It was first prepared by Branconnot in 1839. It is a finely divided white solid similar in chemical and explosive properties to NC. It finds some use in commercial blasting explosives but the substance is unstable for propellant manufacture as, unlike NC, it does not yield tough colloids when gelatinized.

(b) Manitol Hexanitrate (nitromannite): It was first prepared by Domonte and Menart in 1847. It is prepared by nitrating manitol with nitric acid, then adding sulphuric acid to precipitate the product.

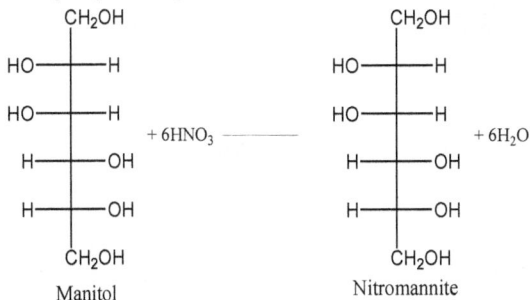

Manitol   Nitromannite

(c) Pentaerythriol tetranitrate (PETN): It was first prepared by Tollens and Wigand in 1891 but was not used until First World War.

It is made by nitrating pentaerythriol with 99% nitric acid (5 to 6 parts) at 15-20^{0}C, followed by crystallization from acetone.

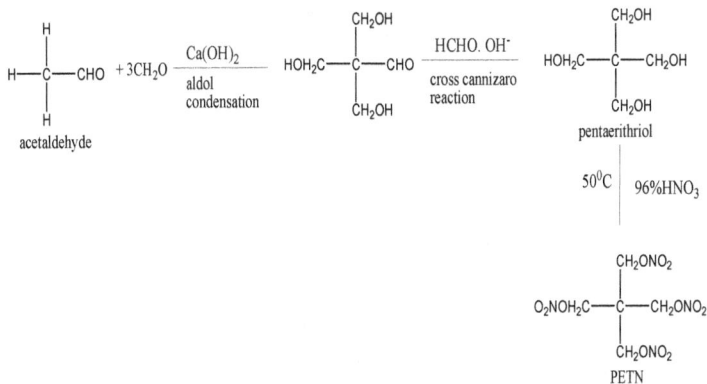

Pentaerithriol required for the above reaction is prepared from acetaldehyde and formaldehyde by Cannizaro's reaction. It is very stable chemically but is sensitive to impact and friction and is used in admixture, for example with TNT in pentolites. It is rather too costly to complete with NG in commercial explosives, and is more sensitive and less stable than RDX. PETN decomposes as follow:

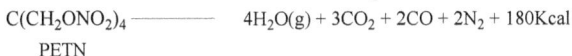

$$C(CH_2ONO_2)_4 \longrightarrow 4H_2O(g) + 3CO_2 + 2CO + 2N_2 + 180Kcal$$
PETN

(iv) Nitramides

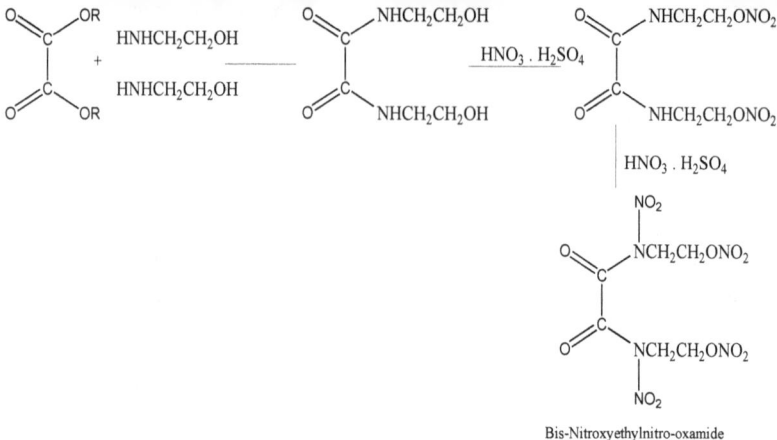

Bis-Nitroxyethylnitro-oxamide

(a) Bis-Nitroxyethylnitro-oxamide: This explosive is obtained from oxalic acid ester and monoethanolamine in the following way:

(b) 1, 3-Dinitraimidazolodone: This explosive is prepared by the nitration of imidazolidone-2. The latter compound is obtained from ethylene and diethyl carbonate or carbonylchloride.

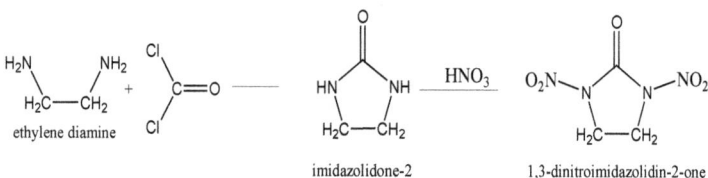

ethylene diamine   imidazolidone-2   1,3-dinitroimidazolidin-2-one

(v) Nitramines

(a) Cyclonite or RDX or s-Trimethylene trinitramine: It was first prepared by Hemmings in 1899. However, Von, Hertz discovered its value as an explosive in 1920. It was extensively used in 2nd World War.

Cyclonite is manufactured by the following methods:

(i) Hale method: Cyclonite was first of all synthesized by a British chemist G.C. Hale in 1925. He synthesized it by the destructive nitration (nitrolysis) of hexamethylene tetramine with conc. nitric acid. The yield of this method is 80%.

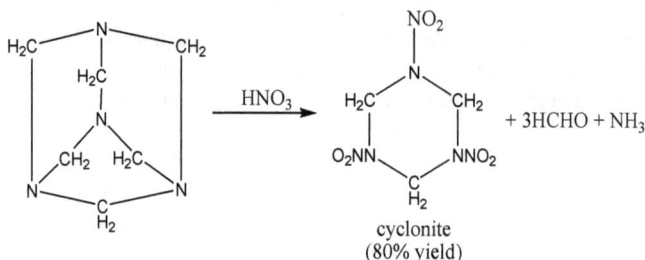

cyclonite
(80% yield)

The nitration of hexamethylene tetramine is carried out in a nitrator which is a stainless steel vessel filtered with stirrer and cooling Jacket. 99% nitric acid is employed for the reaction and the temperature is maintained at 20^0C. As soon as the nitration is completed, the nitrator product is sent to decomposer. The decomposition is initiated by 50 to 55% $NaNO_2$ solution. Cyclonite formed is separated from nitric acid by centrifuging. Recrystallization of cyclonite is done from acetone. Recrystallised cyclonite contains 15% moisture.

(ii) Ebele-Schiessler-Rose method: According to this method, cyclonite is prepared by the addition of paraformaldehyde and ammonium nitrate to acetic anhydride at 0^0C.

$$3CH_2O + 3NH_4NO_3 + 6(CH_3CO)_2O \text{———} (CH_3N.NO_2)_3 + 12CH_3COOH$$

cyclonite

When above method was studied in detail, it was found that this involves in following two steps:

(a) In the first step, there occurs synthesis of hexamethylene tetramine in acetic anhydride from formaldehyde and ammonium nitrate.

$$6CH_2O + 4NH_4NO_3 + 3(CH_3CO)_2O \text{———} (CH_2)_6N_4 + 4HNO_3 + 6CH_3COOH$$

(b) In the second step, there occurs Hale nitrolysis.

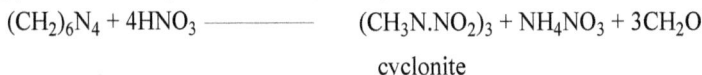

$$(CH_2)_6N_4 + 4HNO_3 \text{———} (CH_3N.NO_2)_3 + NH_4NO_3 + 3CH_2O$$

cyclonite

From the second step, there occurs the regeneration of ammonium nitrate and formaldehyde which are reused in step (a). It means that this process must be quantitative. But the yield is only 60%. This is attributed to the fact that some formaldehyde has been converted to unreactive methylene and polymethelyne diactates.

(i) Bachmann method: This method is considered to be the combination of the above two methods. In this method, cyclonite is prepared by mixing hexamethylene tetramine with nitric acid, ammonium nitrate and acetic anhydride.

$$(CH_2)_6N_4 + 4HNO_3 + 4HNO_3 + Ac_2O \text{———} 2(CH_3N.NO_2)_3 + 12CH_3COOH$$

cyclonite

The yield of this method is 80%.

(ii) Wolfram method: This was developed in Germany by Wolfram which involves the following two steps:

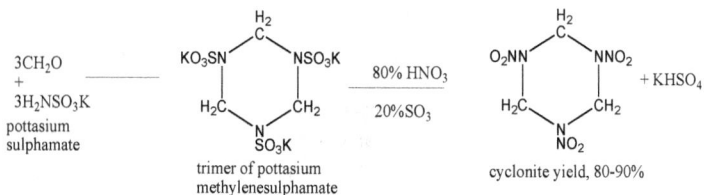

$$3CH_2O + 3H_2NSO_3K$$
pottasium
sulphamate

trimer of pottasium
methylenesulphamate

80% HNO_3
20% SO_3

cyclonite yield, 80-90%

$+ KHSO_4$

Use: Cyclonite is very powerful (1.55×TNT) explosive with high rate of detonation. But it is quite sensitive to impact. Therefore, it is used in combination with TNT. For example, Torpex is a mixture, of cyclonite, TNT and aluminium which is used for mines, depth charges and torpedo warheads.

(b) HMX or 1, 3, 5, 7-Tetranitro 1, 3, 6, 7-tetra-aza-cyclo-octane: This explosive synthesized by the nitrolysis of hexamine tetranitrate.

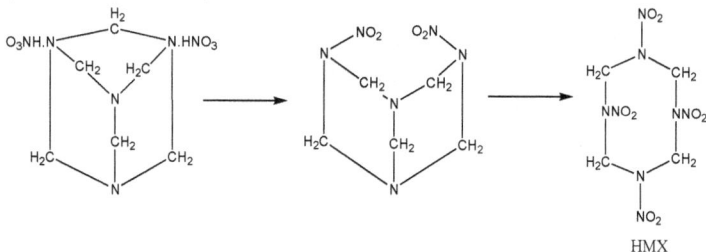

HMX

Similar to cyclonite, it is very powerful explosive with high rate of detonation. But it is a very sensitive to impact. Therefore, it is generally used in combination with TNT. This mixture is used for mines, depth charges and torpedo, warheads.

(c) Tetryl (2, 4, 6-trinitrophenyl methyl nitramine): It was first prepared by Meetens in 1877.

Preparation: (i) Tetryl is generally prepared from the nitration of dimethylamine sulphate with a mixture of nitric acid, sulphuric acid and water in a stainless steel nitrator at 100 ^{0}C.

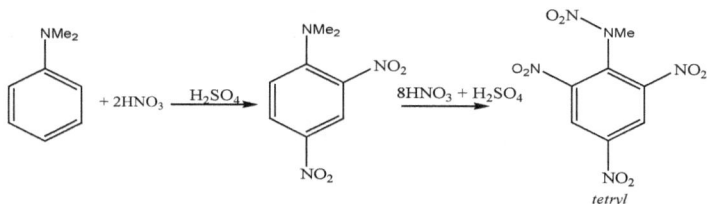

tetryl

Dimethylaniline required for the above reaction is prepared by the catalytic vapour-phase dehydration of a mixture of aniline and methyl alcohol.

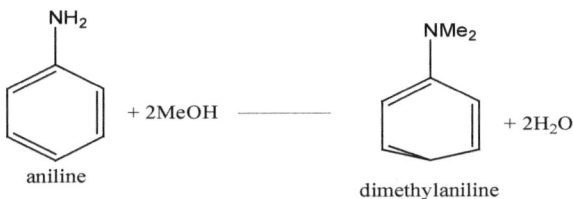

aniline + 2MeOH ──────── dimethylaniline + 2H₂O

Process: Dimethylaniline is dissolved in 96% sulphuric acid. This mixture is added to a mixture of nitric acid (67%) and sulphuric acid (16%) which is taken in nitrator. The temperature of the reaction vessel is maintained at about 70^0C by cooling coils. The ppt. of tetryl is run down on stainless filters and acid is separated for recovery. Crude tetryl is then boiled with water to remove acidity and other water soluble impurities. Finally, tetryl may be recrystallising from benzene or acetone.

(ii) Tetryl may also be prepared by alkylating 2,4-dinitrochlorobenzene with methylamine followed by the nitration of the product.

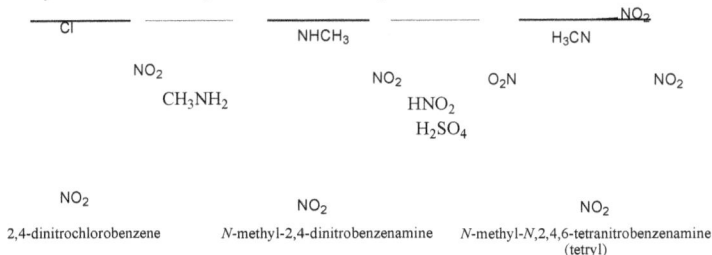

2,4-dinitrochlorobenzene → *N*-methyl-2,4-dinitrobenzenamine → *N*-methyl-*N*,2,4,6-tetranitrobenzenamine (tetryl)

Use:-It is pale yellow crystalline substance. It is used as the booster explosive in high explosive shells. Its velocity of detonation in 7500m/sec. it occurs the skin and cause dermatitis. It is not hydroscopic but reacts with aqueous sodium hydroxide giving picric acid.

(d) Pentryl (s-Trinitro phenol nitraminomethyl nitrate): It may be prepared from 2, 4-dinitrochlorobenzene in the following manner.

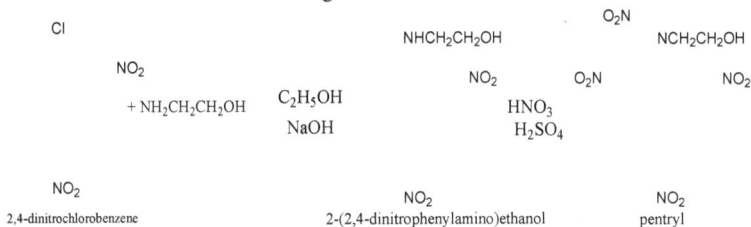

2,4-dinitrochlorobenzene → 2-(2,4-dinitrophenylamino)ethanol → pentryl

Pentryl may also be prepared by nitration of N-hydroxyethylaniline.

NHCH$_2$CH$_2$OH

O$_2$N

NCH$_2$CH$_2$OH

O$_2$N   NO$_2$

HNO$_3$
H$_2$SO$_4$

N-hydroxyethylaniline

NO$_2$
pentryl

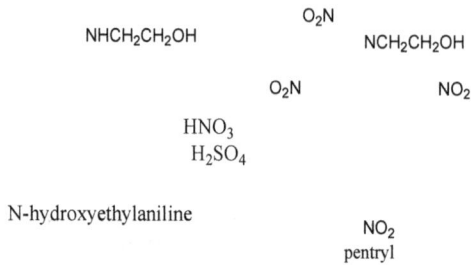

Pentryl is less sensitive than tetryl.

(e) Hexyl (Hexanitrodiphenylamine): It is being prepared from aniline and dinitrochloro-benzene in the following way:

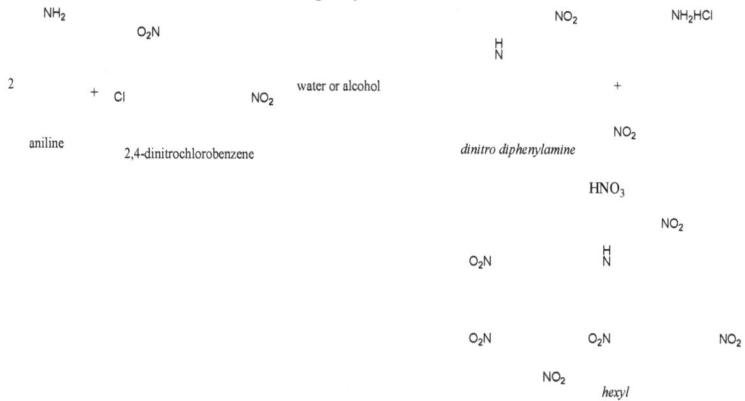

NH$_2$ NO$_2$ NH$_2$HCl

O$_2$N $\overset{H}{N}$

2   +   Cl   NO$_2$ water or alcohol   +

aniline   2,4-dinitrochlorobenzene   dinitro diphenylamine   NO$_2$

HNO$_3$

NO$_2$

O$_2$N $\overset{H}{N}$

O$_2$N   O$_2$N   NO$_2$

NO$_2$
hexyl

Hexyl is generally used in combination with other explosives as a filling bombs, mines and warheads for torpedos.

(f) Ethylenedinitramine (EDNA, halite): It was first described by Franchimont and Klobbie in 1887. It is made from ethyleneurea by nitration, then hydrolysis. The formula of helalite is [(CH$_2$NHNO$_2$)$_2$].

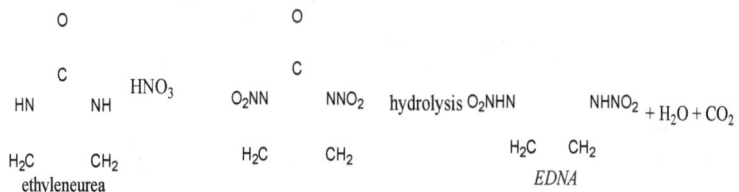

O   O

C   C
HN   NH $\xrightarrow{HNO_3}$ O$_2$NN   NNO$_2$ $\xrightarrow{hydrolysis}$ O$_2$NHN   NHNO$_2$ $+ H_2O + CO_2$

H$_2$C   CH$_2$ H$_2$C   CH$_2$ H$_2$C   CH$_2$
ethyleneurea   EDNA

It was used during the Second World War in some bursting charges. This explosive is more powerful than tetryl and slightly less sensitive to impact but forms sensitive salts with some metals, notably lead. It is an constituent of the binary explosive, edatol.

(vi) Guanidine Explosives

(a) Guanidine Nitrate:

(i) On a large scale, guanidine nitrate is synthesized from calcium carbide in the following way:

$$CaC_2 + N_2 \xrightarrow{1000^0C} CaNCN + C$$
$$\text{calcium cynamide}$$

$$CaNCN + H_2O \longrightarrow NH_2CN + Ca(OH)_2$$
$$\text{cynamide}$$

$$Ca(OH)_2 + CO_2 \longrightarrow CaCO_3 + H_2O$$

When cynamide is treated with alkali, dicynamide is obtained. The latter product is first dissolved with ammonium nitrate at 160^0C under 1500lbs pressure when guanidine nitrate is obtained. The dry crystalline from of guanidine nitrate is obtained after evaporating the liquid ammonia.

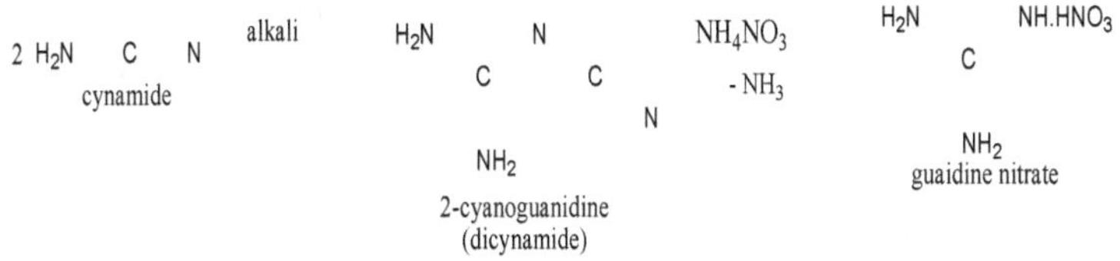

2-cyanoguanidine (dicynamide) → guaidine nitrate

The yield of this method is 85%.

(ii) Guanidine nitrate is also prepared by treating the dry calcium cyanamide with ammonium nitrate and urea at 120^0C under atmospheric pressure.

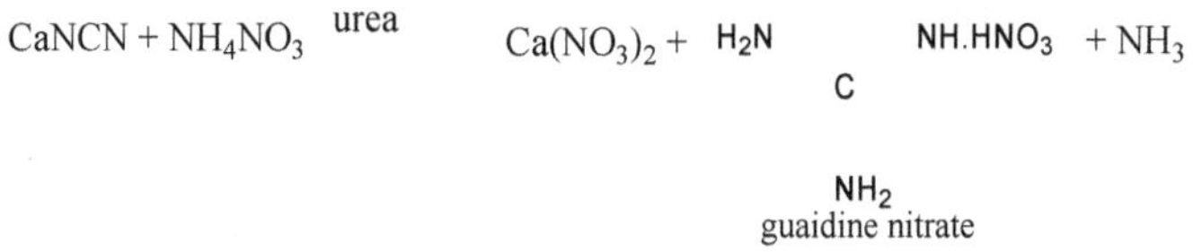

guaidine nitrate

Use: Due to non-hygroscopic nature, guanidine nitrate has replaced ammonium nitrate as a composite explosive.

(b) Nitroguanidine (picrite): Nitroguainidine was the first prepared by Joussel in 1870. However, it was used during Second World War. It is prepared by treating guanidine nitrate with conc. sulphuric acid at 10^0C.

$$H_2N \qquad NH.HNO_3 \qquad\qquad\qquad H_2N \qquad NNO_2$$
$$C \qquad\qquad \xrightarrow[H_2O]{conc, H_2SO_4} \qquad\qquad C$$
$$NH_2 \qquad\qquad\qquad\qquad\qquad NH_2$$

guaidine nitrate   nitro guaidine

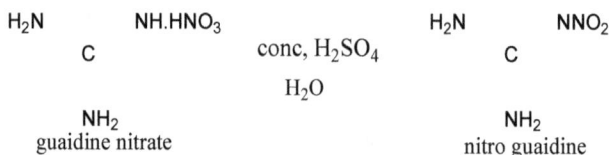

Nitroguanidine is used for making fishless powders. It is introduced into gun cordites as an anti-flash agent and coolant. Its decomposition liberates a large volume of nitrogen which cools and prevents ignition of the gases issuing from the gun, low temperature combustion markedly prolongs the life of gun barrels. It is insensitive and has good thermal stability.

(vii) Diazo and Miscellaneous Compouds

(a) Mercury fulminate [Hg(ONC)$_2$]: It was first made in the seventeenth century and was used in Nobel's aloneior in admixture with potassium chlorate for many years. However, it has poor thermal stability, particular under damp conditions and it has been replaced by lead azide as an initiatory and by mixtures containing lead styphnate for priming purposes. Mercury furminate is prepared in small batches by reaction of mercury with nitric acid and ethanol in an all-glass apparatus. It is a grayish substance. It is not hygroscopic but in the presence of water may react with metals, particularly aluminium.

(b) Lead azide: It is a salt of hydrazoic acid (N_2H). In 1893, Lonze studied the possibility of lead azide as military explosive. Lead azide is prepared by treating sodium azide with lead acetate or nitrate. The sodium azide in turn is prepared by heating sodamide in a current of nitrous acid.

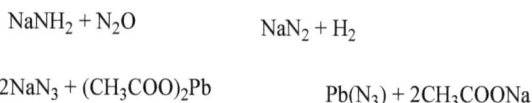

$$NaNH_2 + N_2O \qquad\qquad NaN_2 + H_2$$

$$2NaN_3 + (CH_3COO)_2Pb \qquad\qquad Pb(N_3) + 2CH_3COONa$$

Procedure: A 5% aqueous solution of lead acetate having a small amount of dextrin is taken in vessel. To this, a 2% aqueous solution of sodium azide is added. The ppt. of lead azide is formed. It is separated, washed first with water and then with rectified spirit. Finally, the ppt. of lead azide is dried at room temperature by passing dry air. The dry powder is then sieved through silken sieve by remote control. Finally it is stored in appropriate magazines.

It is important to remember that operations of precipitation, separation, drying, and sieving are hazardous in nature and are therefore carried out from behind blast proof walls.

Use: Lead azide is white to pinkish white powder. It has replaced mercury fulminate as an initiating or primary explosive for blasting caps due to the following reasons:

(i) Mercury fulminate is less stable than lead azide.

(ii) Mercury fulminate requires more costly raw material than lead azide which requires the cheaper raw material. Both lead azide and mercury fulminate undergo decomposition exothermically on being struck as follows:

$$Pb(N_3)_2 \qquad Pb + 3H_2 + 100.6\ Kcal$$

$$Hg(ONC)_2 \qquad Hg\ (g) + N_2 + 2CO + 117Kcal$$

The velocity of detonation of lead azide is 4500m/sec at 3.5 density. In purpose of moisture, lead azide may react with metals, notably copper to give extremely sensitive azides; initially systems must either constructed of metals such as aluminium, which do not give dangerous azides or be hermetically scale against moisture.

Lead azide is less sensitive to ignition by flash, percussion or stab than mercury fulminate, and in some types of composite detonator lead styphnate is added to increase flash sensitiveness.

(c) Lead styphnate (lead 2,4, 6-trinitroresorcinate): It was first made in 1914 by Von Hertz. It is obtained as small yellow-brown needles by reaction of acidified magnesium styphnate with lead nitrate or acetate in hot aqueous solution.

magnesium styphnate   lead styphnate

Lead styphnate is relatively poor initiating agent but finds application because of its ease of ignition. It is used in considerable quantity as an ingredient of cap and priming compositions, and as a cover charge above lead azide.

Normal lead styphnate presents considerable static rise during handling. A monobasic form of lead styphnate has some limited application because of its improved compatibility and stability at high temperatures.

(d) Tetrazene or Tetracene: It was first prepared in 1910 but its molecular structure has been established with certainly only recently. The starting compound for the manufacture of tetrazene is nitroguanidine (I). The latter compound on reduction yields amino guanidine (II) which on treatment with sodium nitrite in the absence of acid yields nitroaminoguiadine (III). A part of nitroaminoguanidine loses a molecule of water to form guanyl azide (IV). The (IV) finally combines with the rest of (III) to form tetrazene.

NH₂ ... chemical structure scheme

NH_2 | NH_2 | NH_2 | NH_2 | NH_2

C NH₂   C NH 4[H]   C NH   C NH   C NH
                                        -H₂O
N   HN   HN   HN   N

NO₂   NO₂   NH₂   NHNO   N   NH₂

2-nitroguanidine (I)   (II)   (III)   (IV)

NH

C   H   N   H
     N   N
(III) + (IV)   H₂N   N   N   C   NHNO
               H

NH

tetrazene

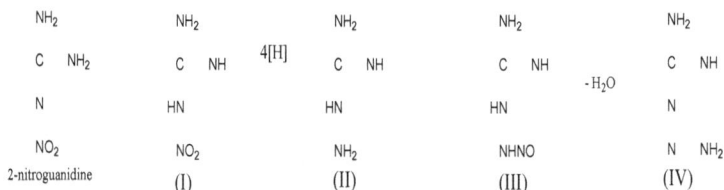

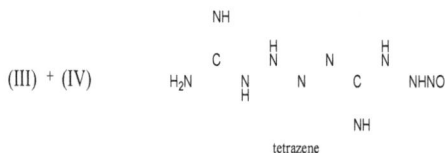

It has been found by X-ray examination that tetrazene is in the form of 1-amino-a[(1H-tetrazol-5-yl)azo]guanidine hydrate.

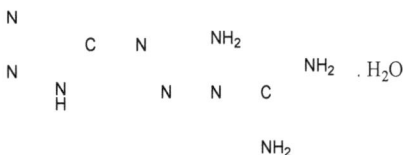

N

C   N NH_2

N NH_2 . H_2O

N   N   N   C

H

NH_2

Two polymorphic modifications of tetrazene are known; the β-form has a crystal density of 1.64 g/cm^3. Tetrazene is not used alone but small amounts are mixed (3-5%) with other intiators notably lead styphnate, to enhance their sensitiveness towards initiation by mechanical stimuli. This effect may be associated with the temperature of ignition (135°C) of tetrazene. Tetrazene has partially replaced lead azide or mercury fulminate.

(d) Dinol (Diazodinitrophenol or 4,6-dinitrobenzene-1,2-diazoxide): It was first diazo compound discovered by Griess in 1858. Its use as a primary explosive has been discovered recently.

Griess prepared Dinol by passing nitrous gas into an alcoholic solution of picramic acid. Now days it is dyeing prepared more conveniently by carrying out the diazotization of picramic acid in aqueous solution with sodium nitrite and hydrochloric acid at 0°C.

OH   OH   O N

O_2N NH_2 O_2N N_2Cl O_2N N

   diazotisation

   0^0C

NO_2 NO_2 NO_2

picramic acid   unstable   dinol

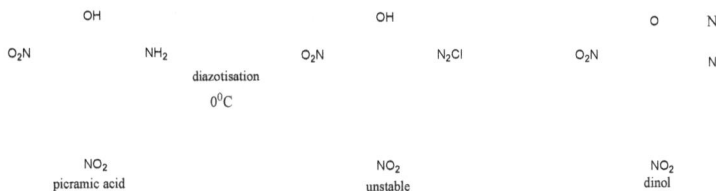

Procedure: Picramic acid is prepared by evaporating of picric acid in alcohol solution with ammonium sulphide. Picramic acid forms red needle shaped crystsls, m. p. 169^0C.

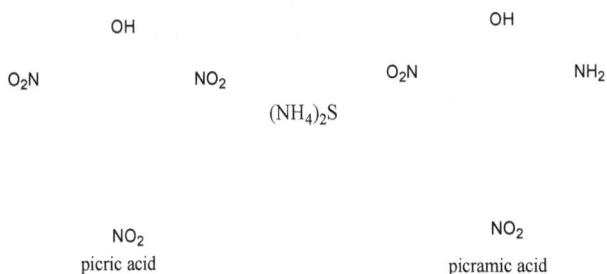

OH   OH

O_2N NO_2 O_2N NH_2

    $(NH_4)_2S$

NO_2 NO_2

picric acid   picramic acid

Picramic acid obtained as above is suspended in 5%HCl in a vessel. This mixture is cooled to 0^0C by the cooling mixture. To the mixture, about 3 to 4% aqueous solution of sodium nitrite is added and the whole mixture is stirred for 20 minutes. The product obtained is filtered and washed thoroughly with ice water. A dark brown granular crude material is obtained. This can be recrystallised from acetone when a pale yellow amorphous powder of dinol is obtained. Its colour gets darkened on exposure to sunlight. It does not react with water.

Use: Dinol is less sensitive to impact than mercury fulminate and lead azide. However, it detonates when struck with a sharp blow. It burns without quick flash. It is common primary explosive used in detonators and fuses.

4. Food colours

4.1 Introduction

Food color is any substance that is added to food or drink to change its color. Food coloring is used both in commercial food production and in domestic cooking. Due to its safety and general availability, food coloring is also used in a variety of non-food applications, for example in home craft projects and educational settings etc.

4.2 Purpose of food coloring

People associate certain colors with certain flavors, and the color of food can influence the perceived flavor in anything from candy to wine. For this reason, food manufacturers add dyes to their products. Sometimes the aim is to simulate a color that is perceived by the consumer as natural, such as adding red coloring to glace cherries (which would otherwise be beige), but sometimes it is for effect, like the green ketchup that Heinz launched in 2000.While most consumers are aware that food with bright or unnatural colors (such as the green ketchup, or children's cereals such as Froot Loops) likely contain food coloring, far fewer people know that seemingly "natural" foods such as oranges and salmon are sometimes also dyed to mask natural variations in color. Color variation in foods throughout the seasons and the effects of processing and storage often make color addition commercially advantageous to maintain the color expected or preferred by the consumer. Some of the primary reasons include

(i) Offsetting of color loss due to light, air, extremes of temperature, moisture, and storage conditions.

(ii) Masking natural variations in color.

(iii) Enhancing naturally occurring colors.

(iv) Providing identity to foods.

(v) Protecting flavors and vitamins from damage by light.

(vi) Decorative or artistic purposes such as cake icing.

4.3 Regulation

Food colorings are tested for safety by various bodies around the world and sometimes different bodies have different views on food color safety. In the United States, an FD and C number (which generally indicates that the FDA has approved the colorant for use in foods, drugs and cosmetics) are given to approved synthetic food dyes that do not exist in nature, while in the European Union, E numbers are used for all additives, both synthetic and natural, that are approved in food applications.

Most other countries have their own regulations and list of food colors which can be used in various applications, including maximum daily intake limits. Natural colors are not required to be tested by a number of regulatory bodies throughout the world, including the United States FDA. The FDA lists "*color additives exempt from certification*" for food in subpart A of the Code of Federal Regulations - Title 21 Part

73.However, this list contains substances which may have synthetic origins. There are **26** colors permitted to be used in food and **28** to be used in cosmetics and pharmaceuticals.

There are major three categories of food colors;

1) Natural colors
2) Synthetic colors
3) Lakes and dyes

4.4 Natural Food Color

Natural Food Color is any dye, pigment or any other substance obtained from vegetable, animal, mineral, or source capable of coloring food drug, cosmetic or any part of human body, colors come from variety of sources such as seeds, fruits, vegetables, algae & insect. A growing number of natural food dyes are being commercially produced, partly due to consumer concerns surrounding synthetic dyes. Some examples include:

(a) Caramel coloring (E150), made from caramelized sugar, used in cola products and also in cosmetics.

(b) Annatto (E160b), a reddish-orange dye made from the seed of the Achiote.

(c) A green dye made from chlorella algae (chlorophyll, E140)

(d) Cochineal (E120), a red dye derived from the cochineal insect, Dactylopius coccus.

(e) Betanin extracted from beets.

(f) Turmeric (curcuminoids, E100)

(g) Saffron (carotenoids, E160a)

(h) Paprika (E160c)

(i) Elderberry juice

To ensure reproducibility, the colored components of these substances are often provided in highly purified form, and for increased stability and convenience, they can be formulated in suitable carrier materials (solid and liquids)According to the application a suitable Natural Color can be achieved by keeping in mind the factors such as PH. heat, light storage and the other ingredients of the formula or recipe. The storage conditions for natural colors depend on the particular need of the product. A tight sealed container is best to store he product in a cool storage to preserve color strength and quality, along with its degree of cooling point.

Here is a list of few natural food colors

Table:1 natural food colours

Natural colour	EEC Number	Description
Annatto	E 160 b	Liqiud /Powder (os) Liquid/Powder(ws)
Turmeric	E100	Paste 35% curcumin Liquid(os) up to 12% curcumin Liquid(ws) up to 10% curcumin Powder up to 95% curcumin
Cochineal Carmini	E120	Powder, N. L. T. 50% carminic acid

		Powder, 30-95% carminic acid
		Powder, up to 40% carminic acid
		Liquid(ws), 2-10% carminic acid
Paprika	E 160 c	Liquid(os),10,000 to 1,60,000 Cu
		Liquid(ws),10,000 to 50,000 Cu
		Powder, spray dried
Anthocyanins	E 163	Liquid/powder(ws)
β-Carotenes Carotenoids	E 160 a	Powder(ws)up to 7.5% mix carotenes
		Liquid(os) up to 2.5% mix carotinoids
Gardenia	-	Liquid(ws)
Iron oxides	E 172	Powder
Marigold extract	E 161 b	Paste(os) up to 10 % xanthophylls
Lutein	-	Powder up to 51 % lutein
Chlorophyll	E 140	Liquid(ws/os)
	E 141	Liquid(ws/os), Powder (ws)
Titanium dioxide	E 172	Powder, paste (ws/os)
Carbon black	E 153	Powder, paste, emulsion in oil
Beet root	E 162	Liquid/powder(ws)
Safflower	-	Liquid/powder(ws)
Caramel	E 160 d	Powder(os) 2-20%
Monascus	-	Liquid/powder(ws)
Red cabbage	-	Powder
Raddish	-	Powder

4.5 Detail of few major colors

(i) Annatto:

Annatto, also called *Raucous*, is a derivative of the *achiote trees* of tropical regions of the Americas, used to produce *a red food coloring* and also as a flavoring. Its scent is described as "slightly peppery with a hint of nutmeg" and flavor as *slightly sweet and peppery*. Annatto is produced from the reddish pulp which surrounds the seed of the achiote (*Bixaorellana L.*). It is used in many cheeses (e.g., Cheddar, Red Leicester, and Brie), margarine, butter, rice, smoked fish, and custard powder. Annatto is commonly found in Latin America and Caribbean cuisines as both a coloring agent and for flavoring. Central and South American natives use the seeds to make a body paint, and lipstick. For this reason, the achiote is sometimes called the lipstick-tree. Achiote originated in South America and has spread in popularity to many parts of Asia. The heart shaped fruits are brown or reddish brown at maturity, and are covered with short, stiff hairs. When fully mature, the fruits split open exposing the numerous dark red seeds. While the fruit itself is not edible, the orange-red pulp that covers the seed is used as a commercial food coloring and dye. The achiote dye is prepared by stirring the seeds in water or oil.

History

Annatto has long been used by indigenous Caribbean and South American cultures. It is believed to originate in Brazil. It was probably not initially used as a food additive but for other reasons, such as body painting, to ward off evil, and as an insect repellent. The ancient Aztecs called it *achiotl*, and it was used for Mexican manuscript painting in the sixteenth century, as a food coloring, as a food additive, Annatto has the E number *E160b*. The fat soluble part of the crude extract is called *bixin,* the water soluble part is called *norbixin,* and both share the same E number as annatto. Annatto seed contains 4.5-5.5% pigments, which consists of 70-80% bixin. In the United States, annatto extract is listed as a color additive *"exempt from certification"* and is commonly considered to be a natural color. The yellowish orange color is produced by the chemical compounds bixin and norbixin, which are classified as xanthophylls, a type of carotenoid. However, unlike beta-carotene, another well-known carotenoid, they do not have the correct chemical structures to be vitamin A precursors. The more norbixin in an Arnatto color, the more yellow it is; a higher level of bixin gives it a more reddish shade. Unless an acid-proof version is used, it takes on a pink shade at low P^H.

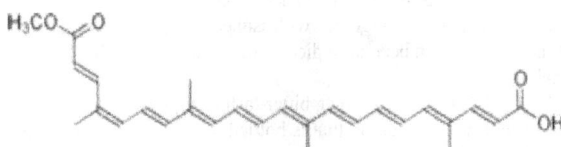

Cheddar cheese is often colored as an allergic, annatto has been linked with many cases of food-related allergies, and is the only natural food coloring believed to cause as many allergic type reactions as artificial food coloring.

(ii) Betanin

IUPAC name: 4-(2-(2-carboxy-5-(β-D-glucopyranosyloxy)-2,3-dihydro-6-hydroxy-1H-indol-1-yl)ethenyl) -2, 6-Pyridinedicarboxylic acid

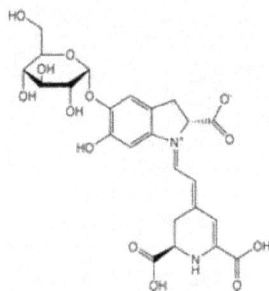

Molecular formula: $C_{24}H_{27}N_2O_{13}$
Molar mass: 551. 48 g/mol
Betanin, or Beetroot Red, is a *red glycosidic food dye* obtained from beets; its aglycone, obtained by hydrolyzing away the glucose molecule, is betanidin. As a food additive, its E number is *162*. Betanin degrades when subjected to light, heat, and oxygen; therefore, it is used in frozen products, products with short shelf life, or products sold in dry state. Betanin can survive pasteurization when in products with high sugar content. Its sensitivity to oxygen is highest in products with high content of water and/or containing metal cations (e.g. iron and copper); antioxidants like ascorbic acid and sequestrants can slow this process down, together with suitable packaging. In dry form Betanin is stable in presence of oxygen.

Betanin is usually obtained from the extract of beet juice; the concentration of Betanin in red beet can reach 300-600 mg/kg. Other dietary sources of Betanin and other Betanins include the *opuntia cactus, Swiss chard,* and the leaves of some strains of Amaranth. The color of Betanin depends on pH; between *four* and *five* it is *bright bluish-red* becoming *blue-violet* as the pH increases. Once the pH reaches *alkaline* levels Betanin degrades by hydrolysis, resulting in a *yellow-brown color.*

(i) Betanin can be also used for coloring meat and sausages.

(ii) The most common uses of Betanin are in coloring ice cream and powdered soft drink beverages; other uses are in some sugar confectionery, e.g. fondants, sugar strands, sugarcoating, and fruit or cream fillings. In hot processed candies, it can be used if added at the final part of the processing. Betanin is also used in soups as well as tomato and bacon products.

(iii) Betanin absorbs well from the gut and acts as an antioxidant.

Betanin is a betalain pigment, together with isobetanin, probetanin, and neobetanin. Other pigments contained in beet are indicaxanthin and vulgaxanthins.

(iii) Caramel

Caramel color (E150) is a dark, rather bitter-tasting liquid, the highly concentrated product of near total caramelization that is bottled for commercial and industrial use. Beverages such as cola use caramel coloring, and it is also used as food coloring. Caramel color or caramel coloring is a soluble food coloring. It is made by a carefully controlled heat treatment of carbohydrates, generally in the presence of acids, alkalis, or salts, in a process called caramelization. It is more fully oxidized than caramel candy and has an odour of burnt sugar and a somewhat bitter taste. Its color ranges from *pale yellow* to *amber* to *dark brown.*

Caramel color is one of the oldest and most widely-used food colorings, and is found in almost every kind of industrially produced food, including: batters, beer, brown bread, buns, chocolate, cookies, cough drops, dark liquor such as brandy, rum, and whisky, chocolate flavored flour based confectionery, coatings, custards, decorations, fillings and toppings, potato chips, dessert mixes, doughnuts, fish and shellfish spreads, frozen desserts, Fruit preserves, glucose tablets, gravy browning, ice cream, pickles, sauces and dressings, soft drinks (especially colas), sweets, vinegar, and wines.

4.6 Production

Caramel color is made by the *controlled heat treatment* of carbohydrates (nutritive sweeteners which are the monomers glucose and fructose or their polymers, *e.g.* glucose syrups, sucrose, invert syrups, and dextrose, generally in the presence of food-grade acids, alkalis, and salts, in a process called caramelization. Antifoaming agents may be used as processing aids during its manufacture. Unlike caramel candy, it tends towards maximum oxidation of the sugar to produce a caramel concentrate that has an odor of burnt sugar and a somewhat bitter taste in its raw liquid form. Its color ranges from pale yellow to amber to dark brown. Caramel color molecules carry either a positive or negative charge depending upon the reactants used in their manufacture. Problems such as precipitation, flocculation or migration can be eliminated with the use of a properly charged caramel color for the intended application.

4.7 Classification

Internationally, the Joint FAO/WHO Expert Committee on Food Additives (JECFA) recognizes *four* classes of caramel color, differing by intended application and in the reactants used in their manufacture, each with its own INS and E number that are listed in the table below.

Table :2 classification and E number

Class	INS No.	E No.	Description	Restriction in reaction	Uses
I	150a	E150a	Plain caramel, caustic caramel, spirit caramel	No ammonium or sulfite compounds can be used	Whiskey among Many
II	150b	E150b	Caustic sulfite caramel	In the presence of sulfite compounds but no ammonium compounds can be used	Beer and confectionary
III	150c	E150c	Ammonia caramel, baker's, caramel, confectioner's caramel, beer caramel	In the presence of Ammonium compo-unds but no sulfite compounds can be used	Beer, synthetic soya sauce, and confectionery
IV	150d	E150d	Sulfite ammonia caramel, acid-proof caramel, soft drink caramel	In the presence of both sulfite and ammonium compounds	Acidic environ- ments such as soft drinks

Physical properties

Caramel color is a *colloid.* Although the primary function of caramel color is for coloring, it also serves additional functions. In soft drinks, it functions as an emulsifier to help retard the formation of certain types of *"floc"* and its light protective quality can aid in preventing oxidation of the flavoring components in bottled beverages.

Caramel color has excellent microbiological stability. it is manufactured under very high temperature, high acidity, high pressure, and high specific gravity, it is essentially sterile as it will not support microbial growth unless in a dilute solution. When reacted with sulfites, caramel color may retain traces of sulfite after processing. However, in finished food products, labeling is usually only required for sulfite levels above 10 ppm.

(i) Carmine

Carmine also called Crimson Lake, Cochineal, Natural Red 4, C I 75470, or E120, is a pigment of a *bright red color* obtained from the carminic acid produced by some scale insects, such as the cochineal and the Polish cochineal, and is used as a general term for a particularly deep red color of the same name.

Carmine is used in the manufacture of artificial flowers, paints, crimson ink, rouge, and other cosmetics, and is routinely added to food products such as yogurt and certain brands of juice, most notably those of the ruby-red variety.

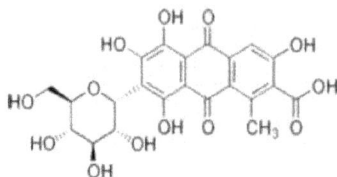

Production

Carmine may be prepared from cochineal, by boiling dried insects in water to extract the carminic acid and then treating the clear solution with alum, cream of tartar, stannous chloride, or potassium hydrogen oxalate; the coloring and animal matters present in the liquid are thus precipitated. Other methods are in use in which egg white, fish glue, or gelatin are sometimes added before the precipitation.

The quality of carmine is affected by the temperature and the degree of illumination during its preparation, sunlight being requisite for the production of a brilliant hue. It also differs according to the amount of alumina present in it. It is sometimes adulterated with cinnabar, starched other materials; from these the carmine can be separated by dissolving it in ammonia. Good carmine should crumble readily between the fingers when dry. Carmine lake is a pigment obtained by adding freshly precipitated alumina to decoction of cochineal.

Carmine can be used as a staining agent in microbiology, as a *Best's carmine* to stain glycogen, *mucicarmine* to stain *acidic mucopolysaccharides,* and *carmalum* to stain *cellnuclei.* In these applications, it is applied together with a mordant, usually an Al (III) salt.

Allergy

Carmine is used as a food dye in many different products such as juices, ice cream, yogurt, and candy, and as a dye in cosmetic products such as eye shadow and lipstick. Although principally a red dye, it is found in many foods that are shades of red, pink, and purple. As a food dye it has been known to cause severe allergic reactions and anaphylactic shock in some people. Food products containing carmine-based food dye may prove to be a concern for people who are allergic to carmine, or people who choose not consume any or certain animals, suchas vegetarians, vegans.

(ii) Carotene

The term **carotene** is used for several related hydrocarbon substances having the formula$C40Hx$, which are synthesized by plants but cannot be made by animals. Carotene is an orange photosynthetic pigment important for photosynthesis. Carotenes are all colored to the human eye. They are responsible for the orange color of the carrot, for which this class of chemicals is named and for the colours of many other fruits and vegetables (for example, sweet potatoes and orange cantaloupe melon). Carotenes are also responsible for the orange (but not all of the yellow) colours in dry foliage. They also (in lower concentrations) impart the yellow coloration to milk-fat and butter. Omnivorous animal species which are relatively poor converters of colored dietary carotenoids to colorless retinoid have yellow colored body fat, as a result of the carotenoids retention from the vegetable portion of their diet. The typical yellow-colored fat of humans and chickens is a result of fat storage of carotenes from their diets.

Carotenes contribute to photosynthesis by transmitting the light energy they absorb from chlorophyll. They also protect plant tissues by helping to absorb the energy from singlet oxygen, an excited form of the oxygen molecule O_2 which is formed during photosynthesis. β-Carotene is composed of two retinyl groups, and is broken down in the mucosa of the human small intestine by beta-carotene 15,15'-monooxygenase to retinal, a form of vitamin A. β-Carotene can be stored in the liver and body fat and converted to retinal as needed, thus making it a form of vitamin A for humans and some other mammals. The carotenes α-carotene and γ-carotene, due to their single retinyl group (beta-ionone ring), also have some vitamin-A activity (though less than β-carotene), as does the xanthophylls carotenoids β-cryptoxanthin. All other carotenoids, including lycopene, have no beta-ring and thus no vitamin A activity (although they may have antioxidant activity and thus biological activity in other ways). Animal species differ greatly in their ability to convert retinyl (beta-ionone) containing carotenoids to retinals. Carnivores in general are poor converters of dietary ionine containing carotenoids, and pure carnivores such as cats and ferets lack beta-carotene 15, 15'-monooxygenase and cannot convert any carotenoids to retinals at all (resulting in carotenes not being a form of vitamin A for these species).

Molecular structure

Chemically, carotenes are polyunsaturated hydrocarbons containing 40 carbon atoms per molecule, variable numbers of hydrogen atoms, and no other elements. Some carotenes are terminated by hydrocarbon rings, on one or both ends of the molecule. All are colored to the human eye, due to extensive systems of conjugated double bonds. Structurally carotenes are terpenes, synthesized biochemically from eight isoprene units. Carotenes are found in plants in two primary forms designated by characters from the Greek alphabet:

- alpha-carotene (α-carotene)
- Beta-carotene (β-carotene).
- Gamma, delta, epsilon, and zeta (γ, δ, ε, and ζ-carotene) also exist.

Since they are hydrocarbons, and therefore contain no oxygen, carotenes are fatsoluble and insoluble in water (in contrast with other carotenoids, the xanthophylls, which contain oxygen and thus are less chemically hydrophobic).

Dietary sources

The following foods are particularly rich in carotenes;

Sweet potatoes, Carrots, Cantaloupe melon, Mangoes, Apricots, Persimmon, Spinach, Kale, Turnip greens, Beet greens, Mustard greens, Broccoli, Parsley, Romaine lettuce, Ivy gourd, Winter squash, Pumpkin, Cassava.

Absorption from these foods is enhanced if eaten with fats, as carotenes are fat soluble, and if the food is cooked for a few minutes until the plant cell wall splits and the color is released into any liquid. 6 μg of dietary β-carotene supplies the equivalent of 1 μg of retinol, or 1 RE (Retinol Equivalent). This is equivalent to 3⅓ IU of vitamin A.

The multiple forms

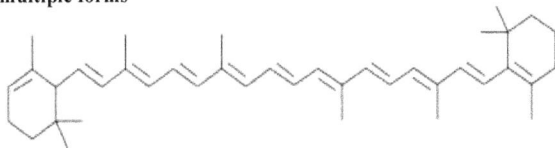

α-carotene

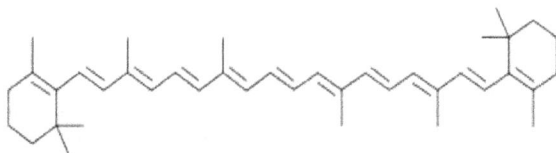

β-carotene

The two primary isomers of carotene, α-carotene and β-carotene, differ in the position of double bonds in the cyclic group at the end. β-Carotene is the more common form and can be found in yellow, orange, and green leafy fruits and vegetables. As a rule of thumb, the greater the intensity of the orange color of the fruit or vegetable, the more β-carotene it contains. Carotene protects plant cells against the destructive effects of ultraviolet light. β-Carotene is an anti-oxidant.

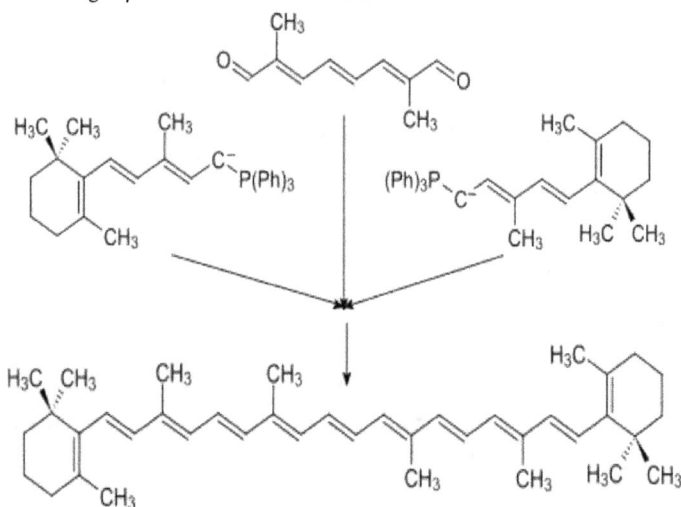

Nomenclature

Carotenes are carotenoids containing no oxygen. Carotenoids containing some oxygen are known as xanthophylls. The two ends of the β-carotene molecule are structurally identical, and are called *β-rings*. Specifically, the group of nine carbon atoms at each end forms a *β-ring*. The α-carotene molecule has a *β-ring* at one end; the other end is called a *ε-ring*. There is no such thing as a "α-ring".

(iii) Carthamin

Carthamin is a natural red pigment derived from safflower (*Carthamus tinctorius*), earlier known as carthamine. It is used as a dye and a food coloring. As a food additive, it is known as Natural Red 26. Safflower has been cultivated since ancient times, and carthamin was used as a dye in ancient Egypt. It was used extensively in the past for dyeing wool for the carpet industry in European countries and to create cosmetics for geisha and kabuki artists in Japan, where the color is called *Beni*. It competed with the early synthetic dye fuchsine as a silk dye after fuchsine's 1859 discovery. It is composed of two chalcones; the conjugated bonds being the cause of the red color. It is derived from precarthamin by a decarboxylase. It should not be confused with carthamidin, other flavonoids.

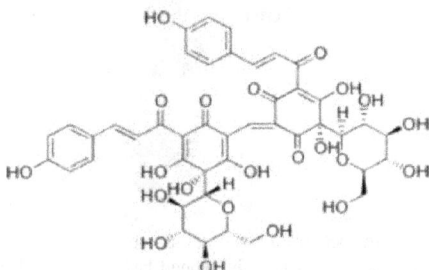

(iv) Curcumin

Curcumin is the principal curcumin oil of the popular Indian spice turmeric, which is a member of the ginger family (Zingiberaceae). The other two curcuminoids are desmethoxycurcumin and bis-desmethoxycurcumin. The curcuminoids are polyphenols and are responsible for the yellow color of turmeric. Curcumin can exist in at least two tautomeric forms, *keto* and *enol*. The enol form is more energetically stable in the solid phase and in solution. Curcumin can be used for boron quantification in the so-called curcumin method. It reacts with boric acid forming a red colored compound, known as rosocyanine. Curcumin is brightly colored and may be used as a food coloring. As a food additive, its E number is E100

Chemistry

Curcumin incorporates several functional groups. The aromatic ring systems, which are polyphenols, are connected by two α, β-unsaturated carbonyl groups. The two carbonyl groups form a diketone. The diketone form stable enols or are easily deprotonated and form enolates, while the α, β-unsaturated carbonyl is a good Michael acceptor and undergoes nucleophilic addition.

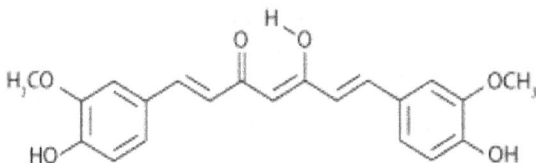

The structure was first identified in 1910 by *Kazimierz Kostanecki, J. Miłobędzka* and *Wiktor Lampe*. Curcumin is used as a reagent for Boron in EPA Method 212.3.Boron by colorimetry.

(v) Turmeric

Turmeric (Curcuma longa) is a rhizomatous herbaceous perennial plant of the ginger family, **Zingiberaceae**. It is native to tropical South Asia and needs temperatures between 20°C and 30°C, and a considerable amount of annual rainfall to thrive. Plants are gathered annually for their rhizomes, and re-seeded from some of those rhizomes in the following season. The rhizomes are boiled for several hours and then dried in hot ovens, after which they are ground into a deep orange-yellow powder commonly used as a spice in curries and other South Asian and Middle Eastern cuisine, for dyeing, and to impart color to mustard condiments. Its active ingredient is curcumin and it has a distinctly earthy, slightly bitter, slightly hot peppery flavor and a mustardy smell. In medieval Europe, turmeric became known as Indian Saffron, since it was widely used as an alternative to the far more expensive saffron spice.

In non-South Asian recipes, turmeric is sometimes used as an agent to impart rich, custard like yellow color. It is used in canned beverages and baked products, dairy products, ice, yogurt, yellow cakes, orange juice, biscuits, popcorn color, sweets, cake icings, cereals, sauces, gelatins, etc. It is a significant ingredient in most commercial curry powders. Turmeric is mostly used in savory dishes, as well as some sweet dishes such as the cake Sfouf. Although usually used in its dried, powdered form, turmeric is also used fresh, much like ginger. It has numerous uses in Far Eastern recipes, such as fresh turmeric pickle which contains large chunks of soft turmeric. Turmeric (coded as E100 when used as a food additive) is used to protect food products from sunlight. The oleoresin is used for oil-containing products. The curcumin/ polysorbate solution or curcumin powder dissolved in alcohol is used for water-containing products. Over coloring, such as in pickles, relishes, and mustard, are sometimes used to compensate for fading. In combination with annatto (E160b), turmeric has been used to color cheeses, yogurt, dry mixes, salad dressings, winter butter and margarine. Turmeric is also used to give a yellow color to some prepared mustards, canned chicken broths and other foods (often as a much cheaper replacement for saffron).

Turmeric is widely used as a spice in South Asian and Middle Eastern cooking. Many Persian dishes use Turmeric, for the coloring of rice bottoms as well as a starter ingredient for almost all Iranian fry ups (which typically consist of oil, onions and turmeric followed by any other ingredients that are to be included). In Nepal, turmeric is widely grown and is extensively used in almost every vegetable and meat dish in the country for its color as well as for its medicinal value. In South Africa turmeric is traditionally used to give boiled white rice a golden color.

Composition

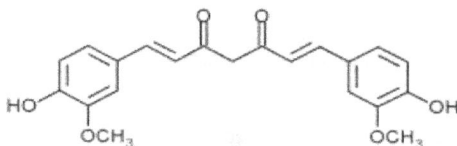

Curcumin Keto form

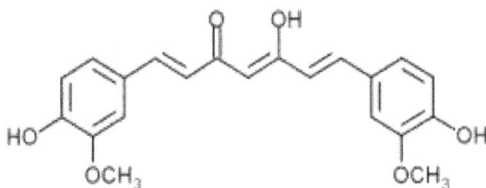

Curcumin enol form

Turmeric contains up to 5% essential oils and up to 3% curcumin, a polyphenol. It is the active substance of turmeric and it is also known as C.I. 75300, or Natural Yellow 3. The systematic chemical name is (1*E*, 6*E*)-1, 7-bis (4-hydroxy-3-methoxyphenyl)-1,6-heptadiene-3,5-dione.It can exist at least in two tautomeric forms, keto and enol. The keto form is preferred in solid phase and the enol form in solution. Curcumin is a pH indicator. In acidic solutions (pH < 7.4) it turns yellow whereas in basic (pH > 8.6) solutions it turns bright red.

(vi) Anthocyanin

Anthocyanins (from Greek: (anthos) = flower + (kyanos) = blue) are water-soluble vacuolar pigments that may appear red, purple, or blue according to pH. They belong to apparent class of molecules called flavonoids synthesized via the phenylpropanoid pathway; they are odorless and nearly flavorless, contributing to taste as a moderately astringent sensation. Anthocyanins occur in all tissues of higher plants, including leaves, stems, roots, flowers, and fruits. Anthoxanthins are their clear, white to yellow counterparts occurring in plants. Anthocyanins are derivatives of anthocyanidins which include pendant sugars.

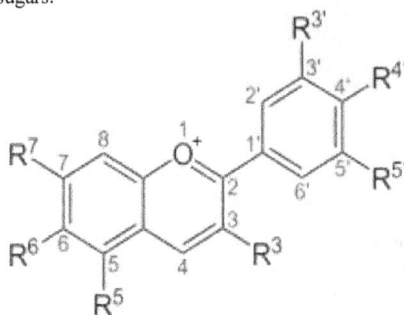

Function

In flowers, bright reds and purples are adaptive for attracting pollinators. In fruits, the colorful skins also attract the attention of animals, which may eat the fruits and disperse the seeds. In photosynthetic tissues (such as leaves and sometimes stems), anthocyanins have been shown to act as a "sunscreen", protecting cells from high-light damage by absorbing blue-green and UV light, thereby protecting the tissues from photo inhibition, or high-light stress. This has been shown to occur in red juvenile leaves, autumn leaves, and broad-leaved evergreen leaves that turn red during the winter. It has also been proposed that red coloration of leaves may camouflage leaves from herbivores blind to red wavelengths, or signal un palatability, since anthocyanin synthesis often coincides with synthesis of unpalatable phenolic compounds. In addition to their role as light-attenuators, anthocyanins also act as powerful antioxidants. However, it is not clear whether anthocyanins can significantly contribute to scavenging of free-radicals produced through metabolic processes in leaves, since they are located in the vacuole, and thus, spatially separated from metabolic reactive oxygen species. Some studies have shown that hydrogen peroxide produced in other organelles can be neutralized by vacuolar anthocyanin.

Occurrence

Anthocyanins are found in the cell vacuole, mostly in flowers and fruits but also in leaves, stems, and roots. In these parts they are found predominantly in outer cell layers such as the epidermis and peripheral mesophyll cells. Most frequent in nature are the glycosides of cyanidin, delphinidin, malvidin, pelargonidin, peonidin and petunidin. Roughly 2% of all hydrocarbons fixated in photosynthesis are converted into flavonoids and their derivatives such as the anthocyanins. There is no less than109 tons of anthocyanins produced in nature per year. Not all land plants contain anthocyanin; in the Caryophyllales (including cactus, beets, and amaranth), they are replaced by betalains. Plants rich in anthocyanins are Vaccinium species, such as blueberry, cranberry and bilberry, Rubus berries including black rasp berry, red raspberry and blackberry, blackcurrant, cherry, egg plant peel, black rice, Concord grape and muscadine grape, red cabbage and violet petals. Anthocyanins are less abundant in banana, asparagus, pea, fennel, pear and potato, and maybe totally absent in certain cultivars of green gooseberries.The highest recorded amount appears to be specifically in the seed coat of black soybean(Glycine max L. Merr.) containing some 2,000 mg per 100 g and in skins and pulp of black chokeberry (Aronia melanocarpa L.) . However, the Amazonian palmberry, acai, having about 320 mg per 100 g of which cyanidin-3-glucoside is the most prevalent individual anthocyanin (approximately 10 mg per 100 g), is also a high-content source for which only a small fraction of total anthocyanins has been determined to date.

(vii) Chlorophyll

Chlorophyll is a green pigment found in most plants, algae, and cyanobacteria. Its name is derived from the Greek (*chloros "green"*) and (*phyllon "leaf"*). Chlorophyll absorbs light most strongly in the blue portion of the electromagnetic spectrum, followed by the red portion. However, it is a poor absorber of green and near-green portions of the spectrum, hence the green color of chlorophyll-containing tissues. Chlorophyll was first isolated by Joseph Bienaime Caventou and Pierre Joseph Pelletier in 1817.

Chlorophyll and photosynthesis

Chlorophyll is vital for photosynthesis, which allows plants to obtain energy from light. Chlorophyll molecules are specifically arranged in and around photo systems which are embedded in the thylakoid membranes of chloroplasts. In these complexes, chlorophyll serves two primary functions. The function of the vast majority of chlorophyll (up to several hundred molecules per photo system) is to absorb light and transfer that light energy by resonance energy transfer to a specific chlorophyll pair in the reaction center of the photo systems. Because of chlorophyll's selectivity regarding the wavelength of light it absorbs, areas of a leaf containing the molecule will appear green.

The two currently accepted photo system units are Photo system II and Photo system I, which have their own distinct reaction center chlorophylls, named P680 and P700, respectively. These pigments are named after the wavelength (in nanometers) of their red-peak absorption maximum. The identity, function and spectral properties of the types of chlorophyll in each photo system are distinct and determined by each other and the protein structure surrounding them. Once extracted from the protein into a solvent (such as acetone or methanol), these chlorophyll pigments can be separated in a simple paper chromatography experiment, and, based on the number of polar groups between chlorophyll a and chlorophyll b, will chemically separate out on the paper. The function of the reaction center chlorophyll is to use the energy absorbed by and transferred to it from the other chlorophyll pigments in the photo systems to undergo a charge separation, a specific redox reaction in which the chlorophyll donates an electron into a series of molecular intermediates called an electron transport chain. The charged reaction center chlorophyll (P680+) is then reduced back to its ground state by accepting an electron. In Photo system II, the electron which reduces P680+ ultimately comes from the oxidation of water into O_2 and H+ through several intermediates. This reaction is how photosynthetic organisms like plants produce O_2 gas, and is the source for practically all the O_2 in Earth's atmosphere. Photo system I typically work in series with Photo system II, thus the P700+ of Photo system I is usually reduced, via many intermediates in the thylakoid membrane, by electrons ultimately from Photo system II. Electron transfer reactions in the thylakoid membranes are complex, however, and the source of electrons used to reduce P700+ can vary.

Chemical structure

Chlorophyll is a chlorin pigment, which is structurally similar to and produced through the same metabolic pathway as other porphyrin pigments such as heme. At the center of the chlorin ring is a magnesium ion. The chlorin ring can have several different side chains, usually including a long phytol chain. There are a few different forms that occur naturally, but the most widely distributed form in terrestrial plants is chlorophyll a. The general structure of chlorophyll a was elucidated by Hans Fischer in 1940, and by 1960, when most of the stereochemistry of chlorophyll a was known, Robert Burns Woodward published a total synthesis of the molecule as then known. In 1967, the last remaining stereo chemical elucidation was completed by Ian Fleming, and in 1990 Woodward and co-authors published an updated synthesis. The different structures of chlorophyll are summarized below:

Table:3 different structure of chlorophyll

Structure	Chlorophyll a	Chlorophyll b	Chlorophyll C_1	Chlorophyll C_2	Chlorophyll d
Molecular formula	$C_{55}H_{72}O_5N_4Mg$	$C_{55}H_{70}O_6N_4Mg$	$C_{35}H_{30}O_5N_4Mg$	$C_{35}H_{28}O_5N_4Mg$	$C_{54}H_{70}O_6N_4Mg$
C_3 Group	-CH=CH$_2$	-CH=CH$_2$	-CH=CH$_2$	-CH=CH$_2$	-CHO
C_7 Group	-CH$_3$	-CHO	-CH$_3$	-CH$_3$	-CH$_3$
C_8 Group	-CH$_2$-CH$_3$	-CH$_2$-CH$_3$	-CH$_2$-CH$_3$	-CH=CH$_2$	-CH$_2$-CH$_3$
C_{17} Group	-CH$_2$-CH$_2$-COO-	-CH$_2$-CH$_2$-COO-	-CH=CH-COOH	-CH=CH-COOH	-CH$_2$-CH$_2$-COO-
C_{17}-C_{18} bond	Single	Single	Double	Double	Single
Occurrence	Universal	Mostly plant	Various Algae	Various Algae	Cyanobacteria

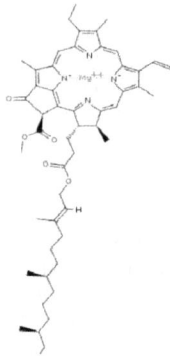

Chlorophyll a   Chlorophyll b   Chlorophyll d

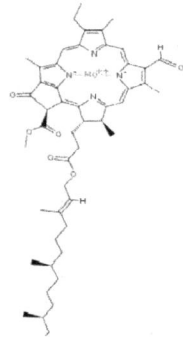

When leaves degreen in the process of plant senescence chlorophyll is converted to a group of colourless tetrapyrroles known as nonfluorescent chlorophyll catabolites (NCC's). These compounds have also been identified in several ripening fruits. General structure of NCC's:

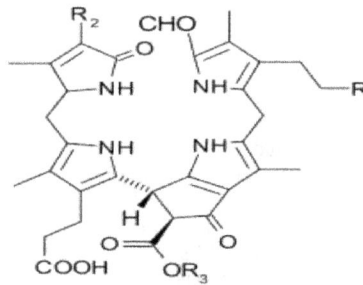

Spectrophotometry

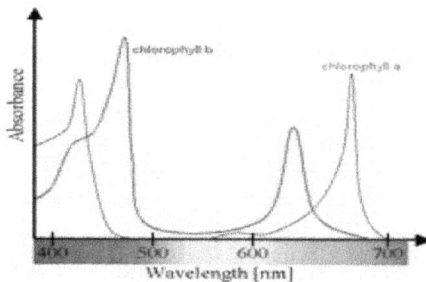

Measurement of the absorption of light is complicated by the solvent used to extract it from plant material, which affects the values obtained, In diethyl ether, chlorophyll a has approximate absorbance maxima of 430 nm and 662 nm, while chlorophyll b has approximate maxima of 453 nm and 642 nm. The absorption peaks of chlorophyll a

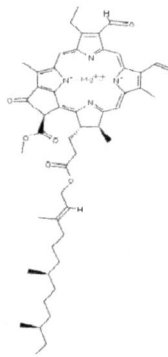

are at 665 nm and 465 nm. Chlorophyll *a* fluorescence at 673 nm (maximum) and 726 nm. The peak molar absorption co-efficient of chlorophyll *a* exceeds 10^5 M^{-1} cm^{-1}, which is among the highest for small-molecule organic compounds.

Biosynthesis

In plants, chlorophyll may be synthesized from succinyl-Co A and Glycine, although the immediate precursor to chlorophyll *a* and *b* is protochlorophyllide. In Angiosperms, the last step, conversion of protochlorophyllide to chlorophyll, is light-dependent and such plants are pale (etiolated) if grown in the darkness. Non-vascular plants and green algae have an additional light-independent enzyme and grow green in the darkness as well Chlorophyll itself is bound to proteins and can transfer the absorbed energy in the required direction. Protochlorophyllide, differently, mostly occur in the free form and under light conditions act as photosensitizer, forming highly toxic free radicals. Hence plants need an efficient mechanism of regulating the amount of chlorophyll precursor. In angiosperms, this is done at the step of aminolevulinic acid (ALA), one of the intermediate compounds in the biosynthesis pathway. Plants that are fed by ALA accumulate high and toxic levels of protochlorophyllide, so do the mutants with the damaged regulatory system.

(viii) Betalain

Betalains are a class of red and yellow indole-derived pigments found in plants of the Caryophyllales, where they replace anthocyanin pigments, as well as some higher order fungi. They are most often noticeable in the petals of flowers, but may color the fruits, leaves, stems, and roots of plants that contain them. They include powerful antioxidant pigments such as those found in beets.

Description

The name "betalain" comes from the Latin name of the common beet (β- vulgaris), from which betalains were first extracted. The deep red color of beets, bougainvillea, amaranth, and many cacti results from the presence of betalain pigments. The particular shades of red to purple are distinctive and unlike that of anthocyanin pigments found in most plants. Betalains may occur in any part of the plant, including the petals of flowers, fruits, leaves, stems, and roots.

There are two categories of betalains:

(a) Betacyanins include the reddish to violet betalain pigments.

(b) Betaxanthins are those betalain pigments which appear yellow to orange.

Among the betaxanthins present in plants include vulgaxanthin, miraxanthin and portulaxanthin, and indicaxanthin. Plant physiologists are uncertain of the function that betalains serve in those plants which possess them, but there is some preliminary evidence that they may have fungicidal properties.

Chemistry

It was once thought that betalains were related to anthocyanins, the reddish pigments found in most plants. Both betalains and anthocyanins are water-soluble pigments found in the vacuoles of plant cells. However, betalains are structurally and chemically unlike anthocyanins. For example, betalains contain nitrogen where as anthocyanins do not. It is now known that betalains are aromatic indole derivatives synthesized from tyrosine. They are not related chemically to the anthocyanins and are not even flavonoids. Each betalain is a glycoside, and consists of a sugar and a colored portion. Their synthesis is promoted by light.

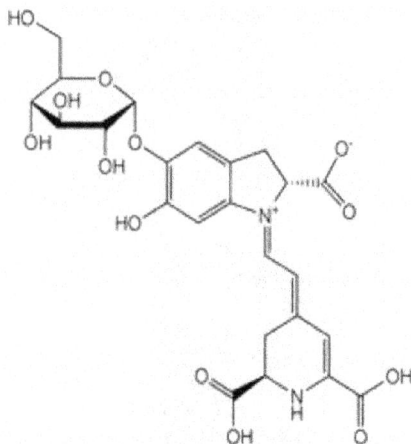

The most heavily studied betalain is betanin, also called beetroot red after the fact that it may be extracted from red beet roots. Betanin is a glucoside, and hydrolyzes into the sugar glucose and betanidin. It is used as a food coloring agent, and the color is sensitive to pH. Other betalains known to occur in beets are isobetanin, probetanin, and neobetanin. Other important betacyanins are amaranthine and isoamaranthine, isolated from species of Amaranthus.

(ix) Myglobin

Myoglobin is an oxygen-binding protein found in the muscle tissue of vertebrates in general and in almost all mammals. It is the heme portion of iron that gives both red blood cells and red muscle their color. In blood, the hemoglobin contains iron, whereas in muscle cells it is the myoglobin that contains the iron. Red blood cells contain hemoglobin, not myoglobin, and so it is hemoglobin that gives red blood cells their color. The only time myoglobin is found in the bloodstream is following muscle injury, when it is released as the result. It is an abnormal finding, and can be diagnostically relevant when found in blood.

Myoglobin (abbreviated Mb) is a single-chain globular protein of 153 or 154 amino acids, containing a heme (iron-containing porphyrin) prosthetic group in the center around which the remaining a pro-protein folds. It has eight alpha helices and a hydrophobic core. It has a molecular weight of and is the primary oxygen-carrying pigment of muscle tissues. Unlike the blood-borne hemoglobin, to which it is structurally related, this protein does not exhibit cooperative binding of oxygen, since positive co-operativity is a property of multimeric/oligomeric proteins only. Instead, the binding of oxygen by myoglobin is unaffected by the oxygen pressure in the surrounding tissue. Myoglobin is often cited as having an "instant binding tenacity" to oxygen given its hyperbolic oxygen dissociation curve. High concentrations of myoglobin in muscle cells allow organisms to hold their breaths longer. Diving mammals such as whales and seals have muscles with particularly high myoglobin abundance.

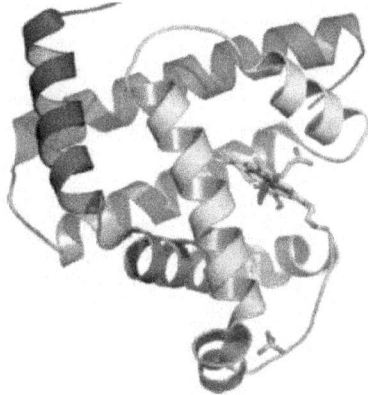

Myoglobin was the first protein to have its three-dimensional structure revealed. In 1958, John Kendrew and associates successfully determined the structure of myoglobin by high resolution X-ray crystallography. For this discovery, John Kendrew shared the 1962 Nobel Prize in chemistry with Max Perutz.

4.8 Meat color

Myoglobin forms pigments responsible for making meat red. The color that meat takes is partly determined by the charge of the iron atom in myoglobin and the oxygen attached to it. When meat is in its raw state, the iron atom is in the +2 oxidation state, and is bound to a dioxygen molecule (O_2). Meat cooked well done is brown because the iron atom is now in the +3 oxidation state, having lost an electron, and is now coordinated by a water molecule. Under some conditions, meat can also remain pink all through cooking, despite being heated to high temperatures. If meat has been exposed to nitrites, it will remain pink because the iron atom is bound to NO, nitric oxide (true of, e.g., corned beef or cured hams). Grilled meats can also take on a pink "smoke ring" that comes from the iron binding a molecule of carbon monoxide to give metmyoglobin. Raw meat packed in a carbon monoxide atmosphere also shows this same pink "smoke ring" due to the same coordination chemistry. Notably, the surface of the raw meat also displays the pink color, which is usually associated in consumers' minds with fresh meat. This artificially-induced pink color can persist in the meat for a very long time, reportedly up to one year. Hormel and Cargill are both reported to use this meat-packing process, and meat treated this way has been in the consumer market since 2003. Myoglobin is found in Type I muscle, Type II A and Type II B, but most texts consider myoglobin not to be found in smooth muscle.

Structure and bonding

Myoglobin contains a porphyrin ring with an iron center. There is a proximal histidine group attached directly to the iron center, and a *distal histidine* group on the opposite face, not bonded to the iron. Many functional models of myoglobin have been studied. One of the most important is that of picket fence porphyrin by James Collman. This model was used to show the importance of the distal prosthetic group. It serves three functions:

(i) To form hydrogen bonds with the dioxygen moiety, increasing the O_2 binding constant.

(ii) To prevent the binding of carbon monoxide, whether from within or without the body. Carbon monoxide binds to iron in an end-on fashion, and is hindered by the presence of the distal histidine, which forces it into a bent conformation. CO binds to heme 23,000 times better than O_2, but only 200 times better in hemoglobin and myoglobin. Oxygen binds in a bent fashion, which can fit with the distal histidine.

(iii) To prevent irreversible dimerization of the oxymyoglobin with another deoxymyoglobin species.

(i) Paprika

Paprika is a spice made from the grinding of dried fruits of *Capsicum annuum* (e.g., bell peppers or chili peppers). In many European languages, the word paprika refers to bell peppers themselves. The seasoning is used in many cuisines to add color and flavor to dishes. Paprika can range from sweet (mild, not hot) to spicy (hot). Flavors also vary from country to country.

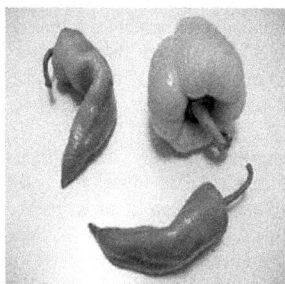

Usage

Paprika is used as an ingredient in a broad variety of dishes throughout the world. Paprika is principally used to season and color rices, stews, and soups, such as goulash, and in the preparation of Sausages as an ingredient that is mixed with meats and other spices. Paprika was first produced in Spain, as that country was also responsible for the introduction of the bell pepper into Europe. The highest quality paprika and most expensive paprika come from Spain.

Nutrition

Capsicum peppers used for paprika are unusually rich in vitamin C, a fact discovered in 1932by Hungary's 1937 Nobel prize-winner Albert Szent-Györgyi. Much of the vitamin C content is retained in paprika, which contains more vitamin C than lemon juice by weight. Paprika is also high in other antioxidants, containing about 10% of the level found in acai berries. Prevalence of nutrients, however, must be balanced against quantities ingested, which are generally negligible for spices.

4.9 Synthetic Food Colors

Synthetic Food Colors also known as Artificial Food Colours, are manufactured chemically and are the most commonly used dyes in the food, pharmaceutical and cosmetic industries. Seven dyes were initially approved under the Pure Food and Drug Act of 1906, but several have been delisted and replacements have been found.

Current seven

In the USA, the following seven artificial colorings are permitted in food under act of 2007:

(i) FD&C Blue No. 1 – Brilliant Blue FCF, E133 (Blue shade)

(ii) FD&C Blue No. 2 – Indigotine, E132 (Dark Blue shade)

(iii) FD&C Green No. 3 – Fast Green FCF, E143 (Bluish green shade)

(iv) FD&C Red No. 40 – Allura Red AC, E129 (Red shade)

(v) FD&C Red No. 3 – Erythrosine, E127 (Pink shade)

(vi) FD&C Yellow No. 5 – Tartrazine, E102 (Yellow shade)

(vii) FD&C Yellow No. 6 – Sunset Yellow FCF, E110 (Orange shade)

Delisted

(i) FD&C Red No. 2 – Amaranth (dye)

(ii) FD&C Red No. 4

(iii) FD&C Red No. 32 was used to color Florida oranges.

(iv) FD&C Orange No. 1, was one of the first water soluble dyes to be commercialized, and one of seven original food dyes allowed under the Pure Food and Drug Act of June 30,1906.

(v) FD&C Orange No. 2 was used to color Florida oranges.

(vi) FD&C Yellows No. 1, 2, 3, and 4

(vii) FD&C Violet No. 1

Synthetic colors are of two types;

(a) Primary Food Colors

Primary colors are those when they are mixed produce other colors.

A list of primary colours is given below:

Product/Colour Shade	C.I.No.	F.D. & C.No	E.No.
QUINOLINE YELLOW	47005	-	E 104
(Di sodium salt of disulfonates of 2-(2quinolyl) - 1, 3 indandione.)			
TARTRAZINE	19140	Yellow 5	E 102
(Tri sodium salt of 5-hydroxy (1-p-sulphophenyl 4- (p-sulphophenylazo) pyrazol -3-carboxylic acid			
SUNSET YELLOW FCF	15985	Yellow 6	E 110
(Di sodium salt of disulfonates of 2-(2quinolyl) - 1, 3 indandione.)			
ERYTHROSINE	45430	Red 3	E 127
(Di sodium salt of disulfonates of 2-(2quinolyl) - 1, 3 indandione.)			
ERYTHROSINE	45430	Red 3	E 127
(Di sodium salt of disulfonates of 2-(2quinolyl) - 1, 3 indandione.)			
PONCEAU 4R	16255	-	E 124
(Di sodium salt of disulfonates of 2-(2quinolyl) - 1, 3 indandione.)			
ALLURA RED	16035	Red 40	E 129
(Di sodium salt of disulfonates of 2-(2quinolyl) - 1, 3 indandione.)			
CARMOISINE	14720	-	E 122
(Di sodium salt of disulfonates of 2-(2quinolyl) - 1, 3 indandione.)			
AMARANTH	16185	Red 2	E 123
(Di sodium salt of disulfonates of 2-(2quinolyl) - 1, 3 indandione.)			
CHOCOLATE BROWN HT	20285	-	E 155
(Di sodium salt of disulfonates of 2-(2quinolyl) - 1, 3 indandione.)			

BRILLIANT BLUE FCF	42090	Blue 1	E 133

(Di sodium salt of disulfonates of 2-(2quinolyl) - 1, 3 indandione.)

PATENT BLUE V	42015	-	E 131

(Di sodium salt of disulfonates of 2-(2quinolyl) - 1, 3 indandione.)

INDIGO CARMINE	73015	Blue 2	E 132

(Di sodium salt of disulfonates of 2-(2quinolyl) - 1, 3 indandione.)

BLACK PN	28440	-	E 151

(Di sodium salt of disulfonates of 2-(2quinolyl) - 1, 3 indandione.)

FAST RED E	16045	-	-

(Di sodium salt of disulfonates of 2-(2quinolyl) - 1, 3 indandione.)

GREEN S	44090	-	E 142

(Di sodium salt of disulfonates of 2-(2quinolyl) - 1, 3 indandione.)

RED 2G	18050	-	E 128

(Di sodium salt of 8-actamido-1-hydroxy-2- phenylazonaphthalene-3,6 disulponate)

(b) Blended Food Colors

Blended Colors are prepared from mixing of previously certified batches of primary colors. Blends can be made available to meet specific requirements of a customer in terms of shade and strength. Widely used blends are as below:

EGG YELLOW	VC 10
YOLK YELLOW	VC 11
ORANGE RED	VC 12
STRAWBERRY RED	VC 13
ROSE PINK	VC 14
RASPBERRY RED	VC 15
GRAPE	VC 16
VIOLET	VC 17
COFFEE BROWN	VC 18
CHOCOLATE BROWN	VC 19
DARK CHOCOLATE	VC 20
LIME GREEN	VC 21
APPLE GREEN	VC 22
PEA GREEN	VC 23
BLACK CURRANT	VC 24

Detail of few synthetic colors
1. Tatrazine

Tartrazine (E number E102 or C I 19140) is a synthetic lemon yellow azodye used as a food coloring .It is water soluble and has a maximum absorbance in an aqueous solution at427±2 nm. Tartrazine is a commonly used color all over the world, mainly for yellow, but can also be used with Brilliant Blue FCF (FD & C Blue 1, E133) or Green S (E142) to produce various green shades.

IUPAC name: Trisodium (4E)-5-oxo- 1-(4-sulfonatophenyl)- 4-[(4-sulfonatophenyl)hydrazono]- 3-pyrazole carboxylate.

Other name: FD & C Yellow 5

Molecular formula: $C_{16}H_9N_4Na_3O_9S_2$

Molar mass: 534.3 g/ mol

NaOOC $\diagdown$

NaO_3S ... N=N ... N ... SO_3Na

OH

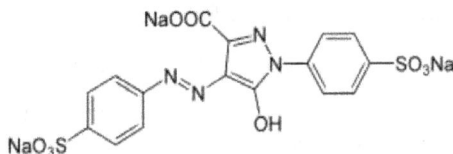

Products including tartrazine commonly include confectionery, cotton candy, soft drinks (Mountain Dew), energy drinks, instant puddings, flavored corn chips, cereals (cornflakes, muesli, etc .), cake mixes, pastries, custard powder, soups (particularly instant or "cube" soups), sauces, some rice's (like paella, risotto, etc.), powdered drink mixes, sports drinks, ice cream, ice pops, candy, chewing gum, marzipan, jam, jelly, gelatins, marmalade, mustard, yogurt, noodles such as Kraft Dinner, pickles and other pickled products, certain brands of fruit squash, fruit cordial, potato chips, Biscuits, and many convenience together with glycerin, lemon and honey products.

2. Sunset yellow FCF

IUPACname: Disodium 6-hydroxy-5-[(4-sulfophenyl) azo]-2-naphthalenesulfononate
Other names: Orange Yellow S; FD & C Yellow 6; C I 15985; E110
Molecular formula: $C_{16}H_{10}Na_2O_7S_2N_2$
Molar mass: 452.37 g/mol
Melting point: 300 °C

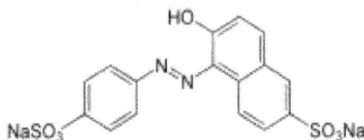

HO

$NaSO_3$ ... N=N ... SO_3Na

Sunset Yellow FCF (also known as Orange Yellow S, FD & C Yellow 6 or C I 15985) is colorant that may be added to foods to induce a color change. It is denoted by E Number E110, and has the capacity for inducing an allergic reaction. It is a synthetic coal tar and azo yellow dye useful in fermented foods which must be heat treated. It may be found in orange squash, orange jelly, marzipan, Swiss roll, apricot jam, citrus marmalade, lemon curd, sweets, hot chocolate mix and packet soups, trifle mix, breadcrumbs and cheese sauce mix and soft drinks. Specifically it can be found in the capsules of Dayquil (in high concentrations), some extra strength Tylenol, Astropeach yogurt , fortune cookies, some red sauces, certain pound cakes, Cheetos, snack chips, and other yellow, orange, and red food products.

Sunset Yellow is often used in conjunction with E123, Amaranth, in order to produce a brown colouring in both chocolates and caramel. At high concentrations, Sunset Yellow in solution with water undergoes a phase change from anisotropic liquid to a nematic liquid crystal. This occurs between 0.8 M and 0.9 M at room temperature.

Sunset Yellow is a sulfonated version of Sudan I, a possible carcinogen, which is frequently present in it as an impurity. Sunset Yellow itself may be responsible for causing an allergic reaction in people with an aspirin in tolerance, resulting in various symptoms including gastric upset, diarrhoea, vomiting, nettle rash (urticaria) and swelling of the skin (angioedema). The coloring has also been linked to hyperactivity in young children. As a result of these problems, there have been repeated calls for the total withdrawal of Sunset Yellow from food use.

3. Allura red AC

IUPAC name: Disodium 6-hydroxy-5-((2-methoxy-5-methyl-4- sulfophenyl) azo)-2-naphtha- lene -sulfonate

Other names: Allura Red, Food Red 17, C I 16035, FD&C Red 40, E129, 2-naphthalene sulfonic acid disodium salt

Molecular formula: $C_{18}H_{14}N_2Na_2O_8S_2$

Molar mass: 496.42 g mol^{-1}

Appearance: dark red powder

Melting point : >300 °C

Allura Red AC is a red azo dye that goes by several names including: Allura Red, Food Red17, C I 16035, FD & C Red 40 , 2-naphthalenesulfonic acid, 6-hydroxy-5-((2-methoxy-5-methyl-4-sulfophenyl)azo)-, disodium salt, and disodium 6-hydroxy-5-((2-methoxy-5-methyl-4-sulfophenyl)azo)-2-naphthalene-sulfonate. It is used as a food dye and has the E number E129. Allura Red AC was originally introduced in the United States as a replacement for the use of amaranth as a food coloring. It has the appearance of a dark red powder. It usually comes as a sodium salt but can also be used as both calcium and potassium salts.

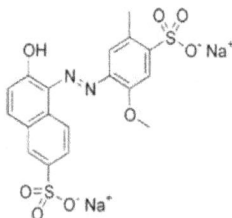

It is soluble in water. In water solution, its maximum absorbance lies at about 504 nm. Its melting point is at >300 degrees Celsius. Allura Red AC is one of many High Production Volume Chemicals. Red AC was originally manufactured from coal tar but is now mostly made from petroleum. Despite the popular misconception, Allura Red AC is not derived from any insect, unlike the food coloring carmine which is derived from the female cochineal insect. Related dyes include Sunset Yellow FCF, Scarlet GN, tartrazine, and Orange B. Allura Red AC has fewer health risks associated with it in comparison to other azo dyes. However, some studies have found some adverse health effects that may be associated with the dye.

Regulation

In Europe, Allura Red AC is not recommended for consumption by children. It is banned in Denmark, Belgium, France, Switzerland, and Sweden. The European Union approves Allura Red AC as a food colorant, but EU countries' local laws banning food colorants are preserved. In Norway, it was banned between 1978 and 2001, a period in which azo dyes were only legally used in alcoholic beverages and some fish products. In the United States, Allura Red AC is approved by the Food and Drug Administration for use in cosmetics, drugs, and food. It is used in some tattoo inks and is used in many products, such as soft drinks, children's medications, and cotton candy.

4. Brilliant blue FCF

Other names: FD & C Blue No.1, Acid Blue 9, D & C Blue No. 4, Alzen Food Blue No. 1, Atracid Blue FG, Erioglaucine Eriosky blue, Patent Blue AR, Xylene Blue VSG

Molecular formula: $C_{37}H_{34}N_2Na_2O_9S_3$

Brilliant Blue FCF, also known under commercial names, is a colorant for foods and other substances to induce a color change. It is denoted by E number E133 and has a color index of 42090. It has the appearance of a reddish-blue powder. It is soluble in water, and the solution has a maximum absorption at about 628 nm.

Chemistry

It is a synthetic dye derived from coal tar. It can be combined with tartrazine (E102) to produce various shades of green. It is usually a disodium salt. The diammonium salt has CAS number [2650-18-2]. Calcium and potassium salts are also permitted. It can also appear as an aluminium lake. The dye is poorly absorbed from the gastro-intestinal tract and 95% of the absorbed dye can

be found in the feces. It also reacts with certain bile pigments to form green feces.

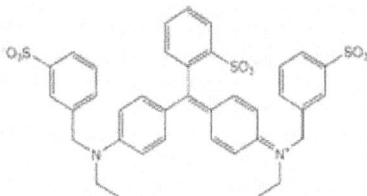

Applications

As a blue color, Brilliant Blue FCF is often found in ice cream, tinned processed peas, dairy products, sweets and drinks. It is also used in soaps, shampoos, and other hygiene and cosmetics applications. In soil science, Brilliant Blue is applied in tracing studies to visualize in filtration and water distribution in the soil. It is used as the stain in Listerine's Agent Cool Blue mouthwash.

Health and safety

Brilliant Blue FCF has previously been banned in Austria, Belgium, Denmark, France, Germany, Greece, Italy, Norway, Spain, Sweden, and Switzerland among others but has been certified as a safe food additive in the EU and is today unbanned in most of the countries. It has the capacity for inducing an allergic reaction in individuals with pre-existing moderate asthma.

5. Brilliant black BN

IUPAC name: Tetra sodium (6Z)-4-acetamido-5-oxo-6-[[7-sulfonato-4-(4-sulfonatophenyl) azo-1-naphthyl] hydrazono] naphthalene-1,7-disulfonate.

Other names: C I Food Black 1; 1743 Black; Black PN; Blue Black BN; Brilliant Acid Black; C I 28440; Certicol Black PNW; Cilefa Black B; E 151; Edicol Supra Black BN; Hexacol Black PN; L Black 8000;Melan Black; Xylene Black F.

Molecular formula: $C_{28}H_{17}N_5Na_4O_{14}S_4$

Molar mass: 867.68 g/mol.

Brilliant Black BN is a synthetic black diazo dye. It is soluble in water. It usually comes as tetra sodium salt. It has the appearance of solid, fine powder or granules. Calcium and potassium salts are allowed as well.

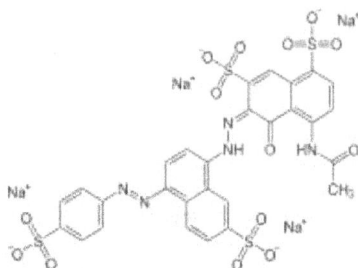

When used as a food dye, its E number is E151. It is used in food decorations and coatings, desserts, sweets, ice cream, mustard, red fruit jams, soft drinks, flavored milk drinks, fish paste, lumpfish caviar and other foods. It appears to cause allergic or intolerance reactions, particularly amongst those with aspirin intolerance. It is a histamine liberator, and may worsen the symptoms of asthma. It is one of the colorants that the Hyperactive Children's Support Group recommends be eliminated from the diet of children. It is banned in Canada, United States, Finland and Japan.

6. Citrus red 2

IUPAC name: 1-(2, 5-Dimethoxy-phenylazo)-naphthalen-2-ol.

Other names: Citrus Red No. 2, C I Solvent Red 80, C I 12156, E121.

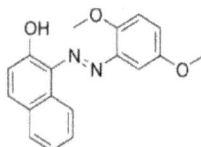

Citrus Red 2, Citrus Red No. 2, C I Solvent Red 80, or C I 12156 is an artificial dye. As a food dye, it is permitted by Food and Drug Administration (FDA) since 1956 only for use in the United States on skin on some oranges. While the dye is a carcinogen, it does not penetrate the orange peel into the pulp].It is an orange to yellow solid or a dark red powder with a melting point of 156 °C. Citrus Red 2 is not water-soluble, but readily soluble in many organic solvents.

7. Erythrosine

Erythrosine, also known as Red No. 3, is a cherry-pink synthetic fluorine food coloring. It is the disodium salt of 2, 4, 5, 7-tetraiodofluorescein.

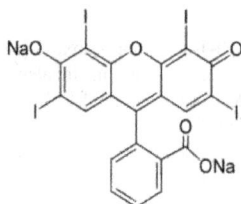

Its maximum absorbance is at 530 nm in an aqueous solution, and it is subject to photo degradation.

Uses
It is used as a food coloring, in printing inks, as a biological stain, a dental plaque disclosing agent and a radio pique medium. Erythrosine is commonly used in sweets such as some candies and popsicles, and even more widely used in cake-decorating gels. It is also used to color pistachio shells. As a food additive, it has the E number E127.

8. Fast green FCF
IUPAC name: ethyl -[4 -[[4 -[ethyl -[(3-sulfophenyl) methyl] amino] phenyl] -(4 - hydroxyl -2 - sulfophenyl) methylidene] -1 -cyclohexa -2, 5 - dienylidene] -[(3 - sulfophenyl) methyl] azanium

Other names: Food green 3, FD &C Green No. 3, Green 1724, Solid Green FCF, and C I 42053

Molecular formula: $C_{37}H_{37}N_2O_{10}S_3$

Fast Green FCF is a sea green triarylmethane food dye. Its E number is E143.Its absorption maximum is at 625 nm. Fast Green FCF is poorly absorbed by the intestines. Its use as a food dye is prohibited in European Union and some other countries. It can be used for tinned green peas and other vegetables, jellies, sauces, fish, desserts, and dry bakery mixes at level of up to 100 mg/kg.

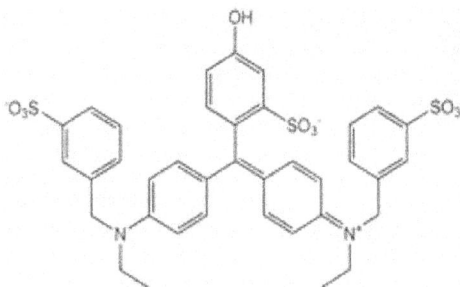

Toxicology
This substance has been found to have tumorigenic effects in experimental animals, as well as Mutagenic effects in both experimental animals and humans. It furthermore risks irritation of eyes, skin, digestive tract, and respiratory tract in its undiluted form.

9. Green S
IUPAC name: Sodium 4-[(4-dimethylaminophenyl)-(4-dimethylazaniumylidene-1-cyclohexa-2, 5-dienylidene) methyl]-3-hydroxynaphthalene-2,7-disulfonate.

Other names: Food Green S; FD&C Green 4; Acid green 50; Lissamine Green B; Wool Green S; C I 44090;E142.

Molecular formula: $C_{27}H_{25}N_2O_7S_2Na$.

Green S is a green synthetic coal tar triarylmethane dye. As a food dye, it has E number E142. It can be used in mint sauce, desserts, gravy granules, sweets, ice creams, and tinned peas. Green S is prohibited as a food additive in Canada, United States, Japan, and Norway. Green S is a vital dye, meaning it can be used to stain living cells. It is used in ophthalmology, among fluoresces in and Rose Bengal, to diagnose various disorders of the eye's surface. Green S may cause allergic reactions and is one of the colorants that the Hyperactive Children's Support Group recommends to be eliminated from the diet of children.

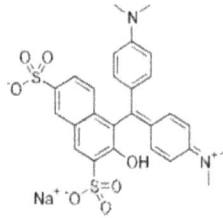

10. Patent blue V

Patent Blue V, also called Food Blue 5 or Sulphan Blue, is a dark bluish synthetic dye used as a food coloring. As a food additive, it has E number E131. It is a sodium or calcium salt of [4-(α-(4-diethylaminophenyl)-5-hydroxy-2,4-disulfophenyl-methylidene)-2, 5-cyclohexadien-1-ylide ne] diethyl ammonium hydroxide inner salt. It has the appearance of a violet powder.

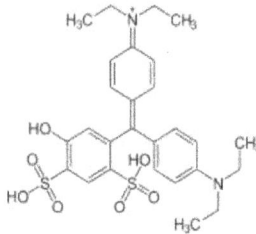

Its CAS number is [3536-49-0]. It is not widely used, but can be found in Scotch eggs and certain jelly sweets. Patent Blue V is banned as a food dye in Australia, USA, and Norway.

11. Ponceau 4R

IUPAC name: Trisodium (8Z)-7-oxo-8-[(4-sulfonatonaphthalen-1-yl) hydrazinylidene] naphtha lene -1, 3-disulfonate.

Ponceau 4R (also known as C I 16255, Cochineal Red A, C I Acid Red 18, Brilliant Scarlet 3R, Brilliant Scarlet 4R, New Coccine, SX purple) is a synthetic colorant that may be added to foods to induce a color change. It is denoted by E Number E124. Its chemical name is trisodium salt of 1-(4-sulpho-1-napthylazo) - 2-napthol- 6, 8-disulphonic acid. Ponceau 4R is a red azo dye usually synthesized from coal tar which can be used in a variety of food products.

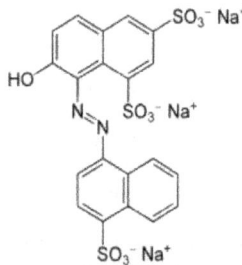

Health effects
Because it is an azo dye, it may elicit intolerance in people allergic to salicylates (aspirin). Additionally, it is a histamine liberator, and may intensify symptoms of asthma. Ponceau 4R is considered carcinogenic in some countries, including the USA, Norway, and Finland, and it is currently listed as a banned substance by the U.S. Food and Drug Administration (FDA).

12. Quinoline yellow WS
Quinoline yellow, Quinoline Yellow WS, C I 47005, or Food Yellow 13, is a yellow food dye. Chemically it is a mixture of disulfonates (principally), monosulfonates and trisulfonates of 2-(2-quinolyl) indan-1, 3-dione.

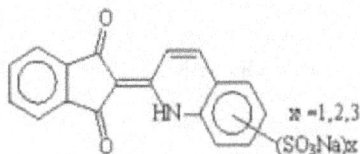

The color Quinoline Yellow SS (Spirit Soluble), which lacks the sulfonate groups, is a related form that is insoluble in water.

Uses
As a food additive with the E number E104, it is used as colorant that induces a dull yellow or greenish yellow color.

13. Red 2G
Red 2G is a synthetic red azo dye. It is soluble in water and slightly soluble in ethanol. It usually comes as a disodium salt of 8-acetamido-1-hydroxy-2-phenylazonaphthalene-3, 6-disulphonate.

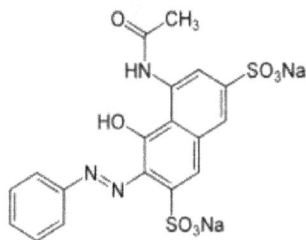

4.10 Food dye
In the European Union, Red 2G is used as a food dye E number E 128. However, it is only permitted for use in breakfast sausages with a minimum cereal content of 6% and burger meat with a minimum vegetable and/or cereal content of 4%.Red 2G is banned in Australia, Austria, Canada, Japan, Norway, Sweden, Malaysia and the United. It was banned in Ireland, Israel and Greece in July 2007.It is relatively insensitive to the bleaching effect of sulfur dioxide (E 220) and sodium metabisulfite (E 223). In the intestines, Red 2G can be converted to the toxic compound aniline, so there are concerns Red 2G may ultimately interfere with blood hemoglobin, as well as cause cancer.

1. Indigo carmine

Indigo Carmine, or 5, 5'-indigodisulfonic acid sodium salt, is a pH indicator with the chemical formula: $C_{16}H_8N_2Na_2O_8S_2$

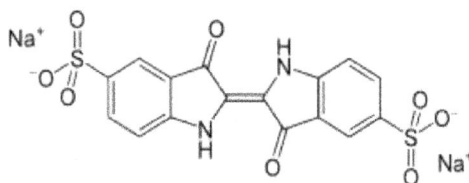

Uses

The primary use of Indigo carmine is as a pH indicator. It is blue at pH 11.4 and yellow at 13.0. Indigo carmine is also a redox indicator, turning yellow upon reduction. Another use is as a dissolved ozone indicator. It is also used as a dye in the manufacturing of capsules.

Health concerns

Indigo carmine is harmful to the respiratory tract if swallowed. It is also an irritant to the skin and eyes. Proper laboratory cautions (lab coat, gloves, and goggles) are advised.

2. Chocolate brown

IUPAC name: Disodium 4-[(2E)-2-[(5Z)-3-(hydroxymethyl)-2,6-dioxo-5-[(4-sulfonatonaph-thalen-1-yl)hydrazinylidene]-1-cyclohex-3-enylidene]hydrazinyl] naph thalene-1-sulfonate.

Other names: Chocolate Brown HT, Food Brown 3, C I 20285, E155
Molecular formula: $C_{27}H_{18}N_4Na_2O_9S_2$
Molar mass: 652.56 g/mol.

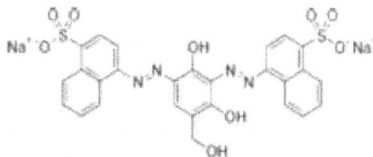

Chocolate Brown is a brown synthetic coal tar diazo dye. When used as a food dye, its E number is E155. It is used to substitute cocoa or caramel as a colorant. It is used mainly in chocolate cakes, but also in milk and cheeses, yoghurts, jams, fruit products, fish, and other products. It may provoke allergic reactions in asthmatics, people sensitive to aspirin, and other sensitive individuals, and may induce skin sensitivity. It is one of the food colorings that the Hyperactive Children's Support Group recommends be eliminated from the diet of children. It is banned in Austria,

Belgium, Denmark, France, Germany, United States, Norway, Switzerland, and Sweden. It is used in nearly every major brand of chocolate flavored milk in Australia.

4.11 Dyes and lakes

Color additives are available for use in food as either "dyes" or lake pigments (commonly known as "lakes").Dyes dissolve in water, but are not soluble in oil. Dyes are manufactured as powders, granules, liquids or other special purpose forms. They can be used in beverages, dry mixes, baked goods, confections, dairy products, pet foods, and a variety of other products. Dyes also have side effects which lakes do not, including the fact that large amounts of dyes ingested can color stools.

Lakes are made by combining dyes with salts to make insoluble compounds. Lakes tint by dispersion. Lakes are not oil soluble, but are oil dispersible. Lakes are more stable than dyes and are ideal for coloring products containing fats and oils or items lacking sufficient moisture to dissolve dyes. Typical uses include coated tablets, cake and doughnut mixes, hard candies and chewing gums, lipsticks, soaps, shampoos, talc, etc.

LAKE QUINOLINE YELLOW	47005:1	E 104
LAKE TARTRAZINE	19140:1	E 102
LAKE SUNSET YELLOW FCF	15985:1	E 110
LAKE ERYTHROSINE	45430:1	E 127
LAKE PONCEAU 4R	16255:1	E 124
LAKE ALLURA RED	16035:1	E 129
LAKE CARMOISINE	14720:1	E 128
LAKE AMARANTH	16185:1	E 123
LAKE CHOCCOLATE BROWN HT	20285:1	E 155
LAKE BRILLIANT BLUE FCF	42090:1	E 133
LAKE INDIGO CARMINE	73015:1	E 132

Lake pigments find wide usage in areas like foodstuffs, pet foods, drugs and pharmaceuticals, cosmetics, plastics, plastic films, can linings, plastic food containers, inks and stationery.

1. Amarnath (dye)

IUPAC name: trisodium (4E)-3-oxo-4-[(4-sulfonato-1-naphthyl) hydrazono]naphthalene-2,7-disulfonate.
Other names: FD & C Red No. 2,E 123,C I Food Red 9,Acid Red 27,Azorubin S,C I 16185.
Molecular formula: $C_{20}H_{11}N_2Na_3O_{10}S_3$
Molar mass: 604.47305 gm/mol
Melting point: 120 °C (decomposes)

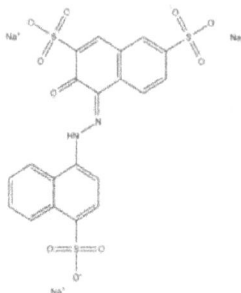

Amaranth is a dark red to purple azo dye in powder form. Once used as a food dye and to color cosmetics, but since 1976 it has been banned in the United States by the Food and Drug Administration (FDA) as it is a suspected carcinogen.

It usually comes as a trisodium salt. It has the appearance of reddish-brown, dark red to purple water-soluble powder that decomposes at 120 °C without melting. Its water solution has absorption maximum at about 520 nm. Amaranth is made from coal tar. Amaranth is an anionic dye. It can be applied to natural and synthetic fibers, leather, paper, and phenol-formaldehyde resins. As a food additive it has E number E123. Amaranth's use is still legal in some countries, notably in the United Kingdom where it is most commonly used to give glace cherries their distinctive color.

2. Aniline yellow

IUPAC name: 4-Phenyldiazenylaniline

Other names: Ceres Yellow, Fast spirit Yellow, Oil Yellow AAB, Oil Yellow AN, Organol Yellow, Solvent Yellow, C I 11000.

Molecular formula: $C_{12}H_{11}N_3$

Molar mass: 197.24 gm/mol

Melting point: 123-126 °C

Boiling point: > 360 °C

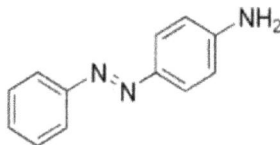

Aniline Yellow is a yellow azo dye and an aromatic amine. It is a derivate of azobenzene. It has the appearance of an orange powder. It is a carcinogen. Aniline Yellow was the first azo dye. it was first produced in 1861 by C. Mene. The second azodye was Bismarck Brown in 1863. Aniline Yellow was commercialized in 1864 as the first commercial azo dye, a year after Aniline Black. It is manufactured from aniline.

Uses

Aniline Yellow is used in microscopy for vital staining. It is also used in insecticides, lacquers, varnishes, waxes, oil stains, and styrene resins. It is also an intermediate in synthesis of other dyes, e. g. chrysoidine, indulines, Solid Yellow, and Acid Yellow.

3. Azo violet

IUPAC name: 4-(4-nitrophenyl) azobenzene-1, 3-diol

Other names: Azoviolet; Magneson

Molecular formula: $C_{12}H_9N_3O_4$
Appearance: dark red to brown crystalline powder
Density: 1. 45 gm/cm^3
Flash point: 261.7 °C

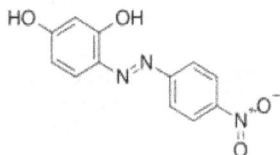

Azo violet is an organic compound with the chemical formula $C_{12}H_9N_3O_4$. It is used as a dye and
a pH indicator. It Forms blue color with magnesium in slightly alkaline medium. This reaction depends on adsorption of the dye on magnesium hydroxide. Ammonium ions reduce sensitivity of reaction. Any salt or salt solution containing Magnesium will give this Violet coloring of solution. However, ammonium compounds such as Ammonium Chloride or Ammonium Hydroxide will give a different color based on amount used. Still, it is used as identification test for Magnesium.

4. Black 7984

IUPAC name: Tetra sodium 6-amino-4-hydroxy-3-[[7-sulfonato-4-[(4-sulfonatophenyl) azo]-1-
naphthyl] azo] naphthalene-2,7-disulfonate
Other names: Food Black 2, C I 27755
Molecular formula: $C_{26}H_{19}N_5Na_4O_{13}S_4$
Molar mass: 733.69 gm/mol

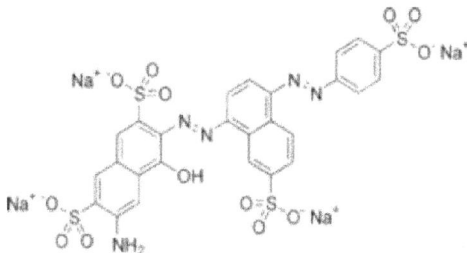

Black 7984, Food Black 2, or C.I. 27755, is a brown-to-black synthetic diazo dye. It is often used as the tetra sodium salt. When used as a food dye, it has E number E152. Its use in food is discontinued in USA and EU since 1984. It is also not permitted in Australia and Japan. Black 7984 is also used in cosmetics. Black 7984 may cause allergic or intolerance reactions, particularly amongst those with an aspirin intolerance. It is a histamine liberator, and may worsen the symptoms of asthma. It is one of the colorants that the Group recommends be eliminated from the diet of children.

5. FD and C Orange Number 1

FD and C Orange Number 1 was one of the first water soluble dyes to be commercialized, and one of seven original food dyes allowed under the Pure of June 30, 1906. [1] In the early1950s, after several cases were reported of sickness in children who had eaten foods with excessive amounts of dye, the FDA conducted new, more thorough and rigorous testing on food dyes. Orange 1 was outlawed for food use in 1956.

6. Fast yellow AB

IUPAC name: 2-amino-5-[(E)-(4-sulfophenyl) diazenyl]- benzenesulfonic acid
Other names: Fast Yellow, Acid Yellow, C I 13015, Food Yellow 2
Molecular formula: $C_{12} H_{11} N_3 O_6 S_2$
Molar mass: 357.36 gm/mol.

Fast Yellow AB is an azo dye denoted by E number E105. It was used as a food dye. It is now delisted in both Europe and USA and is forbidden if used in foods and drinks, as toxicological data has shown it is harmful.

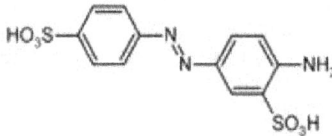

7. Orange B

IUPAC name: Disodium 4-[N'-[3-ethoxycarbonyl-5-oxo-1-(4-sulfonatophenyl)-4-yrazolylidene] hydrazino]-1-naphthalenesulfonate.
Other names: C I Acid Orange 137.

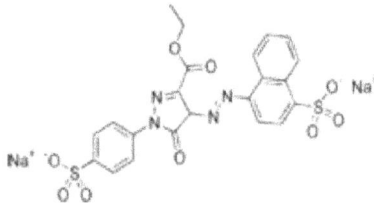

Orange B is a food dye from the azo dye group. It is approved by Food and Drug Administration (FDA) for use only in hot dog and sausage casings or surfaces, only up to 150ppm of the finished food weight. It usually comes as disodium salt.

8. Orange GGN

Orange GGN, also known as alpha-naphthol orange, is a food dye. In Europe it is denoted by the E Number E111. It is the disodium salt of 1-(m-sulfophenylazo)-2-naphthol-6-sulfonic acid.

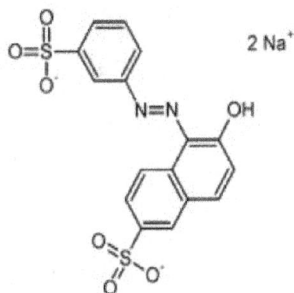

It is currently delisted in Europe and USA, because toxicological data has shown it is harmful. The absorption spectrum of Orange GGN and Sunset Yellow is nearly identical in visible and ultraviolet range, but they can be distinguished by their IR spectra.

9. Ponceau 6R

Ponceau 6R, or Crystal ponceau 6R, Crystal scarlet, Brilliant crystal scarlet 6R, Acid Red 44, or C.I. 16250, is a red azo dye. It is soluble in water and slightly soluble in ethanol. It is used as a food dye, with E number E126. It is also used in histology, for staining fibrin with the MSB Trichromestain. It usually comes as disodium salt. Amaranth is a closely related azo dye, also usable in trichrome staining. Its CAS number is [2766-77-0].

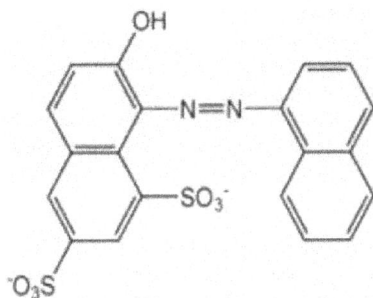

10. Scarlet GN

Scarlet GN or C I Food Red 2, Ponceau SX, FD & C Red No. 4, or C I 14700 is a red azo dye used as a food dye. When used as a food additive, it has the E number E125. It usually comes as a disodium salt. It is permitted in fruit peels and maraschino cherries. Its CAS number is [4548-53-2].

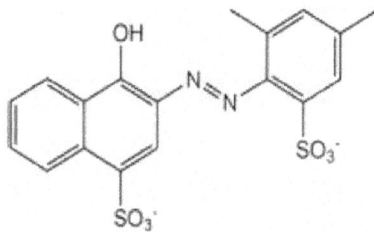

11. Sudan I

Sudan I (also commonly known as C I Solvent Yellow 14 and Solvent Orange R), is alysochrome, a diazo-conjugate dye with a chemical formula of 1-phenylazo-2-naphthol. Sudan I is a powdered substance with an orange-red appearance.

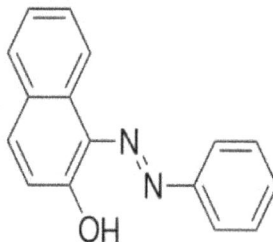

The additive is mainly used to color waxes, oils, petrol, solvents and polishes. Sudan I has also been adopted for coloring various foodstuffs, including particular brands of curry powder and chili powder, although the use of Sudan I in foods is now banned in many countries because Sudan I, Sudan III, and Sudan IV have been classified as category3 carcinogens by the International Agency for Research on Cancer. Sudan I is also present as an impurity in Sunset Yellow, which is its disulfonated water-soluble version.

12. Yellow 2G

Yellow 2G is a colorant in food denoted by E number E107.It has the appearance of a yellow powder, soluble in water. It is a synthetic coal tar and yellow azo dye. It appears to cause allergic or intolerant reactions, particularly amongst those with aspirin intolerance and asthma sufferers. It is one of the colours that the Hyperactive Children's Support Group recommends be eliminated from the diet of children. Currently only the UK in the European Union uses this dye and the EU is proposing a total ban. Its use is banned in Austria, Japan, Norway, Sweden, Switzerland and the United States.

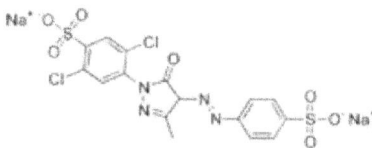

13. Azorubine

Azorubine, carmoisine, Food Red 3, Azorubin S, Brillantcarmoisin O, Acid Red 14, or C I 14720 is a synthetic red food dye from the azo dye group. It usually comes as a disodiumsalt.

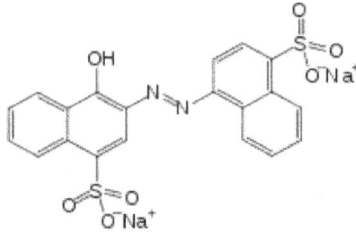

It is a red to maroon powder. It is used for the purposes where the food is heat-treated after fermentation. It has E number E122. Some of the foods it can be present in Swiss roll, jams, preserves, yoghurts, jellies, breadcrumbs, and cheesecake mixes.

4.12 Possible health effects

It appears to cause allergic or intolerance reactions, particularly amongst those with aspirin intolerance. Other reactions can include a rash similar to nettle rash and skin swelling .Asthmatics sometimes react badly to it.

4.13 Criticism and health implications

Though past research showed no correlation between Attention-deficit hyper activity disorder (ADHD) and food dyes, new studies now point to synthetic preservatives and artificial coloring agents as aggravating ADD and ADHD symptoms, both in those affected by these disorders and in the general population. Older studies were inconclusive quite possibly due to inadequate clinical methods of measuring offending behavior. Parental reports were more accurate indicators of the presence of additives than clinical tests.

- Norway banned all products containing coal tar and coal tar derivatives in **1978**. New legislation lifted this ban in 2001 after EU regulations. As such, many FD&C approved colorings have been banned.
- Tartrazine causes hives in less than 0.01% of those exposed to it.
- Erythrosine is linked to thyroid tumors in rats.
- Cochineal, also known as carmine, is derived from insects and therefore is neither vegan nor vegetarian. It has also been known to cause severe, even life threatening allergic reactions in rare cases. This criticism originated during the 1950s. In effect, many foods that used dye became less popular.

Brilliant Blue (BBG) food coloring was cited in a recent study in which rats that had suffered a spinal injury were given an injection of the dye immediately after the injury, and were able to regain or retain motor control. BBG helps protect spine from ATP (adenosine triphosphate), which the body sends to the area after a spinal injury, which further damages the spine by killing motor neurons at the site of the injury.

5. Phytochemistry

5.1 Introduction

Medicinal plants have been the main stay of traditional herbal medicine amongst rural dwellers worldwide since antiquity to date. The therapeutic use of plants certainly goes back to the Sumerian and the Akkadian civilizations in about the third millennium BC. Hippocrates (ca. 460–377 BC), one of the ancient authors who described medicinal natural products of plant and animal origins, listed approximately 400 different plant species for medicinal purposes. Natural products have been an integral part of the ancient traditional medicine systems, e.g. Chinese, Ayurvedic and Egyptian (Sarker & Nahar, 2007). Over the years they have assumed a very central stage in modern civilization as natural source of chemotherapy as well as amongst scientist in search for alternative sources of drugs. About 3.4 billion people in the developing world depend on plant-based traditional medicines. This represents about 88 per cent of the world's inhabitants, who rely mainly on traditional medicine for their primary health care. According to the World Health Organization, a medicinal plant is any plant which, in one or more of its organs, contains substances that can be used for therapeutic purposes, or which are precursors for chemo-pharmaceutical semi synthesis. Such a plant will have its parts including leaves, roots, rhizomes, stems, barks, flowers, fruits, grains or seeds, employed in the control or treatment of a disease condition and therefore contains chemical components that are medically active. These non-nutrient plant chemical compounds or bioactive components are often referred to as phytochemicals ('phyto' term derived from Greek word - *phyto* meance 'plant') or phytoconstituents and are responsible for protecting the plant against microbial infections or infestations by pests (Abo *et al.*, 1991; Liu, 2004; Nweze *et al.*, 2004; Doughari *et al.*, 2009). The study of natural products on the other hand is called phytochemistry. Phytochemicals have been isolated and characterized from fruits such as grapes and apples, vegetables such as broccoli and onion, spices such as turmeric, beverages such as green tea and red wine, as well as many other sources (Doughari & Obidah, 2008; Doughari *et al.*, 2009).

The science of application of these indigenous or local medicinal remedies including plants for treatment of diseases is currently called ethno pharmacology but the practice dates back since antiquity. Ethno pharmacology has been the mainstay of traditional medicines the entire world and currently is being integrated into mainstream medicine. Different catalogues including *De Materia Medica, Historia Plantarum, and Species Plantarum* have been variously published in attempt to provide scientific information on the medicinal uses of plants. The types of plants and methods of application vary from locality to locality with 80% of rural dwellers relying on them as means of treating various diseases. For example, the use of bearberry (*Arctostaphylos uva-ursi*) and cranberry juice (*Vaccinium macrocarpon*) to treat urinary tract infections is reported in different manuals of phytotherapy, while species such as lemon balm (*Melissa officinalis*), garlic (*Allium sativum*) and tee tree (*Melaleuca alternifolia*) are described as broad-spectrum antimicrobial agents

(Heinrich *et al.*, 2004). A single plant may be used for the treatment of various disease conditions depending on the community. Several ailments including fever, asthma, constipation, esophageal cancer and hypertension have been treated with traditional medicinal plants (Cousins and Huffman, 2002; Saganuwan, 2010). The plants are applied in different forms such as poultices, concoctions of different plant mixtures, infusions as teas or tinctures or as component mixtures in porridges and soups administered in different ways including oral, nasal (smoking, snoffing or steaming), topical (lotions, oils or creams), bathing or rectal (enemas). Different plant parts and components (roots, leaves, stem barks, flowers or their combinations, essential oils) have been employed in the treatment of infectious pathologies in the respiratory system, urinary tract, gastrointestinal and biliary systems, as well as on the skin (Rojas *et al.*, 2001; R'ios & Recio, 2005; Adekunle and Adekunle, 2009).

Medicinal plants are increasingly gaining acceptance even among the literates in urban settlements, probably due to the increasing inefficacy of many modern drugs used for the control of many infections such as typhoid fever, gonorrhea, and tuberculosis as well as increase in resistance by several bacteria to various antibiotics and the increasing cost of prescription drugs, for the maintenance of personal health (Levy, 1998; Van den Bogaard *et al.*, 2000; Smolinski *et al.*, 2003). Unfortunately, rapid explosion in human population has made it almost impossible for modern health facilities to meet health demands all over the world, thus putting more demands on the use of natural herbal health remedies. Current problems associated with the use of antibiotics, increased prevalence of multiple-drug resistant (MDR) strains of a number of pathogenic bacteria such as methicillin resistant *Staphylococcus aureus, Helicobacter pylori,* and MDR *Klebsiela pneumonia* has revived the interest in plants with antimicrobial properties (Voravuthikunchai & Kitpipit, 2003). In addition, the increase in cases of opportunistic infections and the advent of Acquired Immune Deficiency Syndrome (AIDS) patients and individuals on immunosuppressive chemotherapy, toxicity of many antifungal and antiviral drugs has imposed pressure on the scientific community and pharmaceutical companies to search alternative and novel drug sources.

5.2. Classes of phytochemicals

5.2.1 Alkaloids

These are the largest group of secondary chemical constituents made largely of ammonia compounds comprising basically of nitrogen bases synthesized from amino acid building blocks with various radicals replacing one or more of the hydrogen atoms in the peptide ring, most containing oxygen. The compounds have basic properties and are alkaline in reaction, turning red litmus paper blue. In fact, one or more nitrogen atoms that are present in an alkaloid, typically as 1^0, 2^0 or 3^0 amines, contribute to the basicity of the alkaloid. The degree of basicity varies considerably, depending on the structure of the molecule, and presence and location of the functional groups (Sarker and Nahar, 2007). They react with acids to form crystalline salts without the production of water (Firn, 2010). Majority of alkaloids exist in solid such as atropine, some as liquids containing carbon, hydrogen, and nitrogen. Most alkaloids are readily soluble in alcohol and though they are sparingly soluble in water, their salts of are usually soluble. The solutions of alkaloids are intensely bitter. These nitrogenous compounds function in the defence of plants against herbivores and pathogens, and are widely exploited as pharmaceuticals, stimulants, narcotics, and poisons due to their potent biological activities. In nature, the alkaloids exist in large proportions in the seeds and roots of plants and often in combination with vegetable acids.

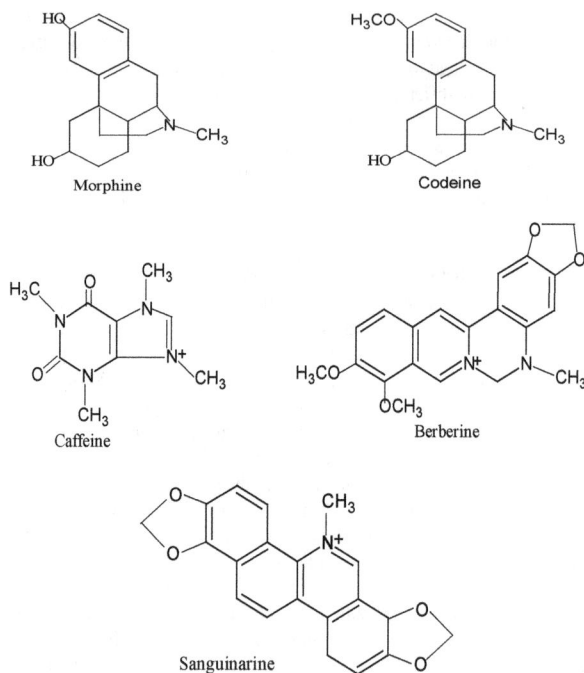

Fig: 1 Basic structures of some pharmacologically important plant derived alkaloids

Alkaloids have pharmacological applications as anesthetics and CNS stimulants (Madziga *et al.*, 2010). More than 12,000-lkaloids are known to exist in about 20% of plant species and only few have been exploited for medicinal purposes. The name alkaloid ends with the suffix *–ine* and plant-derived alkaloids in clinical use include the analgesics morphine and codeine, the muscle relaxant (+)-tubocurarine, the antibiotics sanguinafine and berberine, the anticancer agent vinblastine, the anti-arrythmic ajmaline, the pupil dilator atropine, and the sedative scopolamine. Other important alkaloids of plant origin include the addictive stimulants caffeine, nicotine, codeine, atropine, morphine, ergotamine, cocaine, nicotine and ephedrine (Fig: 1). Amino acids act as precursors for biosynthesis of alkaloids with ornithine and lysine commonly used as starting materials. Some screening methods for the detection of alkaloids are summarized in the following table: 1

Table:1 Methods for detection of alkaloids

Reagent / Test	Composition of the reagent	Result
Meyer's reagent	Potassium mercuric iodide solution	Cream precipitate
Wagner's reagent	Iodine in potassium iodide	Reddish brown precipitate
Tannic acid	Tannic acid	Precipitation
Hager's reagent	A saturated solution of picric acid	Yellow precipitate
Dragendroff's reagent	Solution of potassium bismuth iodide potassium chlorate, a drop of hydrochloric acid, evaporated to dryness, and the resulting	Orange or reddish-brown precipitate (except with caffeine and a few other alkaloids)
Muroxide test for caffeine	residue is exposed to ammonia vapour	Purine alkaloids produce pink colour

5.2.2 Glycosides

Glycosides in general, are defined as the condensation products of sugars (including poly saccharides) with a host of different varieties of organic hydroxyl (occasionally thiol) compounds (invariably monohydrate in character), in such a manner that the hemiacetal entity of the carbohydrate must essentially take part in the condensation. Glycosides are colorless, crystalline carbon, hydrogen and oxygen-containing (some contain nitrogen and sulfur) water-soluble phytoconstituents, found in the cell sap. Chemically, glycosides contain a carbohydrate (glucose) and a non-carbohydrate part (aglycone or ligenin) (Kar, 2007; Firn, 2010). Alcohol, glycerol or phenol represents aglycones. Glycosides are neutral in reaction and can be readily hydrolyzed into its components with ferments or mineral acids. Glycosides are classified on the basis of type of sugar component, chemical nature of aglycone or pharmacological action. The rather older or trivial names of glycosides usually have a suffix 'in' and the names essentially included the source of the glycoside, for instance: strophanthidin from *Strophanthus*, digitoxin from *Digitalis*, barbaloin from *Aloes*, salicin from *Salix*, cantharidin from *Cantharides*, and prunasin from *Prunus*. However, the systematic names are invariably coined by replacing the "ose" suffix of the parent sugar with "oside".This group of drugs are usually administered in order to promote appetite and aid digestion. Glycosides are purely bitter principles that are commonly found in plants of the Genitiaceae family and though they are chemically unrelated but possess the common property of an intensely bitter taste. The bitters act on gustatory nerves, which results in the increased flow of saliva and gastric juices. Chemically, the bitter principles contain the lactone group that may be diterpene lactones (e.g. *andrographolide*) or triterpenoids (e.g.*amarogentin*). Some of the bitter principles are either used as astringents due to the presence of tannic acid, as anti-protozoan, or to reduce thyroxine and metabolism. Examples include cardiac glycosides (acts on the heart), anthracene glycosides (purgative, and for treatment of skin diseases), chalcone glycoside (anticancer), amarogentin, gentiopicrin, andrographolide, ailanthone and polygalin (Fig: 2). Sarker and Nahar (2007) reported that extracts of plants that contain cyanogenic glycosides are used as flavouring agents in many pharmaceutical preparations. Amygdalin has been used in the treatment of cancer (HCN liberated in stomach kills malignant cells), and also as a cough suppressant in various preparations .Excessive ingestion of cyanogenic glycosides can be fatal. Some foodstuffs containing cyanogenic glycosides can cause poisoning (severe gastric irritations and damage) if not properly handled (Sarker and Nahar, 2007). To test for O-glycosides, the plant samples are boiled with HCl/H_2O to hydrolyse the anthraquinone glycosides to respective aglycones, and an aqueous base, e.g. NaOH or NH_4OH solution, is added to it. For C-glycosides, the plant samples are hydrolysed using $FeCl_3/HCl$, and an aqueous base, e.g. NaOH orNH_4OH solution, is added to it. In both cases a pink or violet colour in the base layer after addition of the aqueous base indicates the presence of glycosides in the plant sample.

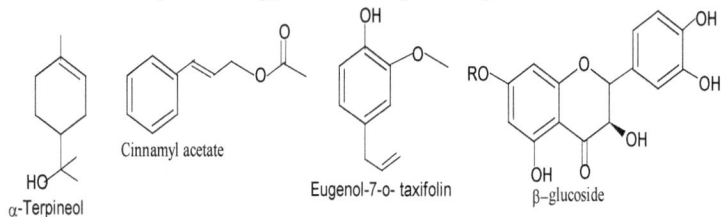

Cinnamyl acetate

α-Terpineol

Eugenol-7-o- taxifolin

β–glucoside

Fig: 2 Basic structures of some pharmacologically important plant derived glycosides

5.2.3 Flavonoids

Flavonoids re important group of poly phenols widely distributed among the plant flora. Structurally, they are made of more than one benzene ring in its structure (a range of C_{15} aromatic compounds) and numerous reports support their use as antioxidants or free radical scavengers (Kar, 2007). The compounds are derived from parent compounds known as flavans. Over four thousand flavonoids are known to exist and some of them are pigments in higher plants. Quercetin, kaempferol and quercitrin are common flavonoidspresent in nearly 70% of plants. Other group of flavonoids includes flavones, dihydroflavons, flavans, flavonols and anthocyanidins (Fig. 3), proanthocyanidins, chalcones and catechin and leucoanthocyanidins.

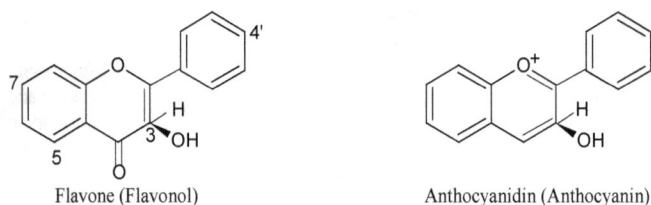

Flavone (Flavonol)

Anthocyanidin (Anthocyanin)

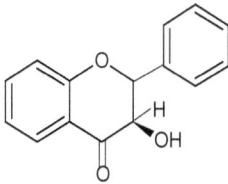
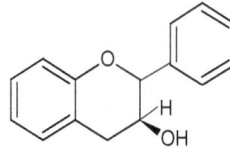

Dihydroflavone (dihydroflavonol)   Flavan (flavanol)
Fig: 3 Basic structures of some pharmacologically important plant derived flavonoids

5.2.4 Phenolics

Phenolics, phenols or polyphenolics (or polyphenol extracts) are chemical components that occur ubiquitously as natural colour pigments responsible for the colour of fruits of plants. Phenolics in plants are mostly synthesized from phenylalanine via the action of phenylalanine ammonia lyase (PAL). They are very important to plants and have multiple functions. The most important role may be in plant defence against pathogens and herbivore predators, and thus are applied in the control of human pathogenic infections (Puupponen-Pimiä *et al.*, 2008). They are classified into (i) phenolic acids and (ii) flavonoids, polyphenolics (flavonones, flavones, xanthones and catechins) and (iii) non-flavonoids polyphenolics. Caffeic acid is regarded as the most common of phenolic compounds distributed in the plant flora followed by chlorogenic acid known to cause allergic dermatitis among humans (Kar, 2007). Phenolics essentially represent a host of natural antioxidants, used as nutraceuticals, and found in apples, green-tea, and red-wine for their enormous ability tocombat cancer and are also thought to prevent heart ailments to an appreciable degree and sometimes are anti-inflammatory agents. Other examples include flavones, rutin, naringin, hesperidin and chlorogenic (Fig: 4).

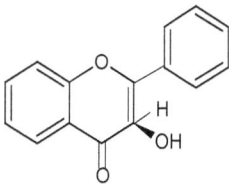
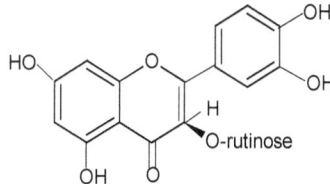

Flavone   Rutin

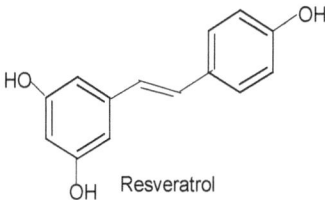
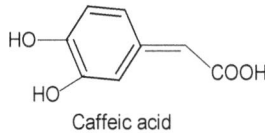

Resveratrol   Caffeic acid

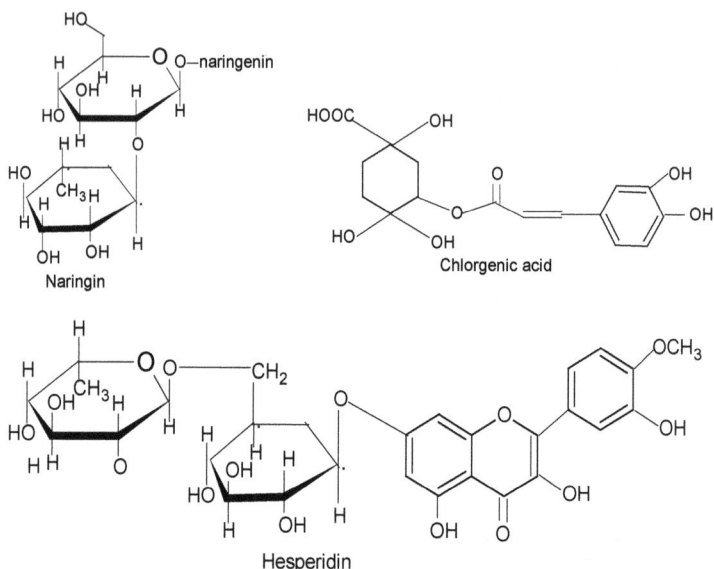

Naringin

Chlorgenic acid

Hesperidin

Fig: 4 Basic structures of some pharmacologically important plant derived phenolics

5.2.5 Saponins

The term saponin is derived from *Saponaria vaccaria* (*Quillaja saponaria*), a plant, which abounds in saponins and was once used as soap. Saponins therefore possess 'soaplike' behaviour in water, i.e. they produce foam. On hydrolysis, an aglycone is produced, which is called sapogenin. There are two types of sapogenin: steroidal and triterpenoidal. Usually, the sugar is attached at C-3 in saponins, because in most sapogenins there is a hydroxyl group at C-3. *Quillaja saponaria* is known to contain toxic glycosides quillajic acid and the sapogenin senegin. Quillajic acid is strenutatory and senegin is toxic. Senegin is also present in *Polygala senega*. Saponins are regarded as high molecular weight compounds in which, a sugar molecule is combined with triterpene or steroid aglycone. There are two major groups of saponins and these include: steroid saponins and triterpene saponins. Saponins are soluble in water and insoluble in ether, and like glycosides on hydrolysis, they give a glycones. Saponins are extremely poisonous, as they cause heamolysis of blood and are known to cause cattle poisoning (Kar, 2007). They possess a bitter and acrid taste, besides causing irritation to mucous membranes. They are mostly amorphous in nature, soluble in alcohol and water, but insoluble in non-polar organic solvents like benzene and n-hexane.

5.2.6 Tannins

These are widely distributed in plant flora. They are phenolic compounds of high molecular weight. Tannins are soluble in water and alcohol and are found in the root, bark, stem and outer layers of plant tissue. Tannins have a characteristic feature to tan, i.e. to convert things into leather. They are acidic in reaction and the acidic reaction is attributed to the presence of phenolics or carboxylic group (Kar, 2007).

They form complexes with proteins, carbohydrates, gelatin and alkaloids. Tannins are divided into hydrolysable tannins and condensed tannins. Hydrolysable tannins, upon hydrolysis, produce gallic acid and ellagic acid and depending on the type of acid produced, the hydrolysable tannins are called gallotannins or gallitannins. On heating, they form pyrogallic acid. Tannins are used as antiseptic and this activity is due to presence of the phenolic group. Common examples of hydrolysable tannins include the flavins (from tea), daidezein, genistein and glycitein (Fig: 5). Tannin rich medicinal plants are used as healing agents in a number of diseases. In Ayurveda, formulations based on tannin-rich plants have been used for the treatment of diseases like leucorrhoea, rhinnorhoea and diarrhea.

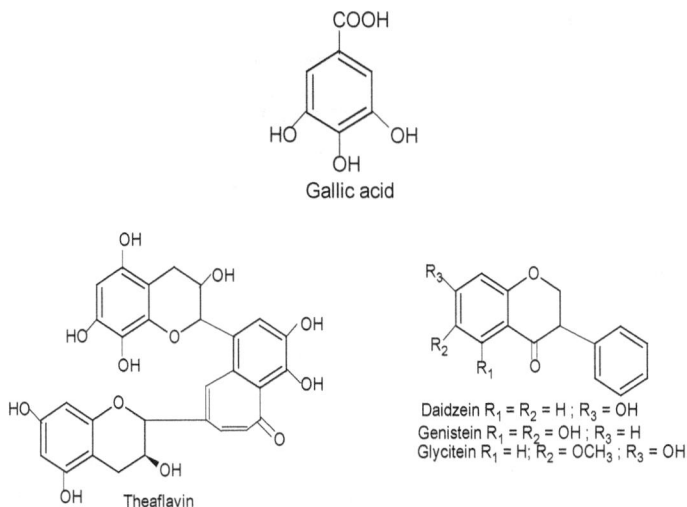

Gallic acid

Theaflavin

Daidzein $R_1 = R_2 = H$; $R_3 = OH$
Genistein $R_1 = R_2 = OH$; $R_3 = H$
Glycitein $R_1 = H$; $R_2 = OCH_3$; $R_3 = OH$

Fig: 5 Basic structures of some pharmacologically important plant derived tannins

5.2.7 Terpenes

Terpenes are among the most widespread and chemically diverse groups of natural products. They are flammable unsaturated hydrocarbons, existing in liquid form commonly found in essential oils, resins or oleoresins (Firn, 2010). Terpenoids includes hydrocarbons of plant origin of general formula $(C_5H_8)n$ and are classified as mono, di, tri and sesquiterpenoids depending on the number of carbon atoms. Examples of commonly important mono terpenes include terpinen-4-ol, thujone, camphor, eugenol and menthol. *Diterpenes (C_{20})* are classically considered to be resins and taxol, the anticancer agent, is the common example. The *triterpenes (C_{30})* include steroids, sterols, and cardiac glycosides with anti-inflammatory, sedative, insecticidal or cytotoxic activity. Common triterpenes: amyrins, ursolic acid and oleanic acid *sesquiterpene (C_{15})* like monoterpenes, are major components of many essential oils (Martinez *et al.,* 2008). The sesquiterpene acts as irritants when applied externally and when consumed internally their action resembles that of gastrointestinal tractirritant. A number of sesquiterpene lactones have been isolated and broadly they have antimicrobial (particularly antiprotozoal) and neurotoxic action. The sesquiterpene lactone, palasonin, isolated from *Butea monosperma* has anthelmintic activity, inhibits glucose up take and depletes the glycogen content in *Ascaridia galli* (Fig: 6). Terpenoids are classified according to the number of isoprene units involved in the formation of these compounds. The major groups are shown in Table: 2.

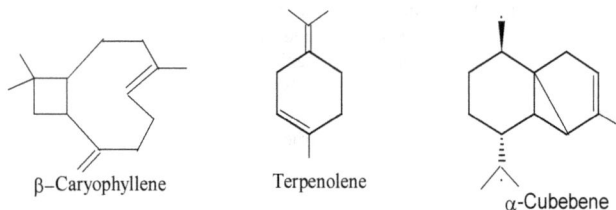

β–Caryophyllene   Terpenolene

α-Cubebene

Fig: 6 Basic structures of some pharmacologically important plant derived terpenes

Table:2 Types of terpenoids according to the number of isopropene units

Type of terpenoids	Number of carbon Atoms	Number of isoprene units	Example
Monoterpene	10	2	Limonene
Sesquiterpene	15	3	Artemisinin
Diterpene	20	4	Forskolin
Triterpene	30	6	α-Amyrin
Tetraterpene	40	8	β-Carotene
Polymeric terpenoid	Several	Several	Rubber

5.2.8 Anthraquinones

These are derivatives of phenolic and glycosidic compounds. They are solely derived from anthracene giving variable oxidized derivatives such as anthrones and anthranols (Maurya *et al.,* 2008; Firn, 2010). Other derivatives such as chrysophenol, aloe-emodin, rhein, salinosporamide, luteolin (Fig: 7) and emodin have in common a double hydroxylation at positionsC-1 and C-8. To test for free anthraquinones, powdered plant material is mixed with organic solvent and iltered, and an aqueous base, e.g. NaOH or NH_4OH solution, is added to it. A pink or violet colour in the base layer indicates the presence of anthraquinones in the plant sample (Sarker and Nahar, 2007).

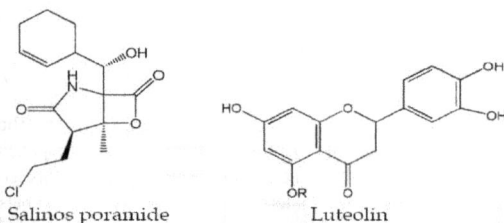

Salinos poramide   Luteolin

Fig: 7 Basic structures of some pharmacologically important plant derived anthraquinones

5.2.9 Essential oils

Essential oils are the odorous and volatile products of various plant and animal species. Essential oils have a tendency evaporate on exposure to air even at ambient conditions and are therefore also referred to as volatile oils or ethereal oils. They mostly contribute to the odoriferous constituents or *'essences'* of the aromatic plants that are used abundantly in enhancing the aroma of some spices (Martinez *et al.*, 2008). Essential oils are either secreted either directly by the plant protoplasm or by the hydrolysis of some glycosides and structures such as directly Plant structures associated with the secretion of essential oils include: Glandular hairs (Lamiaceae e.g. *Lavandula angustifolia*), Oil tubes (or vittae)(Apiaceae e.g. *Foeniculum vulgare* and Pimpinella anisum (Aniseed), modified parenchymal cells (Piperaceae e.g. *Piper nigrum* - Black pepper), Schizogenous or lysigenum passages(Rutaceae e.g. *Pinus palustris* - Pine oil. Essential oils have been associated with different plant parts including leaves, stems, flowers, roots or rhizomes. Chemically, a single volatile oil comprises of more than 200 different chemical components, and mostly the trace constituents are solely responsible for attributing its characteristic flavour and odour (Firn, 2010).

Essential oils can be prepared from various plant sources either by direct steam distillation, expression, extraction or by enzymatic hydrolysis. Direct steam distillation involves the boiling of plant part in a distillation flask and passing the generated steam and volatile oil through a water condenser and subsequently collecting the oil in florentine flasks. Depending on the nature of the plant source the distillation process can be either water distillation, water and steam distillation or direct distillation. Expression or extrusion of volatile oils is accomplished by either by sponge method, scarification, rasping or by a mechanical process. In the sponge method, the washed plant part e.g. citrus fruit (*e.g.*, orange, lemon, grape fruit, bergamot) is cut into halves to remove the juice completely, rind turned inside out by hand and squeezed when the secretary glands rupture. The oozed volatile oil is collected by means of the sponge and subsequently squeezed in a vessel. The oil floating on the surface is separated. For the scarification process the apparatus Ecuelle a Piquer (a large bowl meant for pricking the outer surface of citrus fruits) is used. It is a large funnel made of copper having its inner layer tinned properly. The inner layer has numerous pointed metal needles just long enough to penetrate the epidermis. The lower stem of the apparatus serve two purposes; *first*, as a receiver for the oil; and *secondly*, as a handle. Now, the freshly washed lemons are placed in the bowl and rotated repeatedly when the oil glands are punctured (scarified) thereby discharging the oil right into the handle. The liquid, thus collected, is transferred to another vessel, where on keeping the clear oil may be decanted and filtered. For the rasping process, the outer surface of the peel of citrus fruits containing the oil gland is skillfully removed by a grater. The 'raspings' are now placed in horsehair bags and pressed strongly so as to ooze out the oil stored in the oil glands. Initially, the liquid has a turbid appearance but on allowing it to stand the oil separates out which may be decanted and filtered subsequently. The mechanical process involves the use of heavy duty centrifugal devices so as to ease the separation of oil/water emulsions invariably formed and with the advent of modern mechanical devices the oil output has increased impressively. The extraction processes can be carried out with either volatile solvents (e. g. hexane, petroleum ether or benzene) resulting into the production of 'floral concretes'- oils with solid consistency and partly soluble in 95%

alcohol, or non volatile solvents (tallow, lard or olive oil) which results in the production of perfumes. Examples of volatile oils include amygdaline (volatile oil of bitter almond), sinigrin (volatile oil of black mustard), and eugenol occurring as gein (volatile oil of *Geum urbanum*) (Fig: 8).

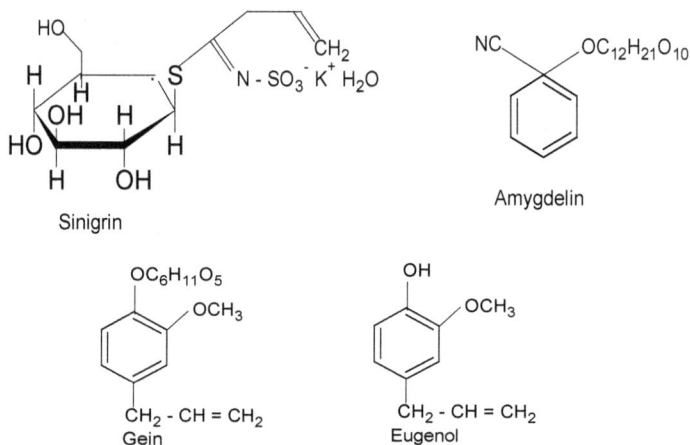

Sinigrin

Amygdelin

Gein

Eugenol

Fig: 8 Basic structures of some pharmacologically important plant derived essential oils

5.2.10 Steroids

Plant steroids (or steroid glycosides) also referred to as 'cardiac glycosides' are one of the most naturally occurring plant phytoconstituents that have found therapeutic applications as arrow poisons or cardiac drugs (Firn, 2010). The cardiac glycosides are basically steroids with an inherent ability to afford a very specific and powerful action mainly on the cardiac muscle when administered through injection into man or animal. Steroids (anabolic steroids) have been observed to promote nitrogen retention in osteoporosis and in animals with wasting illness (Maurya *et al.*, 2008; Madziga *et al.*, 2010). Caution should be taken when using steroidal glycosides as small amounts would exhibit the much needed stimulation on a diseased heart, whereas excessive dose may cause even death. Diosgenin and cevadine (from *Veratrum veride*) are examples of plant steroids (Fig: 9).

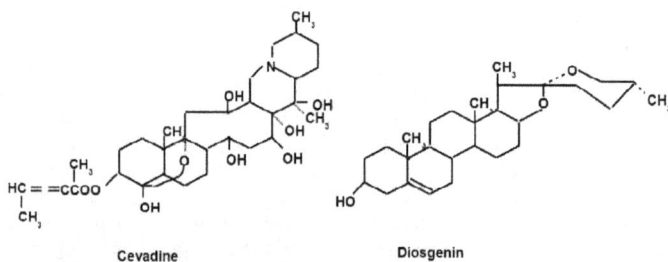

Cevadine

Diosgenin

Fig: 9 Basic structures of some pharmacologically important plant derived steroids

5.3 Mechanism of action of phytochemicals

Different mechanisms of action of phytochemicals have been suggested. They may inhibit microorganisms, interfere with some metabolic processes or may modulate gene expression and signal transduction pathways (Kris-Etherton *et al.*, 2002; Manson 2003; Surh 2003). Phyto chemicals may either be used as chemotherapeutic or chemo preventive agents with Chemoprevention referring to the use of agents to inhibit, reverse, or retard tumorigenesis.In this sense chemo preventive phytochemicals are applicable to cancer therapy, since molecular mechanisms may be common to both chemoprevention and cancer therapy(D'Incalci *et al.*, 2005; Sarkar & Li, 2006). Plant extracts and essential oils may exhibit different modes of action against bacterial strains, such as interference with the phospholipids bilayer of the cell membrane which has as a consequence a permeability increase and loss of cellular constituents, damage of the enzymes involved in the production of cellular energy and synthesis of structural components, and destruction or inactivation of genetic material. In general, the mechanism of action is considered to be the disturbance of the cytoplasmic membrane, disrupting the proton motive force, electron flow, active transport, and coagulation of cell contents (Kotzekidou *et al.*, 2008). Some specific modes of actions are discussed below.

5.3.1 Antioxidants

Antioxidants protect cells against the damaging effects of reactive oxygen species otherwise called, free radicals such as singlet oxygen, super oxide, peroxyl radicals, hydroxyl radicals and peroxynite which results in oxidative stress leading to cellular damage (Mattson and Cheng, 2006). Natural antioxidants play a key role in health maintenance and prevention of the chronic and degenerative diseases, such as atherosclerosis, cardiac and cerebral ischema, carcinogenesis, neurodegenerative disorders, diabetic pregnancy, rheumatic disorder, DNA damage and ageing (Uddin *et al.*, 2008; Jayasri *et al.*, 2009). Antioxidants exert their activity by scavenging the 'free-oxygen radicals' thereby giving rise to a fairly 'stable radical'. The free radicals are meta stable chemical species, which tend to trap electrons from the molecules in the immediate surroundings. These radicals if not scavenged effectively in time, they may damage crucial bio molecules like lipids, proteins including those present in all membranes, mitochondria and, the DNA resulting in abnormalities leading to disease conditions (Uddin et al. 2008). Thus, free radicals are involved in a number of diseases including: tumor inflammation, hemorrhagic shock, atherosclerosis, diabetes, infertility, gastrointestinal ulcerogenesis, asthma, rheumatoid arthritis, cardiovascular disorders, cystic fibrosis, neurodegenerative diseases (e.g. Parkinsonism, Alzheimer's diseases), AIDS and even early senescence (Chen *et al.*, 2006; Uddin *et al.*, 2008). The human body produces insufficient amount of antioxidants which are essential for preventing oxidative stress. Free radicals generated in the body can be removed by the body's own natural antioxidant defenses such as glutathione or catalyses (Sen, 1995). Therefore this deficiency had to be compensated by making use of natural exogenous antioxidants, such as vitamin C, vitamin E, flavones, α-carotene and natural products in plants (Madsen and Bertelsen, 1995; Rice-Evans *et al.*, 1997; Diplock *et al.*, 1998).

Plants contain a wide variety of free radicals scavenging molecules including phenols, flavonoids, vitamins, terpenoids hat are rich in antioxidant activity (Madsen and Bertelsen, 1995; Cai and Sun, 2003). Many plants, citrus fruits and leafy vegetables are the source of ascorbic acid, vitamin E, caratenoids, flavanols and phenolics which possess the ability to scavenge the free radicals in human body. Significant antioxidant properties have been recorded in phytochemicals that are necessary for the reduction in the occurrence of many diseases (Hertog & Feskens, 1993; Anderson & Teuber, 2001). Many dietary polyphenolics constituents derived from plants are more effective antioxidants *in vitro* than vitamins E or C, and thus might contribute significantly to protective effects *in vivo* (Rice-Evans & Miller, 1997; Jayasri *et al.,* 2009). Methanol extract of *Cinnamon* contains a number of antioxidant compounds which can effectively scavenge reactive oxygen species including superoxidants and hydroxyl radicals as well as other free radicals *in vitro.* The fruit of *Cinnamon*, an under-utilized and unconventional part of the plant, contains a good amount of phenolic antioxidants to counteract the damaging effects of free radicals and may protect against mutagenesis.

Antioxidants are often added to foods to prevent the radical chain reactions of oxidation, and they act by inhibiting the initiation and propagation step leading to the termination of the reaction and delay the oxidation process. Due to safety concerns of synthetic compounds, food industries have focused on finding natural antioxidants to replace synthetic compounds. In addition, there is growing trend in consumer preferences for natural antioxidants, all of which has given more impetus to explore natural sources of antioxidants.

5.3.2 Anticacinogenesis

Polyphenols particularly are among the diverse phytochemicals that have the potential in the inhibition of carcinogenesis (Liu, 2004). Phenolics acids usually significantly minimize the formation of the specific cancer-promoting nitrosamines from the dietary nitrites and nitrates. Glucosinolates from various vegetable sources as broccoli, cabbage, cauliflower, and Brussel sprouts exert a substantial protective support against the colon cancer. Regular consumption of Brussel sprouts by human subjects (up to 300 g. day–1) miraculously causes a very fast (say within a span of 3 weeks) an appreciable enhancement in the glutathione-S transferase, and a subsequent noticeable reduction in the urinary concentration of a specific purine meltabolite that serves as a marker of DNA-degradation in cancer. Isothiocyanates and the indole-3-carbinols do interfere categorically in the metabolism of carcinogens thus causing inhibition of procarcinogen activation, and thereby inducing the 'phase-II' enzymes, namely: NAD(P)H quinone reductase or glutathione S-transferase, that specifically detoxify the selected electrophilic metabolites which are capable of changing the structure of nucleic acids. Sulforaphane (rich in broccoli) has been proved to be an extremely potent phase-2enzyme inducer. It predominantly causes specific cell-cycle arrest and also the apoptosis of the neoplasm (cancer) cells. Sulforaphane categorically reduces α-D-gluconolactone which has been established to be a significant inhibitor of breast cancer. Indole-3-carbinol (most vital and important indole present in broccoli) specifically inhibits the Human Papilloma Virus (HPV) that may cause uterine cancer. It blocks the estrogen receptors specifically present in the breast cancer cells as well as down regulates CDK_6, and up regulates p_{21} and p_{27} in prostate cancer cells. It affords G1 cell-cycle arrest and apoptosis of breast and prostate cancer cells significantly and enhances the p 53 expression in cells treated with benzopyrene. It also depresses Akt, NF-kappaB, MAPK, and Bel-2 signaling pathways to a reasonably good extent. Phytosterols block the development of tumors (neoplasms) in colon, breast, and prostate glands. Although the precise and exact mechanisms whereby the said blockade actually takes place are not yet well understood, yet they seem to change drastically the ensuing cell-membrane transfer in the phenomenon of neoplasm growth and thereby reduce the inflammation significantly.

5.3.3 Antimicrobial activity

Phytoconstituents employed by plants to protect them against pathogenic insects, bacteria, fungi or protozoa have found applications in human medicine (Nascimento et al., 2000).Some phytochemicals such as phenolic acids act essentially by helping in the reduction of particular adherence of organisms to the cells lining the bladder, and the teeth, which ultimately lowers the incidence of urinary-tract infections (UTI) and the usual dental caries. Plants can also exert either bacteriostatic or bactericidal activity ob microbes. The volatile gas phase of combinations of *Cinnamon* oil and clove oil showed good potential to inhibit growth of spoilage fungi, yeast and bacteria normally found on IMF (Intermediate Moisture Foods) when combined with a modified atmosphere comprising a high concentration of $CO_2(40\%)$ and low concentration of $O_2(<0.05\%)$ (Jakhetia et al.,2010). *A. flavus,* which is known to produce toxins, was found to be the most resistant microorganism. It is worthy of note that antimicrobial activity results of the same plant part tested most of the time varied from researcher to researcher. This is possible because concentration of plant constituents of the same plant organ can vary from one geographical location to another depending on the age of the plant, differences in topographical factors, the nutrient concentrations of the soil, extraction method as well as method used for antimicrobial study. It is therefore important that scientific protocols be clearly identified and adequately followed and reported.

5.3.4 Anti-ulcer

Plants extracts have been reported to inhibit both growth of *H. pylori in-vitro* as well as its urease activity (Jakhetia et al., 2010). The efficiency of some extracts in liquid medium and at low pH levels enhances their potency even in the human stomach. Their inhibitory effect on the intestinal and kidney Na+/K+ ATPase activity and on alanine transport in rat jejunum has also been reported (Jakhetia et al., 2010).

5.3.5 Anti-diabetic

Cinnamaldehyde, a phytoconstituent extracts have been reported to exhibit significant antihyperglycemic effect resulting in the lowering of both total cholesterol and triglyceride levels and, at the same time, increasing HDL-cholesterol in STZ-induced diabetic rats. This investigation reveals the potential of cinnamaldehyde for use as a natural oral agent, with both hypoglycaemic and hypolipidemic effects. Recent reports indicate that *Cinnamon*ex extract and polyphenols with procyanidin type-A polymers exhibit the potential to increase the amount of TTP (Thrombotic Thrombocytopenic Purpura), IR (Insulin Resistance), andGLUT4 (Glucose Transporter-4) in 3T3-L1 Adipocytes. It was suggested that the mechanism of *Cinnamon*'s insulin-like activity may be in part due to increase in the amounts of TTP, IRβ and GLUT4 and that *Cinnamon* polyphenols may have additional roles as anti-inflammatory and/or anti-angiogenesis agents (Jakhetia et al., 2010).

5.3.6 Anti-inflammatory

Essential oil of *C. osmophloeum* twigs has excellent anti- inflammatory activities and cytotoxicity against HepG2 (Human Hepatocellular Liver Carcinoma Cell Line) cells.

Previous reports also indicated that the constituents of *C. osmophloeum* twig exhibited excellent anti-inflammatory activities in suppressing nitric oxide production by LPS (Lipopolysaccharide)-stimulated macrophages (Jakhetia *et al.*, 2010).

5.3.7 Multifunctional targets

Multiple molecular targets of dietary phytochemicals have been identified, from pro- and anti-apoptotic proteins, cell cycle proteins, cell adhesion molecules, protein kinases, transcription factors to metastasis and cell growth pathways (Awad and Bradford, 2005;Aggarwal & Shishodia, 2006; Choi & Friso, 2006). Phytochemicals such as epigallocatechin-3-gallate (EGCG) from green tea, curcumin from turmeric, and resveratrol from red wine tend to aim at a multitude of molecular targets. It is because of these characteristics that definitive mechanisms of action are not available despite decades of research (Francis *et al.*, 2002). The multi-target nature of phytochemicals may be beneficial in overcoming cancer drug resistance. This multi-faceted mode of action probably hinders the cancer cell's ability to develop resistance to the phytochemicals. It has also been demonstrated that EGCG has inhibitory effects on the extracellular release of verotoxin (VT) from *E. coli* 0157: H_7 (Voravuthikunchai and Kitpipit, 2003). Ethanol pericarp extracts from *Punica granatum* was also reported to inhibited VT production in periplasmic space and cell supernatant. Mechanisms responsible for this are yet to be understood, however the active compounds from the plant are thought to interfere with the transcriptional and translational processes of the bacterial cell (Voravuthikunchai and Kitpipit, 2003). More work is needed to be done in order to establish this assumption. Phytochemicals may also modulate transcription factors (Andreadi *et al.*, 2006), redox-sensitive transcription factors (Surh *et al.*, 2005), redox signaling, and inflammation. As an example, nitric oxide (NO), a signaling molecule of importance in inflammation, is modulated by plant polyphenols and other botanical extracts (Chan and Fong, 1999; Shanmugam *et al.*, 2008). Many phytochemicals have been classified as phytoestrogens, with health-promoting effects resulting in the phytochemicals to be marketed as nutraceuticals (Moutsatsou, 2007).

5.4 Methods of studying phytochemicals

No single method is sufficient to study the bioactivity of phytochemicals from a given plant. An appropriate assay is required to first screen for the presence of the source material, to purify and subsequently identify the compounds therein. Assay methods vary depending on what bioactivity is targeted and these may include antimicrobial, anti-malarial, anticancer, seed germination, and mammalian toxicity activities. The assay method however should be as simple, specific, and rapid as possible. An *in vitro* test is more desirable than a bioassay using small laboratory animals, which, in turn, is more desirable than feeding large amounts of valuable and hard to obtain extract to larger domestic or laboratory animals. In addition, *in vivo* tests in mammals are often variable and are highly constrained by ethical considerations of animal welfare. Extraction from the plant is an empirical exercise in which different solvents are utilized under a variety of conditions such as time and temperature of extraction. The success or failure of the extraction process depends on the most appropriate assay. Once extracted from the plant, the bioactive component then has to be separated from the co extractives. Further purification steps may involve simple crystallization of the compound from the crude extract, further solvent partition of the co extractives or chromatographic methods in order to fractionate the compounds based on their acidity, polarity or molecular size. Final purification, to provide compounds of suitable purity for such structural analysis, may be accomplished by appropriate techniques such as recry- stallization, sublimation, or distillation.

5.4.1 Extraction of phytochemicals

(a) Solvent extraction

Various solvents have been used to extract different phytoconstituents. The plant parts are dried immediately either in an artificial environment at low temperature (50-60 °C) or dried preferably in shade so as to bring down the initial large moisture content to enable its prolonged storage life and . The dried berries are pulverised by mechanical grinders and the oil is removed by solvent extraction. The defatted material is then extracted in a soxhlet apparatus or by soaking in water or alcohol (95% v/v). The resulting alcoholic extract is filtered, concentrated in vacuo or by evaporation, treated with HCl (12N) and refluxed for at least six hours. This can then be concentrated and used to determine the presence of phytoconstituents. Generally, the saponins do have high molecular weight and hence their isolation in the purest form poses some practical difficulties. The plant parts (tubers, roots, stems, leave etc) are washed sliced and extracted with hot water or ethanol (95% v/v) for several hours. The resulting extract is filtered, concentrated *in vacuo* and the desired constituent is precipitated with ether. Exhaustive extraction (EE) is usually carried out with different solvents of increasing polarity in order to extract as much as possible the most active components with highest biological activity.

(b) Supercritical fluid extraction (SFE)

This is the most technologically advanced extraction system (Patil and Shettigar, 2010). Super Critical Fluid Extraction (SFE) involves use of gases, usually CO_2, and compressing them into a dense liquid. This liquid is then pumped through a cylinder containing the material to be extracted. From there, the extract-laden liquid is pumped into a separation chamber where the extract is separated from the gas and the gas is recovered for re-use. Solvent properties of CO_2 can be manipulated and adjusted by varying the pressure and temperature that one works at. The advantages of SFE are, the versatility it offers in appointing the constituents you want to extract from a given material and the fact that your end product has virtually no solvent residues left in it (CO_2 evaporates completely). The downside is that this technology is quite expensive. There are many other gases and liquids that are highly efficient as extraction solvents when put under pressure (Patil and Shettigar, 2010).

(i). Coupled SFE-SFC: System in which a sample is extracted with a supercritical fluid which then places the extracted material in the inlet part of a supercritical fluid chromatographic system. The extract is than chromatographed directly using supercritical fluid.

(ii). Coupled SFE-GC and SFE-LC: System in which a sample is extracted using a supercritical fluid which is then depressurized to deposit the extracted material in the inlet part or a column of gas or liquid chromatographic system respectively. SFE is characterized by robustness of sample preparation, reliability, less time consuming, high yield and also has potential for coupling with number of chromatographic methods.

(c) Microwave-Assisted extraction

Patil and Shettigar (2010) reported an innovative, microwave-assisted, solvent-extraction technology known as Microwave-Assisted Processing (MAP). MAP applications include the extraction of high-value compounds from natural sources including phytonutrients, nutraceutical and functional food ingredients and

pharmaceutical actives from biomass. Compared to conventional solvent extraction methods, MAP technology offers some combination of the following advantages:
1. Improved products, increased purity of crude extracts, improved stability of marker compounds, possibility to use less toxic solvents;
2. Reduced processing costs, increased recovery and purity of marker compounds, very fast extraction rates, reduced energy and solvent usage.

With microwave-derived extraction as opposed to diffusion, very fast extraction rates and greater solvent flexibility can be achieved. Many variables, including the microwave power and energy density, can be turned to deliver desired product attributes and optimize process economics. The process can be customized to optimize for commercial/cost reasons and excellent extracts are produced from widely varying substrates. Examples include, but are not limited to, antioxidants from dried herbs, carotenoids from single cells and plant sources, taxanes from taxus biomass, essential fatty acids from microalgae and oilseeds, phytosterols from medicinal plants, polyphenols from green tea, flavor constituents from vanilla and black pepper, essential oils from various sources, and many more (Patil and Shettigar, 2010).

(d) Solid phase extraction

This involves sorption of solutes from a liquid medium onto a solid adsorbent by the same mechanisms by which molecules are retained on chromatographic stationary phases. These adsorbents, like chromatographic media, come in the form of beads or resins that can be used in column or in batch form. They are often used in the commercially available form of syringes packed with medium (typically a few hundred milligrams to a few grams) through which the sample can be gently forced with the plunger or by vacuum. Solid phase extraction media include reverse phase, normal phase, and ion-exchange media. This is method for sample purification that separates and concentrates the analyses from solution of crude extracts by adsorption onto a disposable solid-phase cartridge. The analyses is normally retained on the stationary phase, washed and then evaluated with different mobile phase. If an aqueous extract is passed down a column containing reverse phase packing material, everything that is fairly non polar will bind, whereas everything polar will pass through (Patil & Shettigar, 2010).

(e) Chromatographic fingerprinting and marker compound analysis

Chromatographic fingerprint of an Herbal Medicine (HM) is a chromatographic pattern of the extract of some common chemical components of pharmacologically active and or chemical characteristics (Patil & Shettigar, 2010). This chromatographic profile should be featured by the fundamental attributions of "integrity" and "fuzziness" or "sameness" and "differences" so as to chemically represent the HM investigated. It is suggested that with the help of chromatographic fingerprints obtained, the authentication and identification of herbal medicines can be accurately conducted (integrity) even if the amount and/or concentration of the chemically characteristic constituents are not exactly the same for different samples of this HM (hence, "fuzziness") or, the chromatographic finger prints could demonstrate both the "sameness" and "differences" between various samples successfully. Thus, we should globally consider multiple constituents in the HM extracts, and not individually consider only one and/or two marker components for evaluating the quality of the HM products. However, in any HM and its extract, there are hundreds of unknown components and many of them are in low amount. Moreover, there usually exists variability within the same herbal materials. Hence it is very important to obtain reliable chromatographic fingerprints that represent pharmacologically active and chemically characteristic components of the HM. In the phytochemical evaluation of herbal drugs, TLC is being employed extensively for the following reasons: (1) it enables rapid analysis of herbal extracts with minimum sample clean-up requirement, (2) it provides qualitative and semi quantitative information of the resolved compounds and (3) it enables the quantification of chemical constituents. Fingerprinting using HPLC and GLC is also carried out in specific cases. In TLC fingerprinting, the data that can be recorded using a highperformanceTLC (HPLC) scanner includes the chromatogram, retardation factor (R_f) values, the colour of the separated bands, their absorption spectra, λ -max and shoulder inflection/s of all the resolved bands. All of these, together with the profiles on derivatives with different reagents, represent the TLC fingerprint profile of the sample. The information so generated has a potential application in the identification of an authentic drug, in excluding the adulterants and in maintaining the quality and consistency of the drug. HPLC fingerprinting includes recording of the chromatograms, retention time of individual peaks and the absorption spectra (recorded with a photodiode array detector)with different mobile phases. Similarly, GLC is used for generating the fingerprint profiles of volatile oils and fixed oils of herbal drugs. Furthermore, the recent approaches of applying hyphenated chromatography and spectrometry such as High-Performance Liquid Chromatography–Diode Array Detection (HPLC–DAD), Gas Chromatography–Mass Spectroscopy (GC–MS), Capillary Electrophoresis-Diode Array Detection (CE-DAD), High-Performance Liquid Chromatography–Mass Spectroscopy (HPLC–MS) and High-Performance Liquid Chromatography–Nuclear Magnetic Resonance Spectroscopy (HPLC–NMR) could provide the additional spectral information, which will be very helpful for the qualitative analysis and even for the on-line structural elucidation.

(f) Advances in chromatographic techniques

(i) Liquid chromatography
a. Preparative high performance liquid chromatography

There are basically two types of preparative HPLC. One is low pressure (typically under 5 bar) traditional PLC, based on the use of glass or plastic columns filled with low efficiency packing materials of large particles and large size distribution. A more recent form PLC is Preparative High Performance Liquid Chromatography (Prep. HPLC) has been gaining popularity in pharmaceutical industry. In preparative HPLC (pressure >20 bar), larger stainless steel columns and packing materials (particle size 10-30 μ m are needed. The examples of normal phase silica columns are Kromasil 10 μ m, Kromasil 16 μ m, Chiralcel AS 20 μ m whereas for reverse phase are Chromasil C_{18}, Chromasil C_8,YMC C_{18}. The aim is to isolate or purify compounds, whereas in analytical work the goal is to get information about the sample. Preparative HPLC is closer to analytical HPLC than traditional PLC, because its higher column efficiencies and faster solvent velocities permit more difficult separation to be conducted more quickly (Oleszek and Marston, 2000; Philipson, 2007). In analytical HPLC, the important parameters are resolution, sensitivity and fast analysis time whereas in preparative HPLC, both the degree of solute purity as well as the amount of compound that can be produced per unit time i.e. throughput or recovery are important. This is very important in pharmaceutical industry of today because new products (Natural, Synthetic) have to be introduced to the market as quickly as possible. Have available such a powerful purification technique makes it possible to spend less time on the synthesis conditions.

b. Liquid Chromatography- Mass Spectroscopy (LC-MS)

In Pharmaceutical industry LC-MS has become method of choice in many stages of drug development. Recent advances includes electro spray, thermo spray, and ion spray ionization techniques which offer unique advantages of high detection sensitivity and specificity, liquid secondary ion mass spectroscopy, later laser mass spectroscopy with 600MHz offers accurate determination of molecular weight proteins, peptides. Isotopes pattern can be detected by this technique (Oleszek and Marston, 2000; Philipson, 2007).

c. Liquid Chromatography- Nuclear Magnetic Resonance (LC-NMR)

The combination of chromatographic separation technique with NMR spectroscopy is one of the most powerful and time saving method for the separation and structural elucidation of unknown compound and mixtures, especially for the structure elucidation of light and oxygen sensitive substances. The online LC-NMR technique allows the continuous registration of time changes as they appear in the chromatographic run automated data acquisition and processing in LC-NMR improves speed and sensitivity of detection (Daffre *et al.,* 2008). The recent introduction of pulsed field gradient technique in high resolution NMR as well as three-dimensional technique improves application in structure elucidation and molecular weight information. These new hyphenated techniques are useful in the field of pharmacokinetics, toxicity studies, drug metabolism and drug discovery process.

(ii) Gas chromatography
a. Gas Chromatography Fourier Transform Infrared spectrometry
Coupling capillary column gas chromatographs with Fourier Transform Infrared Spectrometer provides a potent means for separating and identifying the components of different mixtures.

B. Gas Chromatography-Mass Spectroscopy
Gas chromatography equipment can be directly interfaced with rapid scan mass spectrometer of various types. The flow rate from capillary column is generally low enough that the column output can be fed directly into ionization chamber of MS. The simplest mass detector in GC is the Ion Trap Detector (ITD). In this instrument, ions are created from the eluted sample by electron impact or chemical ionization and stored in a radio frequency field; the trapped ions are then ejected from the storage area to an electron multiplier detector. The ejection is controlled so that scanning on the basis of mass-to-charge ratio is possible. The ions trap detector is remarkably compact and less expensive than quadrapole instruments. GC-MS instruments have been used for identification of hundreds of components that are present in natural and biological system (Oleszek and Marston, 2000; Philipson, 2007; Daffre *et al.,* 2008).

(iii) Supercritical Fluid Chromatography (SFC)
Supercritical fluid chromatography is a hybrid of gas and liquid chromatography that combines some of the best features of each. This technique is an important third kind of column chromatography that is beginning to find use in many industrial, regulatory and academic laboratories. SFC is important because it permits the separation and

determination of a group of compounds that are not conveniently handled by either gas or liquid chromatography. These compounds are either non-volatile or thermally labile so that GC procedures are inapplicable or contain no functional group that makes possible detection by the spectroscopic or electrochemical technique employed in LC. SFC has been applied to a wide variety of materials including natural products, drugs, foods and pesticides.

5.4.2 Other Chromato-Spectrometric studies

The NMR techniques are employed for establishing connectivity between neighboring protons and establishing C-H bonds. INEPT is also being used for long range hetero nuclear correlations over multiple bonding. The application of Thin Layer chromatography (TLC),High Performance Chromatography(HPLC) and HPLC coupled with Ultra violate (UV)photodiode array detection, Liquid Chromatography-Ultraviolet(LC-UV),LiquidChromatography-Mass Spectrophotometry (LCMS), electrospray (ES) and Liquid Chromatography-Nuclear Magnetic Resonance (LC-NMR) techniques for the separation and structure determination of antifungal and antibacterial plant compounds is on the increase frequently (Oleszek and Marston, 2000; Bohlin and Bruhn, 1999). Currently available chromatographic and spectroscopic techniques in new drug discovery from natural products Currently, computer modelling has also been introduced in spectrum interpretation and the generation of chemical structures meeting the spectral properties of bioactive compounds obtained from plants (Vlietinck, 2000). The computer systems utilize ^{1}H, ^{13}C, ^{2}D-NMR, IR and MS spectral properties (Philipson, 2007). Libraries of spectra can be searched for comparison with complete or partial chemical structures. Hyphenated chromatographic and spectroscopic techniques are powerful analytical tools that are combined with high throughput biological screening in order to avoid re-isolation of known compounds as well as for structure determination of novel compounds. Hyphenated chromatographic and spectroscopic techniques include LC–UV–MS, LC–UV–NMR, LC–UV–ES–MS and GC–MS (Oleszek and Marston, 2000; Philipson, 2007).

5.4.3 Simple assay methods

(a) Antimicrobial assay

Common methods used in the evaluation of the antibacterial and antifungal activities of plant extracts and essential oils, include the agar diffusion method (paper disc and well), the dilution method (agar and liquid broth) (Yagoub, 2008; Okigbo et al., 2009; El-Mahmood, 2009; Aiyegoro et al., 2009), and the turbidimetric and impedimetric monitoring of microbial growth (R'ıos and Recio, 2005). These methods are simple to carry out under laboratory conditions.

(b) Antioxidant assays

Most common Spectrophotometry assay method applied is the DPPH radical scavenging system in which the hydrogen or electrons donation ability of plant extracts are measured from bleaching of purple methanol solution of 2, 2'-diphenyl-1-picrylhydrazyl (DPPH) free radical (Changwei et al., 2006). This Spectrophotometry assay uses the stable radical DPPH as a reagent. DPPH absorbs at 517 nm, and as its concentration is reduced by existence of an antioxidant, the absorption gradually disappears with time. A 2-ml aliquot of a suspension of the ethanol extracts is mixed with 1 ml of 0.5 mM 2,2- diphyenyl-1-picrylhydrazyl (DPPH) solution and 2 ml of 0.1 M sodium acetate buffer (pH 5.5). After shaking, the mixture is incubated at ambient temperature in the dark for 30 minutes, following which the absorbance is measured at 517 nm using a UV-160A spectrometer. A solvent such as ethanol can be used as negative control. Radical scavenging activity is often expressed as percentage inhibition and is often calculated using the formula:

$$\% \text{ radical scavenging activity} = [(A_{control} - A_{test})/A_{control}] \times 100$$

Where $A_{control}$ is the absorbance of the control (DPPH solution without test sample) and A_{test} is the absorbance of the test sample (DPPH solution plus antioxidant).Phenolics content and reducing power of extracts is often determined using the Folin-Ciocalteu method. Equal volumes of Folin-Ciocalteu reagent and given quantity (mg) of plant extracts of different concentrations (e.g. 0.4, 0.3, 0.2, 0.1 and 0.05 mg/ml) are often mixed in different sets of test tubes shaken thoroughly, and left to stand for 1 min. Ten percent of $NaHCO_3$ is then added and the mixture once again allowed to stand for 30minutes, after which the absorbance (725 nm) is measured spectrophotometrically. Gallic acid (0.05-0.5 mg/ml) is often used to produce standard calibration curve and the total phenolic content expressed as mg equivalent of Gallic acid (mg GAE) per gram dry weight of the extract by computing with standard calibration curve (Djeridane *et al.*, 2006).For determination of reducing power of plant extracts, the ferric reducing/antioxidant power (FRAP) assay method can be applied. The assay is based on the reducing power of a compound (antioxidant). A potential antioxidant reduces the ferric ion (Fe3+) to the ferrous ion (Fe^{2+}); the later forms a blue complex (Fe^{2+} / 2, 4, 6- tripyridyl-*s*-triazine (TPTZ)), which increases the absorption at 593 nm. Stronger absorption at this wavelength indicates higher reducing power of the phytochemical, thus higher antioxidant activity. Reaction mixture containing test extract sample at different concentrations (10-100µl) in phosphate buffer (0.2M, pH 6.6) and equal amounts of 1% (w/v) potassium ferricyanide are incubated at 50°C for20 minutes and then the reaction terminated by the addition of equal volumes of 10% (w/v)tricarboxyllic acid (TCA) solution and the mixture centrifuged at 3000rpm for 20 minutes. The supernatant is mixed with equal volume of distilled water and 0.1 % (w/v) ferric chloride solution and the absorbance measured at 700 nm. Increased absorbance of the mixture with concentration indicates the reducing power of extract (Jayasri, 2009).

5.5 Toxicological studies

These are often carried out to determine the toxicity of a plant part. Usually animal models such as mice, guinea pigs or rabbits are often employed. In these procedures, the LD_{50} of the extracts in the experimental animal is often determined via either oral or intradermal administration. The toxic response of experimental animals to the administration of plant alkaloids is usually detected by assay of the serum ALT and AST of the animal as sensitive indicators of hepatocellular damage (Chapatwala *et al.*, 1982). Any toxicity usually results in distortion of hepatocytes membrane integrity due to hepatocellular injury and plasma levels rise, as a consequence of high toxin levels present within hepatocytes.

5.6 Safety concerns for phytochemicals

Plants are natural reservoir of medicinal agents almost free from the side effects normally caused by synthetic chemicals (Fennel *et al.*, 2004). The World Health Organization estimates that herbal medicine is still the main stay of about 75-80% of the world population, mainly in the developing countries for primary health care because of better cultural acceptability, better compatibility with the human body, and lesser side-effects (Kamboj, 2000; Yadav & Dixit, 2008). The over use of synthetic drugs with impurities resulting in higher incidence of adverse drug reactions, has motivated mankind to go back to nature for safer remedies. Due to varied locations where these plants grow, coupled with the problem of different vernacular names, the World Health organization published standards for herbal safety to minimize adulteration and abuse (WHO, 1999).

A number of modern drugs have been isolated from natural sources and many of these isolations were based on the uses of the agents in traditional medicine (Rizvi *et al.*, 2009).Antimicrobial properties of crude extracts prepared from plants have been described and such reports had attracted the attention of scientists worldwide (Falodun *et al.*, 2006; El-Mahmood and Amey, 2007; El-Mahmood, 2009). Herbs have been used for food and medicinal purposes for centuries and this knowledge have been passed on from generation to generation (Adedapo *et al.*, 2005). This is particularly evident in the rural areas where infectious diseases are endemic and modern health care facilities are few and far thus, compelling the people to nurse their ailments using local herbs. Herbal treatments have been adjudged to be relatively safe (WHO, 1999). For instance, daily oral doses ofepigallocatechin-3-gallate (EGCG) for 4 weeks at 800 mg/day in 40 volunteers only caused minor adverse effects (Phillipson, 2007). In a 90-day study of polyphenol E (a formulation of green tea extract with 53% EGCG), the oral no effect level (NOEL) values are 90 mg/kg/day for rats and 600 mg/kg/day for dogs (Boocock *et al.*, 2007). For curcumin given to cancer patients at 3600 mg/day for 4 months or 800 mg/day for 3 months, only minor adverse effects are seen. For resveratrol, a single oral dose at 5 g in 10 volunteers only causes minor adverse effects (Boocock *et al.*, 2007). Though herbs are relatively safe to use, their combined use with orthodox drugs should be done with extreme caution. Concomitant use of conventional and herbal medicines is reported to lead to clinically relevant herb–drug interactions (Liu et al., 2009). The two may interact either pharmacokinetically or pharmaco-dynamically resulting into adverse herbal-drug interactions (Izzo, 2005). St John'swort (*Hypericum perforatum*), used for the treatment of mild to moderate depression, interacts with digoxin, HIV inhibitors, theophylline and warfarin. Some medicinal herbs, when ingested, either affect cytochrome P_{450} isoenzymes by which drugs are metabolised or phosphoglycoprotein transporter systems that affect drug distribution and excretion. Concurrent use of some herbal medicines with other medicines may either lower blood plasma concentrations of medicinal drugs, possibly resulting in suboptimal therapeutic amounts, or lead to toxic concentrations in the blood, sometimes with fatal consequences (Phillipson, 2007).Despite this observation however, it has been reported that phytochemicals act in synergy with chemotherapeutic drugs in overcoming cancer cell drug resistance and that the application of specific phytochemicals may allow the use of lower concentrations of drugs in cancer treatment with an increased efficacy (Liu, 2004).Another advantage with phytochemicals is that, among an estimated 10,000 secondary products (natural pesticides), it has been proposed that human ancestors evolved a generalized defense mechanism against low levels of phytochemicals to enable their consumption of many different plant species containing variable levels of natural pesticides(carcinogens) without subsequent ill health (Liu, 2004). Traces of phytochemicals found in fruits and vegetables may potentiate the immune system and help to protect against cancer (Trewavas and Stewart, 2003). Phytochemicals show biphasic dose responses on mammalian cells. Though at high concentrations they can be toxic, sub-toxic doses may induce adaptive stress response (Ames and Gold, 1991). This includes the activation of signaling pathways that result in increased expression of genes encoding cytoprotective proteins. It is therefore suggested that hormetic mechanisms of action may underlie many of the health benefits of phytochemicals including their action against cancer drug resistance (Mattson, 2008).Molecular mechanisms of herb–drug interaction occur, the most notable is the ATP-binding cassette drug transporters such as P-glycoprotein (You & Moris, 2007) and the drug metabolizing enzymes (known as phase I and phase II enzymes), especially cytochromeP450 3A4 (CYP3A4) (Pal & Mitra, 2006; Meijerman *et al.*, 2006).

5.7 Future: phytochemicals as a sources of antimicrobial chemotherapeutic agents

Though there are few disadvantages associated with natural products research. These include difficulties in access and supply, complexities of natural product chemistry and inherent slowness of working with natural products. In addition, there are concerns about intellectual property rights, and the hopes associated with the use of collections of compounds prepared by combinatorial chemistry methods. Despite these limitations, over a100 natural-product-derived compounds are currently undergoing clinical trials and at least100 similar projects are in preclinical development (Phillipson, 2007). Among these products the highest number are from plant origin (Table:3). Most are derived from plants and microbial sources. The projects based on natural products are predominantly being studied for use in cancer or as anti-infectives. There is also, a growing interest in the possibility of developing products that contain mixtures of natural compounds from traditionally used medicines (Charlish, 2008), while, a defined mixture of components extracted from green tea (Veregen TM) has been approved by the US Food and Drug Administration (FDA) and has recently come on the market.

Table: 3 Drugs based on natural products at different stages of development

Development stage	Plant	Bacterial	Fungal	Animal	Semi synthetic	Total
Preclinical	46	12	07	07	27	99
Phase I	14	05	00	03	08	30
Phase I	41	04	00	10	11	66
Phase I	05	04	00	04	13	26
Per-registration	02	00	00	00	02	04
Total	108	25	07	24	61	225

Most of the leads from natural products that are currently in development have come from either plant or microbial sources. Earlier publications have pointed out that relatively little of the world's plant biodiversity has been extensively screened for bioactivity and that very little of the estimated microbial biodiversity has been available for screening (Doughari *et al.*, 2009). Hence, more extensive collections of plants (and microbes) could provide many novel chemicals for use in drug discovery assays. With the growing realization that the chemical diversity of natural products is a better match to that of successful drugs than the diversity of collections of synthetic compounds and with the global emergence of multidrug resistant pathogens (Feher and Schmidt, 2003) the interest in applying natural chemical diversity to drug discovery appears to be increasing once again (Galm and Shen, 2007).

With the advances in fractionation techniques to isolate and purify natural products (e. g. counter-current chromatography (Doughari *et al.*, 2009) and in analytical techniques to determine structures (Singh and Barrett, 2006), screening of natural product mixtures is now more compatible with the expected timescale of high-throughput screening campaigns. Singh and Barrett (2006) point out that pure bioactive compound can be isolated from fermentation broths in less than 2 weeks and that the structures of more than 90% of new compounds can be elucidated within 2 weeks. With advances in NMR techniques, complex structures can be solved with much less than 1 mg of compound. It has recently been demonstrated that it is possible to prepare a screening library of highly diverse compounds from plants with the compounds being pre-selected from an analysis of the Dictionary of Natural Products to be drug-like in their physicochemical properties (Oleszek and Marston, 2000; Doughari *et al.*, 2009). It will be interesting to see if such a collection proves to been richer in bioactive molecules. Several alternative approaches are also being explored in efforts to increase the speed and efficiency with which natural products can be applied to drug discovery. For instance, there is an attraction to screen the mixtures of compounds obtained from extracts of plant material or from microbial broths to select extracts from primary screens that are likely to contain novel compounds with the desired biological activity using the concept of 'differential smart screens'. This approach involves screening extracts of unknown activity against pairs of related receptor sites. By the comparison of the ratios of the binding potencies at the two receptor sites for a known selective ligand and for an extract, it is possible to predict which extract was likely to contain components with the appropriate pharmacological activity (McGaw *et al.*, 2005; Doughari *et al.*, 2009; Okigbo *et al.*, 2009).

Another approach is the use of 'chemical-genetics profiling' (Doughari *et al.*, 2009). In this method, by building up a database of the effects of a wide range of known compounds, it is possible to interrogate drugs with unknown mechanisms or mixtures of compounds such as natural product mixtures. The technique highlighted unexpected similarities in molecular effects of unrelated drugs (e.g. amiodarone and tamoxifen) and also revealed potential anti-fungal activity of crude extracts. This activity was confirmed by isolation and testing of defined compounds, stichloroside and theopalauamide (Fig: 10).

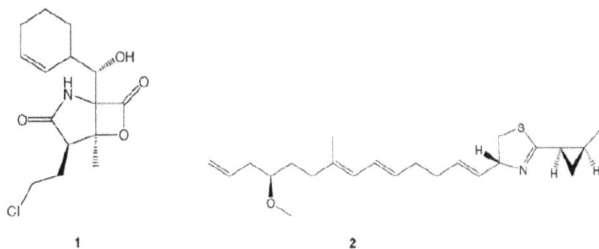

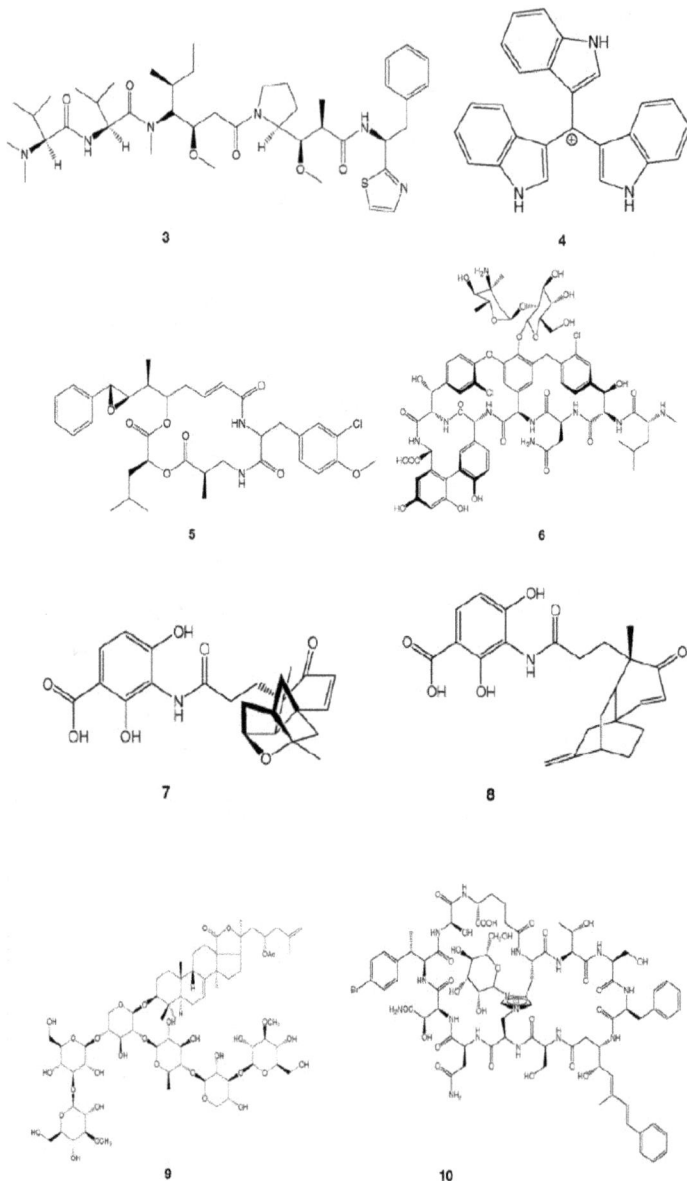

Fig: 10 Natural products – recently discovered and/or in development

(1) SalinosporamideA; (2) curacin A; (3) dolastatin 10; (4) turbomycin A; (5) cryptophicin; (6) vancomycin; (7)platensimycin; (8) platencin; (9) stichloroside; (10) theopalauamide (Source; Doughari *et al.,*2009).

Because these compounds are not structurally similar, they would not have been expected to act via the same biological target, thus providing more chances for a very versatile drug component with high efficacy against antibiotic resistant bacteria. It's been reported that despite the popularity of chemical drugs, herbal medicine in Africa and the rest of the world, continued to be practiced due to richness of certain plants in varieties of secondary metabolites such as alkaloids, flavonoids, tannins and terpenoids (Cowan, 1999; Lewis and Ausubel, 2006; Adekunle and Adekunle, 2009). Stapleton *et al.* (2004) reported that aqueous extracts of tea (*Camellia sinensis*) reversed methicillin resistance in methicillin resistant *Staphylococcus aureus* (MRSA) and also to some extent reduced penicillin resistance in β-lactamase-producing *Staphylococcus aureus*. Also, Betoni *et al.* (2006) reported synergistic interactions between extracts of guaco (*Mikania glomerata*), guava (*Psidium guajava*), clove(*Syzyguim aromaticum*), garlic (*Allium sativum*) lemon grass (*Cymbopogon citratus*) ginger(*Zingiber officinale*) cargueja (*Baccharis trimera*), and mint (*Mentha pieria*) and some antibiotics against *S. aureus*. However, these are preliminary investigations and more works are needed to actually determine the active ingredients in these plants extracts and this may help in improving management of the different infectious diseases that are developing resistance to commonly used antibiotics and possibly to verocytotoxuic bacteria. Furthermore, toxicological studies can also be carried out to determine the reliance on these herbs without many side effects. Researchers have also devised cluster of chemically related scaffolds which are very useful in guiding the synthesis of new compounds. In an attempt to combine the advantages of virtual screening of chemically diverse natural products and their synthetic analogues (scaffolds) with the rapid availability of physical samples for testing, an academic collaboration has established the Drug Discovery Portal (http://www.ddp.strath.ac.uk/).

This brings together a wide variety of compounds from academic laboratories in many different institutions in a database that can be used for virtual screening. Academic biology groups can also propose structures as targets for virtual screening with the Portal's database (and with conventional commercially available databases). Access to the Portal is free for academic groups and the continued expansion of the chemical database means that there is a valuable and growing coverage of chemical space through many novel chemical compounds (Feher and Schmidt, 2003; Galm and Shen, 2007).

Despite all of the advances made by the pharmaceutical industry in the development of novel and highly effective medicines for the treatment of a wide range of diseases, there has been a marked increase in the use of herbal medicines even including the more affluent countries of the world. Germany has the largest share of the market in Europe and it was reported that the sales of herbal medicinal products (HMPs) in 1997 were US$ 1.8 billion (Barnes *et al.*, 2007).Numerous scientific medical/pharmaceutical books have been published in recent years aiming to provide the general public and healthcare professionals with evidence of the benefits and risks of herbal medicines (Barnes *et al.*, 2007; Phillipson, 2007). The pharmaceutical industry has met the increased demand for herbal medicines by manufacturing a range of HMPs many of which contain standardized amounts of specific natural products. In the 1950s, it would not have been possible to predict that in 50 years time there would be a thriving industry producing HMPs based on the public demand for herbal medicines. To date European Pharmacopoeia has even published up to 125monographs on specific medicinal herbs with another 84 currently in preparation (Mijajlovic*et al.*, 2006; Phillipson, 2007. The monographs are meant to provide up-to-date knowledge of phytochemistry for defining the chemical profiles of medicinal herbs and an understanding of analytical tests for identification of the herbs and for the quantitative assessment of any known active ingredients (Phillipson, 2007). Several regulatory bodies including Traditional Medicines Boards (TMBs, in Nigeria and other African Countries), Medicines and Healthcare products Regulatory Agency (MHRA), Herbal Medicines Advisory Committee (HMAC) (Uk) and American Herbal Products Association (AHPA) and several other pharmacopoeia (British, Chinese, German, Japanese) provide guidelines and advice on the safety, quality and utilization of the plant herbal products in several countries (Yadav and Dixit, 2008). Scientific and Research communities are currently engaged in phytochemical research, and pharmacognosy, phytomedicine or traditional medicine are various disciplines in higher institutions of learning that deals specifically with research in herbal medicines. It is estimated that >5000 individual phytochemicals have been identified in fruits, vegetables, and grains, but a large percentage still remain unknown and need to be identified before we can fully understand the health benefits of phytochemicals (Liu, 2004).

5.9 Concluding remarks

With the increasing interest and so many promising drug candidates in the current development pipeline that are of natural origin, and with the lessening of technical drawbacks associated with natural product research, there are better opportunities to explore the biological activity of previously inaccessible sources of natural products. In addition, the increasing acceptance that the chemical diversity of natural products is well suited to provide the core scaffolds for future drugs, there will be further developments in the use of novel natural products and chemical libraries based on natural products in drug discovery campaigns.

6. Synthetic Drugs

6.1 Introduction

To define a drug is not a simple task. According to the most acceptable definition a drug is a substance with an abnormal effect on certain body functions, e.g. strychnine stimulates the action of heart, and aspirin stills its action. Since both of them affects abnormally, the two compounds are known as drugs. in the drug chemistry the first attempt was made in 1920's to find suitable substitutes for morphine; later on in 1930's the therapeutic properties of the sulpha drugs were recognized. After the sulpha drugs, came the vitamins, hormones and antibiotics.

The use of drugs in medicine may be discussed under the following groups:

(a) The drugs which is used in the treatment and cure of specific disease, *viz.* malaria, syphilis, tuberculosis, tuberculosis, microorganisms infection discuses, etc. Such types of drugs are known as chemotherapeutic agents, viz. anti-malarial, sulpha drugs & antibiotics. The chemotherapeutic agents attacks and destroying the cell of the infected host. The first chemo -therapeutic agents introduced by Ehrlich in 1910, Were certain organoasenic compounds against Syphilis.

(b) The drugs which have characteristic effects upon the animal organism but which are not specific remedies for particular diseases. This group includes drugs like analgesics, anesthetics, sedatives, hypnotics, antispasmodics, antihistamines etc. Although, such type of drugs can assist in recovery from a specific bacterial or viral infections, their action is not directed at the pathogenic organism which are the targets in chemotherapy (difference from chemotherapeutic agents).

The medicinal value of the drugs is generally represented by 'therapeutic index' which is described as the ratio of the amount necessary to kill the patient (i.e. the maximum tolerated dose, MTD) to that required for a maximum curative dose, MCD.

$$\text{Therapeutic index} = \frac{\text{Maximum tolerated dose}}{\text{Maximum curative dose}}$$

A therapeutic index of ten means that ten times a dose used for curative purpose would kill the parasite.

6.2 Metabolite antagonisum.

The discovery of sulpha drugs & their uses as antibacterial agent resulted in a principle of drug action. It is well know that p-aminobenzoic acid (PABA) is a normal cellular constituent or metabolite of the bacteria (micro-organism). The bacterium takes up sulphonamide, the structurally similar compound to that of PABA, which can't fulfill the function of PABA, the essential metabolite and hence the growth and reproduction of bacteria stops.

Sulphonamides   PABA

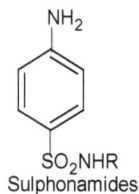

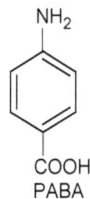

The inhibition or antagonism of the essential metabolite (e.g. PABA) by structurally similar compound (e.g. sulphonamides) is known as metabolite antagonism & the inhibitors are known as metabolite antagonists. Although, the first example of metabolite antagonism was that of PABA antagonism in which the antagonist & metabolite resembled structurally, it is not essential that the structure of all the antagonists resemble with that of the corresponding metabolite, & of course some example of the latter type of the latter type of metabolite antagonisms are given below.

(i) *Histamine and anti-histamines:* Histamine, a essential amino acid is antagonized by all the anti-histamines, many of which has the phenyl, benzyl, alkyl amine structure.

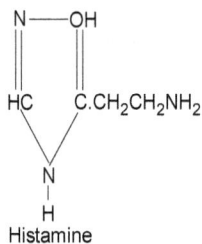

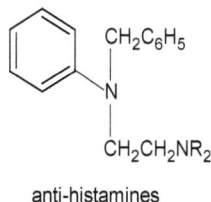

Histamine

anti-histamines

(ii) *Aneurine (vitamin B$_1$) pyrithiamine:* Aneurine, an essential growth factor, is antagonized by pyrithiamine.

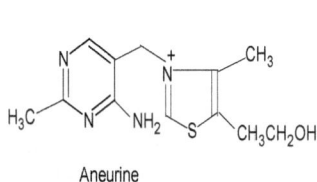

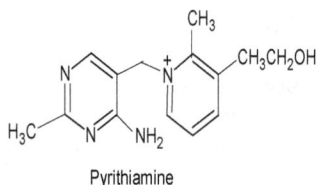

Aneurine

Pyrithiamine

(iii) *Biotine and desthiobiotin:* Biotin is antagonized by desthiobiotin and nordesthiobiotion.

Biotin

Desthiobiotin

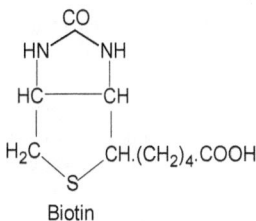

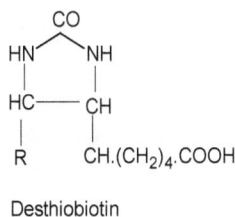

(iv)Some vitamin-B complex antagonists: Pyridine 3-sulphonic acid is antagonistic to nicotinic acid or to nicotinamide, pentoyltaurine is antagonistic to pentothenic acid, and pteroylaspartic acid is antagonistic to pteroylglutamic acid.

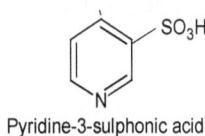

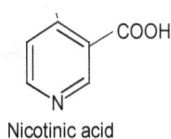

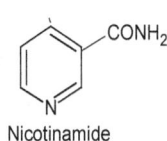

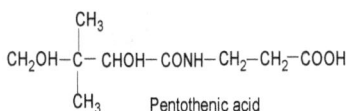

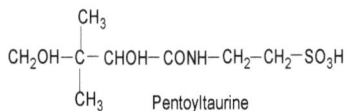

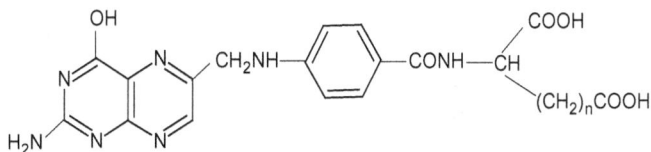

Pyridine-3-sulphonic acid Nicotinic acid Nicotinamide

$CH_2OH-C-CHOH-CONH-CH_2-CH_2-COOH$ Pentothenic acid

$CH_2OH-C-CHOH-CONH-CH_2-CH_2-SO_3H$ Pentoyltaurine

n = 1 pteroylaspartic acid

n = 2 pteroylglutamic acid

From the above examples we see that in some cases the antagonist has identical structure to that of metabolite and hence the search for structurally analogous inhibitors (antagonists) constituted one of the targets in the discovery of new drugs and chemotherapeutic agents. Although, neither all the effective drugs act as direct inhibitors (antagonists) of some structurally identical natural metabolite, and nor all the structural analog are effective drugs, the principle has been very useful in past research and will also in further studies.

6.3 Isosterism

Two molecules or ions containing an identical number and arrangement of electrons are known as isosteres and the phenomenon is known as Isosterism. Most of the physical and chemical properties the two isosteres are almost similar. The most important example of Isosterism is benzene and thiophene; the two isosteres and even their corresponding derivatives like nitro, aldehyde etc. boils at temperatures within very close range.

Since the benzene and thiophene possess a certain degree of parallelism in their physical properties and odour, they and their corresponding derivatives have also similar physiological action, e. g. nitrothiophene and nitrobenzene causes the same degree of toxicity. In general it has been observed that in the majority of cases the benzene ring can be replaced by thiophene ring without much alteration of physiological action. Although exceptions are very rare as the thienylalanine can't take the place of phenylalanine in nutrition.

Friedman applied the term bio-isosterism to those isosteres which have the same type of biological activity also. Some other examples of bio-isosterism are given below.

(i) The 2-thienyl analogue of cocaine has strong local anesthetic activity.

(ii) 2-Methylthienylcinchonic acid and phenylcinchonic acid both almost equally increase the urinary secretion of uric acid.

(iii) Antergan and its 2-thiophene analogue are both active antihistamines.

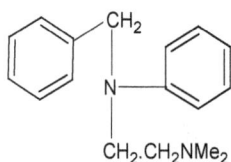

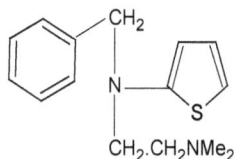

6. 4 Relation of chemical structure and physiological activity

During the 19th century, a number of natural products were isolated and subjected to detailed investigations of their structure and pharmacological action. Some of the compounds are found to possess a definite physiological activity and later on it was observed that the physiological activity and latter on it was observed that the physiological activity of a compound is associated with a particular structural unit or group and hence if this structural unit is present in other compounds the latter also becomes biologically active .such a part of the drug which is responsible for the actual physiological activity is known as pharmacophore group. The pharmacophoric groups is then somewhat modified by the common and simple unit processes to give more active compound with low toxicity. Although there is no broad relationship between the structure and biological activity of the various groups, yet some physiological effect is often associated with particular groups, the more important groups along with the physiological effect exerted by them are described below.

(i) Effect of alkyl groups

Whenever an alkyl groups is introduced in a compound in place of active hydrogen atom, viz.,

$$HCN \rightarrow RCN, \quad Ar\text{-}OH \rightarrow Ar\text{-}OR, \quad RNH_2 \rightarrow RNHR,$$

The biological activity of the alkylated compound is decreased to that of the non-alkylated. For example, the convulsive properties of ammonia are decreased by the introduction of methyl groups and thus trimethylamine becomes free from all these effect. Similarly replacement of the hydrogen of the amino group of aniline lowers the convulsive properties of aniline. Other example in which the introduction of alkyl groups in place of active hydrogen atom (i.e.-OH,-NH$_2$, etc) causes the dimunition of the biological activity are given bellow,

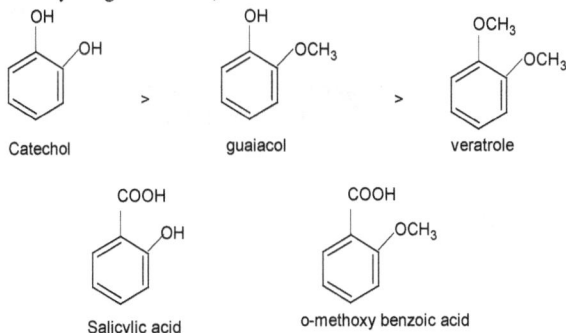

| Catechol | guaiacol | veratrole |

Salicylic acid   o-methoxy benzoic acid

However, there are many exceptions to the above rule, e.g. some simple alkylamines are more toxic than alcohols (R-OH), resorcinol, dimethyl ether is extremely toxic as compared to resorcinol, theobromine(N-dimethyl derivative of xanthine) and caffeine (N-trimethyl derivative of xanthine) are more toxic than the percent compound, xanthenes.

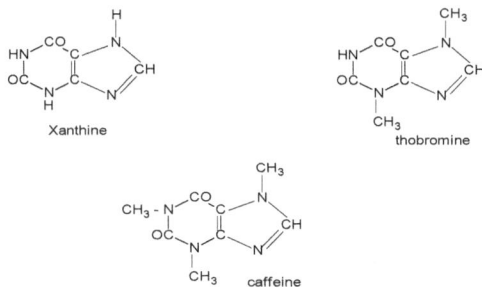

Xanthine

thobromine

caffeine

Similarly in some cases the methylation of a carboxyl, hydroxyl and amino group the full certain masked properties, viz, the cocaine is a strong anesthetic whereas its analogue acid is inert, antipyretic whereas its analogue having only one methyl groups is inert.

(ii) Effect of acidic group

The introduction of acidic groups in the molecule either decreases or completely vanishes the physiological action of the parent compound , e .g. phenol is poisonous, but benzene sulphonic acid is harmless , similarly morphine has a strong physiological action, but morphine sulphuric acid is quite inactive .it might be thought that the above change in physiological action is due to the removal of the hydroxyl group, i. e. anchoring group, but the experimental studies have shown that the entrance of acidic group without altering the active or anchoring group has also the same effects. for example, nitrobenzene is poisonous, but nitrobenzoic acids are harmless; martius yellow (dinitronaphthol) is markedly toxic, but its sulphonic acid (naphthol yellow s) is harmless; aniline is toxic, but m-aminobenzoic acid is harmless; amines are toxic ,while amino-acids are food -stuffs.

However, the physiological properties which have lost by the entrance of acid groups can be restored by esterification. For example, the poisoning action of tyrosin is restored in the hydrochloride of its ethyl ester.

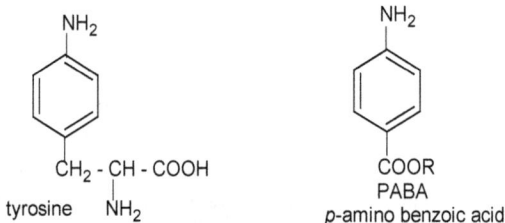

tyrosine $CH_2 - CH - COOH$ with NH_2

$COOR$
PABA
p-amino benzoic acid

Similarly the alkyl esters of p -aminobenzoic acid constituted the important local anesthetics the property which is absent n the acid itself.

The acylation of basic compound by means of organic acids is of great importance in the preparation of synthetic drugs. The phenomenon of acylation reduces the basicity and action of the substance which is then available after hydrolysis in the body and thus exerts its physiological action. The acid group is generally physiologically inactive, and the choice of a particular acid mainly depends upon the solubility and rate of hydrolysis of the derivatives formed. The various acyl derivatives arranged in the decreasing order of solubility are lactyl, acetyl, benzoyl, and salicyl but due to cheapness and ease of hydrolysis generally the acetyl derivatives are the most convenient, because the lactyl derivatives is some time too rapidly hydrolysed , whereas the benzoyl derivative are generally hydrolysed very slowly. However, it must also be remembered that the acyl group may also have its own action quite different to that of acid or base from which it is derived. Moreover, in many cases the acyl group markedly affects the activity of the substance acylated, and so we can't say that its effects are only that of slowing down the rate of action of aromatic compound.

The presence of the benzoyl group is of immense importance to the physiological activity of the various compounds, *viz.,* ecgonine methyl ester is active, whereas its benzoyl derivative is a well known local anesthetic, the cocaine.

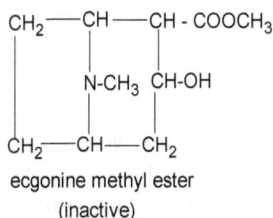

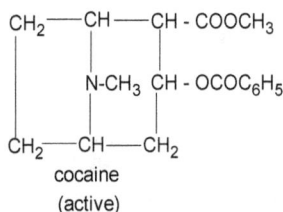

```
CH2———CH———CH - COOCH3
 |       |
 |      N-CH3  CH-OH
 |       |      |
CH2———CH———CH2
```
ecgonine methyl ester
(inactive)

```
CH2———CH———CH - COOCH3
 |       |
 |      N-CH3  CH - OCOC6H5
 |       |      |
CH2———CH———CH2
```
cocaine
(active)

(iii) Effect of halogens

The presence of negative halogen, i.e. when the halogen is present in non-conjugated positions, generally increases both the useful and toxic properties but at the different rates. Generally, it has been observed that the increase in toxicity is negligible whereas the useful properties are considerably increased, especially by means of chlorination. Hence, halogenations are used as a means of stepping up activity and widening the margin of safety in a given series.

On the other hand, the presence of positive halogens' as in acid halides, halogens carbonyl compound, etc., has the opposite effect. The increase in the number of positive halogens takes place with the dimunition of the toxicity of the compound. Compound having available halogens, such as chloramines, are strongly antiseptic depending upon the percentage of hypohalous acid liberated on hydrolysis.

There is no definite difference between the activity of organochlorine, bromine, and iodine compounds, but it is observed that the aliphatic fluorocarbons are much less physiologically active then the corresponding other halogens and even less then the corresponding non- fiuorinated compounds. The less activity of fluorinated compound may be due to their stability.

(iv) Effect of nitro (NO₂) and nitrite (ONO) groups

The entrance of a nitro group in to aromatic compound ,generally, increses the toxicity, e.g. nitrobenzene, nitronaphthol, and nitrothiophene are more toxic than the parent hydrocarbons. however,, the toxicity may be reduced by introducing an easily oxidisable group, sush as methyl and aldehyde,e.g. p-nitrotoluene is less poisonous then nitrobenzene and similarly nitraldehydes are also relatively very little poisonous.

R—N(=O)(→O) nitro compound   R—O—N=O nitrite compound

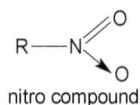

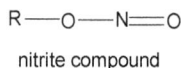

The aliphatic nitrites have the property of dialating the blood vessels and hence they are used to lower blood-pressure. the strength of this effect increases in ascending the series from methyl nitrite to amyl nitrite . since the secondary and tertiary nitrites are easily hydrolysed to alcohol and nitrite, they are more powerful then corresponding primary nitrites.

(v) Effect of amino group

The amino group is toxic and as described earlier the successive alkylation reduces the toxicity (see effect of alkyl groups) . acylation generally, also decreases the physiological action of the parent compound , e.g. aniline is toxic, whereas acetanilide is an important febrifuge (for further details see also effect of acid group). the physiological effect of the amino compound is also decreased by sulphonation and carboxylation . the entrance of a second amino grouop into the benzene nucleus incraaes the toxycity vin all the three phenylene diamines are extrenely poisonous. However, it must be noted that aromatic amines and hydrazines constitute the important examples of antipyretics and analysics.

(vi) Effect of nitrile (-CN) group

The parent compound of all the nitriles, the toxic properties of nitriles is said to be associated with the nitrile group. the nitriles produce coma. in aliphatic series, the lower nitriles are more poisonous than the higher nitriles is said to be associated with the nitrile group. the nitriles produce coma. in aliphatic series, the lower nitriles are more poisonous than the higher nitriles. the isomeric nitrile compounds, i.e. isonitriles (isocyanate) are also poisonous , theybparalyse the respiratory system.

The cyanide ion in inorganic compounds also exert more poisonous effect, e. g. potassium thio cyanate(KCNS), is weakly poisonous, and sodium nitroprusside, $Na_2[Fe(CN)_6(NO)]$, is a strong poison and cause death. But on the same side, the cyanide ion in the sodium ferrocyanide $Na_2[Fe(CN)_6]$ has no physiological action.

(vii) Efeect of unsaturation

The unsaturated compound are usually toxic than their corresponding saturated compounds, e.g. propanol ($CH_3CH_2CH_2OH$) is a mild narcotic whereas allyl alcohol ($CH_2 = CH\text{-}CH_2OH$) has strong poisonous properties, similarly acrolein, crotonaldehyde and carvone are toxic as com- pared to their corresponding saturated compound.

$$CH_2 = CH - CHO$$
acrolein

$$H_3C - CH = CH - CHO$$
crotonaldehyde

carvone
(more toxic)

menthone
(less toxic)

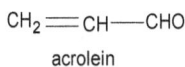

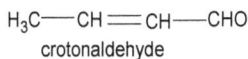

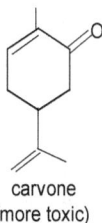

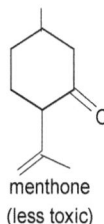

Furthermore, the toxicity of a compound increases with increasing unsaturation, and also produces toxicity in the compounds other than carbon, *viz.*, trivalent arsenic compounds are generally more toxic than those of quinquevalent arsenic, the poisonous mustard gas, $(ClCH_2CH_2)_2SO_2$, on oxidation (which increases the valency of sulphar).

(viii) Effect of isomerism
There are two types of effect of isomerism, *viz.*,

(a) *strutural isomerism:* The difference in the ortho , meta, and para derivatives in the aromatic series is a well known fact, a very interesting example in favour of this is a well known fact. a very interesting example in favour of this statement is that sulphanilamide is very active as drug, whereas the two other isomers are inactive . similarly , we can explain the failure of cocaine as an anaesthetic whereas the ordinary cocaine is a well known local anaesthetic.

(b) *Stereoisomerism:* Sometimes there is a surprisingly great difference in the activity of stereoisomerides, e. g.,maleic acid is toxic whereas its another geometrical isomer the fumaric acid is harmless. Similarly in the optical isomer, one isomer may be very less active or completely inactive than the other. A few example may be quoted in favour of this:(i) atropine(*dl-hyocyanamaine*) is more active than *l-hyocyanamaine*. (ii) *l-nicotine* is two times more poisonous than the *d-nicotine*. (iii) the natural *l-adrenaline* is twelve times as active as the *l-adrenaline.*, etc.

6.5 Antimalarials
Malaria is perheps the most widespread of all the disease. In human being it is caused by the few species of *plasmodium* when an infected female anopheles bites to the man *i.e.*actually first of all the plasmodium protozuan infects female anopheles which then infect the human being. Earlierly (since 1640 AD) cinchona bark was found to be due to the presence of certain alkaloids, *viz. quinine, cinchonidine, cinchonine and quinidine.* Out of which quinine is the most important. Since that time malaria was treated by quinine extracted from the cinchonine bark.

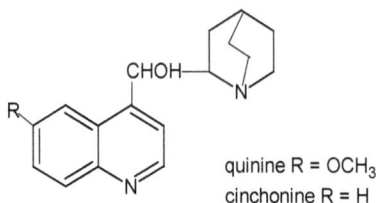

quinine R = OCH$_3$
cinchonine R = H

Apart from quinine, roots of the chinese plant *chang shan* has been used as an antimalarial in Chinese medicine from the time immemorial. The antimalarial activity of the root is due to the presence of two alkaloids, *viz.*, febrifugine and isofebrifugine.

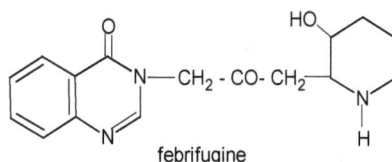

febrifugine

Before World War II It was estimated that nearly ¼ th of the world population suffered from malaria, where as the total output of quinine (the only anti-malarial at that time) was only 800 tons per annum. Moreover during World War II the cinchona cultivated area (*Java*) was captured by Japanese and hence Germany felt an acute shortage of quinine. Now because of the great demand for anti-malarial for armies in World War II, efforts were made to find the synthetic substitutes for quinine, and during that time thousands of synthetic compounds were prepared and tested against malarial parasites, *i.e.* Plasmodium species.

The important and most widely used synthetic anti-malarials can be classified into the following four groups.

(i) 8-Aminoquinoline derivatives
(ii) 4-Aminoquinoline derivatives
(iii) Acridine derivatives
(iv) Pyrimidine and biguanide derivatives

(a) 8-Aminoquinolines

(i)Pemaquine (Plasmoquin or Plasmochin): It was the first synthetic anti-malarial. It is prepared by the condensation of 6-methoxy-8-aminoquinoline (A) with 1-diethylamino-4-bromo pantane (B).

(A) *Synthesis of 6-methoxy-8-aminoquinoline*

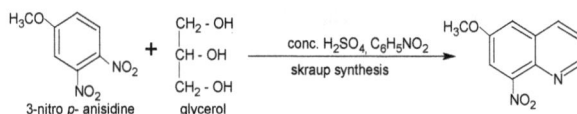

Reduction

H_3CO — 6-methoxy-8-aminoquinoline with NH_2 group

6-methoxy-8-aminoquinoline

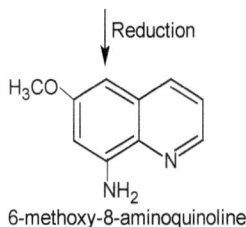

(B) Synthesis of 1-diethylamino-4-bromopentane

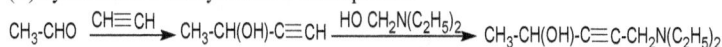

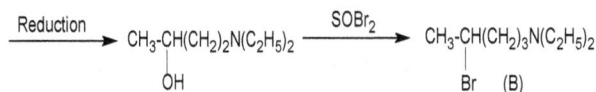

$$CH_3\text{-}CHO \xrightarrow{CH\equiv CH} CH_3\text{-}CH(OH)\text{-}C\equiv CH \xrightarrow{HO\ CH_2N(C_2H_5)_2} CH_3\text{-}CH(OH)\text{-}C\equiv C\text{-}CH_2N(C_2H_5)_2$$

$$\xrightarrow{Reduction} \underset{OH}{CH_3\text{-}CH(CH_2)_2N(C_2H_5)_2} \xrightarrow{SOBr_2} \underset{Br\ \ (B)}{CH_3\text{-}CH(CH_2)_3N(C_2H_5)_2}$$

(C) Condensation of A and B to give pemaquine.

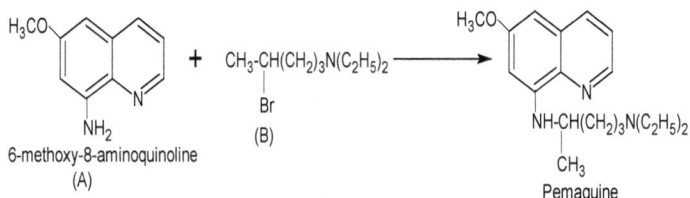

6-methoxy-8-aminoquinoline
(A)

$+$ $\underset{Br\\(B)}{CH_3\text{-}CH(CH_2)_3N(C_2H_5)_2}$ $\longrightarrow$

Pemaquine with $NH\text{-}CH(CH_2)_3N(C_2H_5)_2$ and CH_3

(ii) Primaquine: It is unalkylated amine analogue to pemaquine. It is prepared by condensing 6-methoxy-8-aminoquinoline with 4-bromopentylamine hydrobromide.

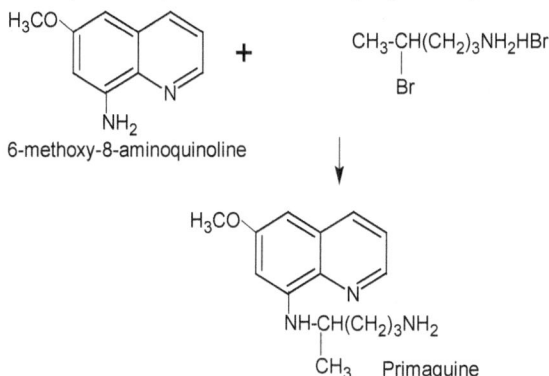

6-methoxy-8-aminoquinoline $+$ $\underset{Br}{CH_3\text{-}CH(CH_2)_3NH_2HBr}$

↓

Primaquine with $NH\text{-}CH(CH_2)_3NH_2$ and CH_3

(iii) Pentaquine: It is obtained by the condensing 6-methoxy-8-amino quinoline with 1-chloro-5-isopropyl aminopentane (C). The compound (C) is obtained by the following method;

teterahydro furfuryl
alcohol

$(CH_3)_2CHNH_2$ → $(CH_3)_2CHN=C$ CH_2OH $\xrightarrow{H_2}$ $CH_2OH(CH_2)_4NHCH(CH_3)_2$ $\xrightarrow{SOCl_2}$

$CH_2Cl(CH_2)_4NHCH(CH_3)_2$
1-chloro-5-isopropyl amino pentane

+ $\underset{\underset{Cl}{|}}{CH_2\text{-}CH_2(CH_2)_3NHCH(CH_3)_2}$

6-methoxy-8-aminoquinoline

$NH\text{-}CH_2(CH_2)_4NHCH(CH_3)_2$

Pentaquine

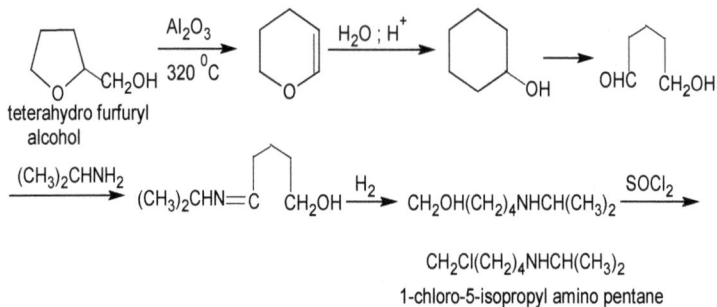

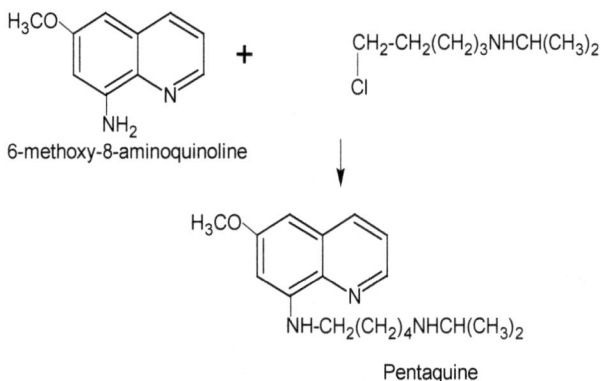

(b) 4-Aminoquinolinenes

(i) Chloroquine: It is prepared by condensing 4, 7-dichoroquinoline with 1-diethylamino-4-aminopentane.

(a) Synthesis of 4, 7- dichloroquinoline

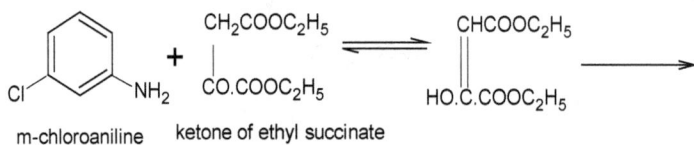

m-chloroaniline   ketone of ethyl succinate

156

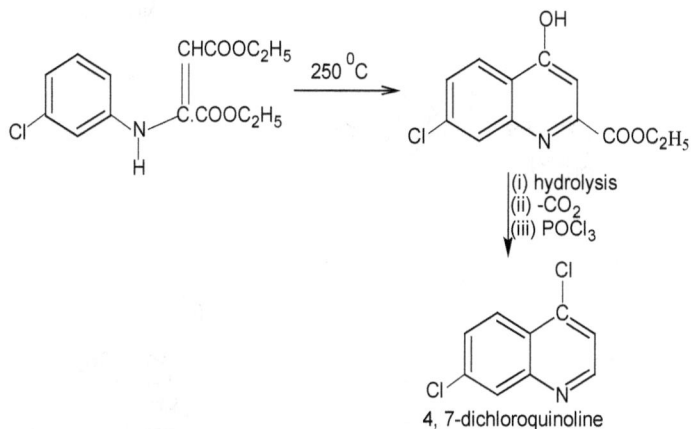

4, 7-dichloroquinoline

Condensation of 4, 7-dichloroquinoline and 1-diethylamino-4-aminopentane

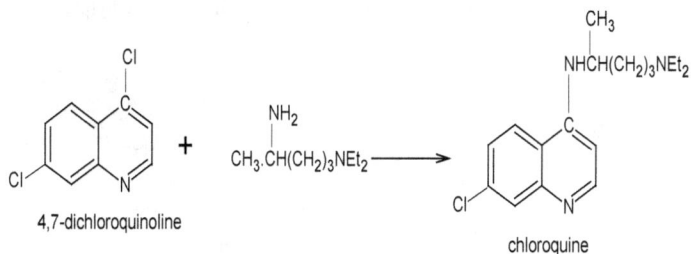

4,7-dichloroquinoline

chloroquine

The sulphate of chloroquine is known as *nivaquine*.

(ii) Santoquine: Santoquine is having one additional methyl group in position 3 in the quinoline ring of chloroquine. It is less reactive then chloroquine.

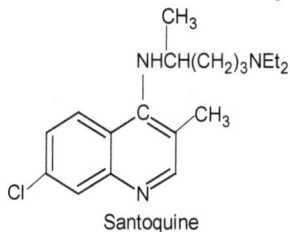

Santoquine

(iii) Camaquine: It is prepared by the condensation of 4,7-dichloroquinoline with 4-amino-2-diethylaminomethyl phenol (E) . The latter compound can be synthesized by the following two methods.

(a) [reaction scheme showing NHAc / OH compound + HCHO, Et$_2$NH → NHAc / OH / CH$_2$NEt$_2$ compound → hydrolysis → NH$_2$ / OH / CH$_2$NEt$_2$ compound]

(b) [reaction scheme showing NO$_2$ / CH$_2$Cl / OH compound + Et$_2$NH → NO$_2$ / CH$_2$NEt$_2$ / OH compound → reduction → NH$_2$ / CH$_2$NEt$_2$ / OH compound]

4-amino-2-diethylaminomethyl phenol

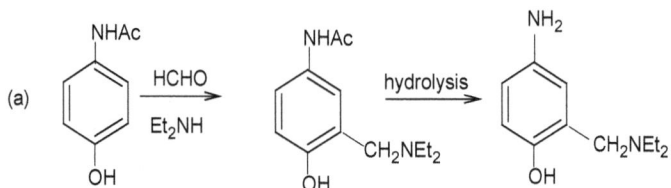

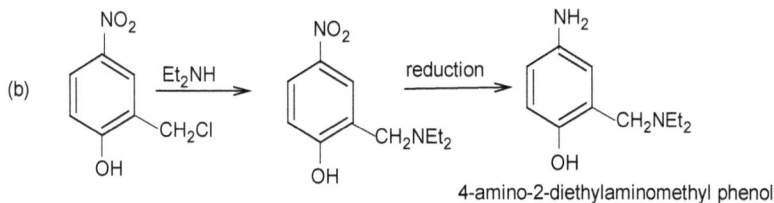

COndensationof 4,7-dichloroquinoline and 4-amino-2-diethylaminomethylphenol to from camaquine.

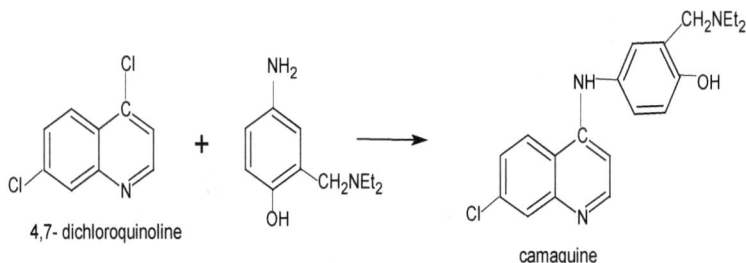

4,7- dichloroquinoline

camaquine

Camaquine is three to four times more active than quinine, especially against the malaria caused by *vivax* and *flaciparum* species of plasmodium. It must be noted that the phenolic group is essential for its anti-malarial activity since its removal depresses and its methylation completely destroys the anti-malarial activity.

(c) Acridine derivative

(i) Mepacrine (atabrin, atebrin, quinacrine or chinacrine): It is quite related to chloroquine in its structure. It is used in the form of water soluble salts, *viz.*, dihydrochloride or bismethyl sulphonates. But its bitter taste and skin colouration are quite objectionable properties. It is prepared by the condensation of 4-methoxy-6, 9-dichloroacridine.

(a) Synthesis of 4-methoxy-6, 9-dichloroacridine.

2, 6-dimethyl toluene   2, 6-dichlorobenzoic acid   p-anisidine

4-methoxy-6,9-dichloroacridine

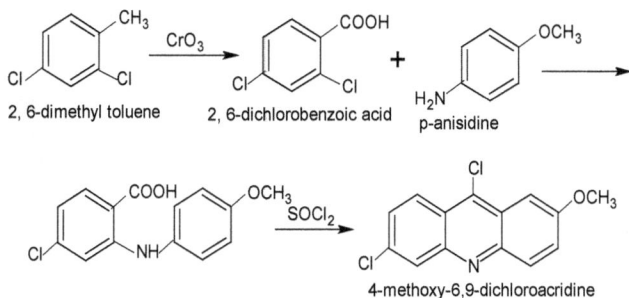

(b) Synthesis of 1-diethylamino-4-aminopentan ().

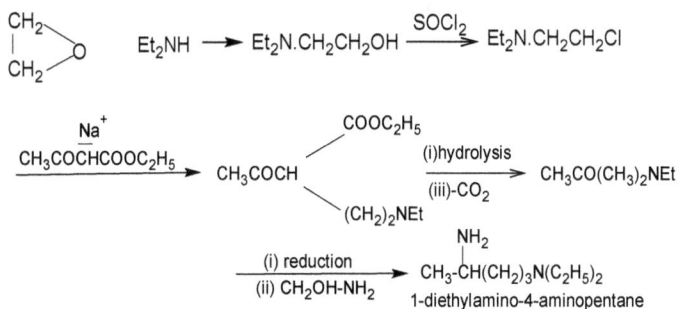

$$Et_2NH \longrightarrow Et_2N.CH_2CH_2OH \xrightarrow{SOCl_2} Et_2N.CH_2CH_2Cl$$

$$\xrightarrow[\text{(iii)-CO}_2]{\text{(i)hydrolysis}} CH_3CO(CH_3)_2NEt$$

$$\xrightarrow[\text{(ii) CH}_2\text{OH-NH}_2]{\text{(i) reduction}} \underset{\text{1-diethylamino-4-aminopentane}}{CH_3\text{-}\overset{\overset{\displaystyle NH_2}{|}}{C}H(CH_2)_3N(C_2H_5)_2}$$

Condensation of 4-methoxy-6,9-dichloroacridine and 1-diethylamino-4-aminopentane.

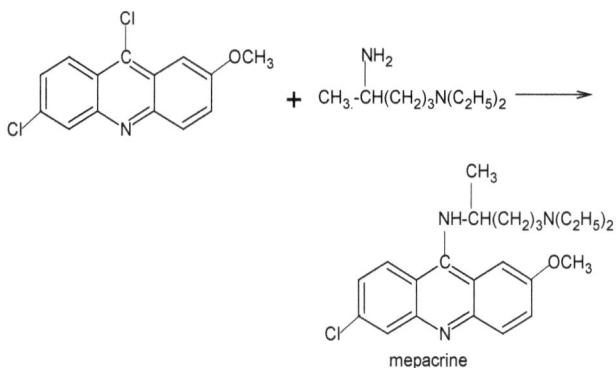

$$+ \quad CH_3\text{-}\overset{\overset{\displaystyle NH_2}{|}}{C}H(CH_2)_3N(C_2H_5)_2 \longrightarrow$$

mepacrine

(ii)   *Azacrin:* It is more active than quinacrine. Its synthesis resembles that of quinacrine, the essential steps are given below.

2,4-dichlorobenzoic acid

2-methoxy-5-amino pyridine

SOCl$_2$

$CH_3.CH(CH_2)_3NEt_2$ with NH_2

Azacrin

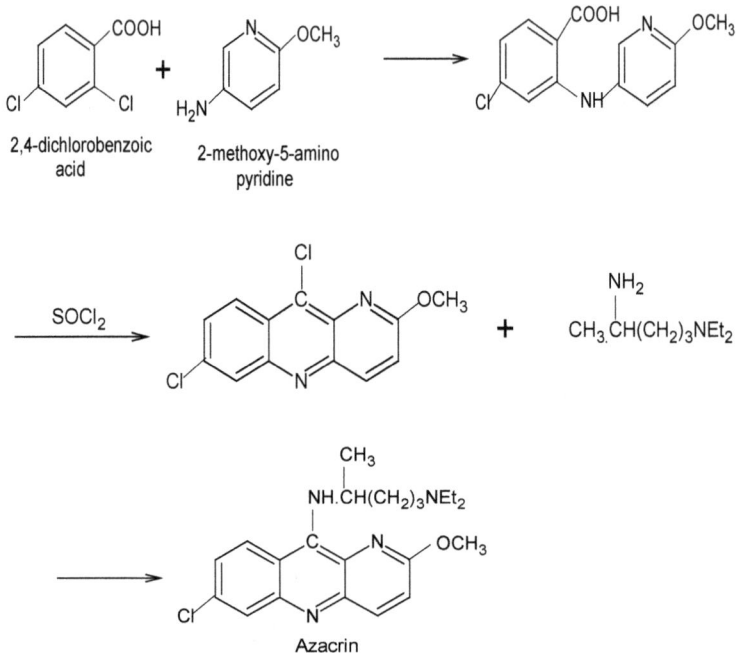

(d) Pyrimidine and biguanide derivatives

(i) Paludrine (proguanil or N^1-p-chlorophenyl $-N^5$- isopropyl diguanide): It is superior to all the anti-malarias known. It is colourless, tasteless, least toxic compounds and has activities on the malarials parasite in as many of its phases as possible. It was discovered by Curd Davey and Rose prepared in the following manner.

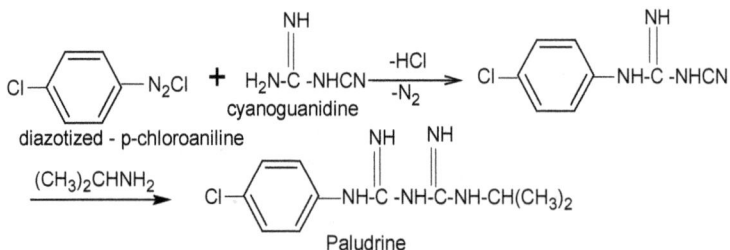

diazotized - p-chloroaniline

cyanoguanidine

Paludrine

Curd et al. (1948)have also synthesized pludrine by heating the calcium or sodium salts of dicyanimide, obtained from calcium cyanamide and chlorocyanogen, with p-chloroaniline.

$$CaCN_2 + ClCN \longrightarrow (C_2N_3)Ca + Cl-\underset{}{\bigcirc}-NH_2 \longrightarrow Cl-\bigcirc-\underset{NH}{\overset{NH}{NH-C}}-NH-\overset{NH}{C}-NH-CH(CH_3)_2$$

Paludrine

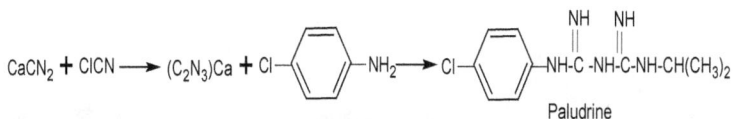

(ii)Pyrimethamine (deraprim or2,4-diamino-5-p-chlorophenyl-6-ethylpyrimidine): It is comparatively non toxic substance, and has a great advantage as a drug due to its tasteless nature. It is synthesized from *p*-chlorobenzyl chloride by the following method.

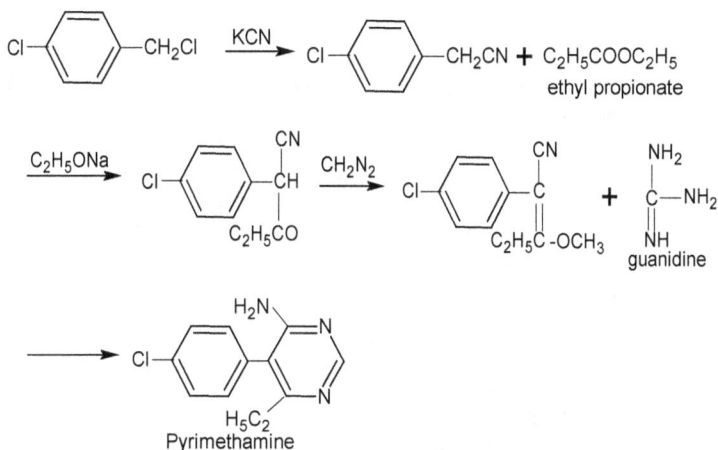

$$Cl-\bigcirc-CH_2Cl \xrightarrow{KCN} Cl-\bigcirc-CH_2CN + C_2H_5COOC_2H_5$$

ethyl propionate

$$\xrightarrow{C_2H_5ONa} Cl-\bigcirc-\underset{C_2H_5CO}{\overset{CN}{CH}} \xrightarrow{CH_2N_2} Cl-\bigcirc-\underset{C_2H_5C-OCH_3}{\overset{CN}{C}} + \underset{NH}{\overset{NH_2}{C-NH_2}}$$

guanidine

$$\longrightarrow Cl-\bigcirc-\underset{H_5C_2}{\overset{H_2N}{\bigcirc}}\underset{N}{\overset{N}{}}$$

Pyrimethamine

6.6 Sulphonamides, Sulpha drugs or anti-bacterials.

Historical: It was Paul Ehrlich idea on dyestuff and their stanning properties to the bacterial cell that stimulated Gerhard Domagk in 1930 for testing certain dyes for their effect on germs in ice. His daughter suffered an infection in her finger that could not be checked by any means and even the surgery attempts failed to stop the spread of deadly infection. Luckily, Domagk gave his girl oral doses of one of his dyes being tested which recovered his girl quickly. It was prontosil (an azo dye) later on known as prontosil which was found to be active against hemolytic streptococci (Domagk, 1935). The dye was used to be prepared by diazotising sulphanilamide followed by coupling with m-phenylenediamine.

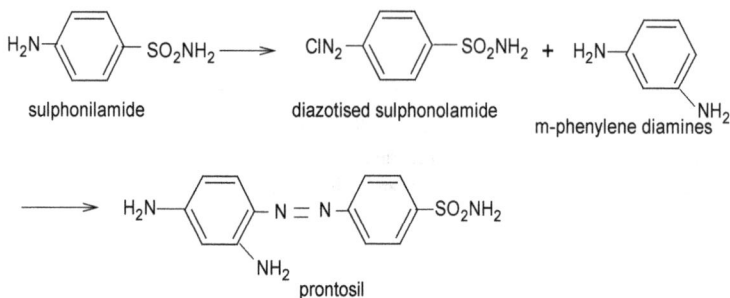

sulphonilamide → diazotised sulphonolamide + m-phenylene diamines

prontosil

Due to law solubility prontosil was generally used as its hydrochloride as a soluble prontosil. Later prontosil was replaced by prontosil soluble due to its superior bacterial activity and greater solubility.

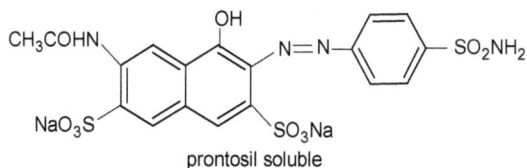

prontosil soluble

Prontosil soluble the disodium salt of 4'-sulphonamidophenylazo-7-acetylamino-1-naphthalene-3,6-disulphonic acid, was used to be prepared by coupling diazotized sulphanilamide with 7-acetylamino-1-naphthol-3,6-disulphonic acid. The Domagk's discovery of prontosil resulted in the discovery of several other prontosils by various groups of scientists.

Discovery of sulphonamide: It was soon learned that the prontosils are degraded *in vivo* to sulphonamide which led to the idea that most probably the activity a prontosil as drugs is due to presence of *Trefouel.*

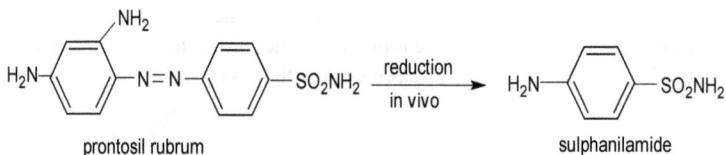

prontosil rubrum → sulphanilamide

This idea led to the testing of sulphonamide and other compounds containing this active, or pharmacophore, group which finally proved that indeed sulphanilamide part was the real active group of prontosils. Since that time, the field of discovery has turned from the azo dyes to the substituted sulphonamide and today thousands of substituted sulphanilamides have been prepared and tested of their antibacterial activity. The sulphanilamides were found to be active against several types of bacteria and has used in the treatment of various dieses, viz, pneumonia, gonorrhea, sinus infections, etc. Moreover, the sulpha drugs are the more effective compounds with little toxicity. Although, nearly 6000 substituted sulphanilamides have been synthesized and tested in a search for drugs superior to sulphanilamides, the letter has steel retained amongst place in chemotherapy because of its low price. For many years sulphonamides were the chief medical weapons against many common Gram-positive bacterial diseases, but now days they have been super seeded, in part, by the much toxic and more broadly effective clinical antibiotics.

But however, because of the simple structure of sulphonamide, the chemical and bacteriostatic studies of sulphonamides have been made in most details among the whole group of related drugs; the various sulphanilamides are discussed below:

(i) sulphanilamide (sulphonamide or p-aminobenzenesulphonamide) : It is the parent compound of all the sulpha drugs. Industrially, it is prepared from acetanilide by the following method.

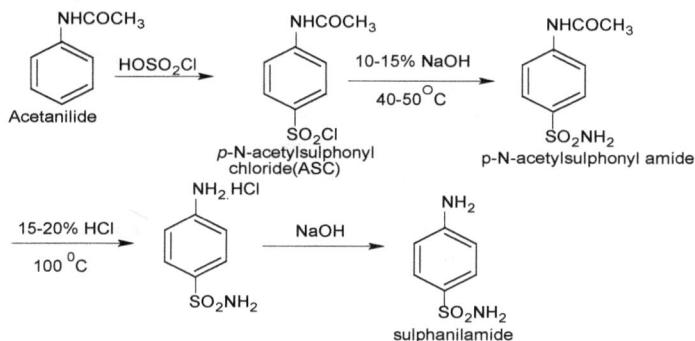

A potentially cheaper method then the above starting from chlorobenzene has not been developed on the industrial scale because of purification difficulties.

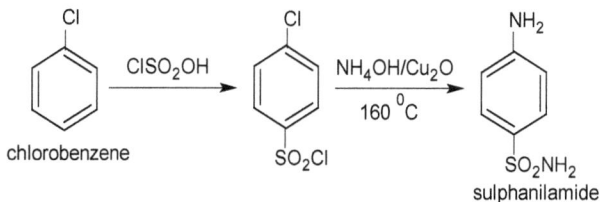

Sulphanilamide is in the control of 'cocci infections', Such as *pneumococci, streptococci, menningcocci* and *gonococci*. But now a day's sulphanilamide is largely replaced by its various derivatives which are less toxic, more effective and in some cases more specific. As described earlier thousands of substituted sulphanilamides have been synthesized and tested for their activity against the common bacterial infections, but only a few of them had been found the posses properties which rendered them useful as drugs.

All the substituted sulphanilamides are derived from the following general structure.

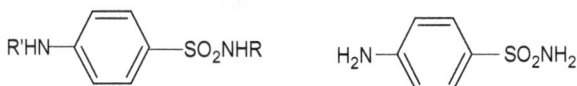

$$R'HN-\!\!\bigcirc\!\!-SO_2NHR \qquad H_2N-\!\!\bigcirc\!\!-SO_2NH_2$$

It must be noted that o and m derivatives are valueless, and substitution on the aromatic ring of sulphanilamide destroys or reduces the activity of the drug, on the other hand, the R' can be changed only within comparatively close limits, but r can be activity of the drug. so now we can say that all the important substituted sulphanilamides are either the 1 or 4- substituted sulphanilamides. in the present book we will be dealing with only the most important sulphanilamides.

(ii) Sulphapyridine (N^1-2- pyridylsulphanimide, M and B 693): It was the first N^1-heterocyclic sulphanilamides to be used as a chemotherapeutic agent. It was found to be very active against pneumonia causing bacteria and lowered the molarity from 2.5-4%.It was also used in staphylococcal and meningococcal infections. Because of its high toxicity in man, it is no longer used now days.

It is obtained by the condensation of the N^4- acetylsulphanilyl chloride with 2-amino pyridine as a solvent followed by alkaline hydrolysis of the product.

2-aminopyridine

$+$ ClSO$_2$—⟨benzene⟩—NHCOCH$_3$

↓ pyridine

⟨pyridine⟩—NH-SO$_2$—⟨benzene⟩—NHCOCH$_3$

↓ NaOH

⟨pyridine⟩—NH-SO$_2$—⟨benzene⟩—NH$_2$

sulphpyridine

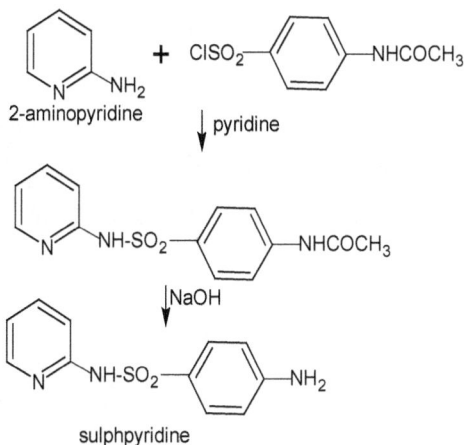

(iii)Sulphathiazole or *cibazole(N^1-2-thiozolyl sulphanilamide):* It is the most highly bacteriostatic sulphonamide drug. It is generally used but particularly useful against staphylococcal infection and in bubonic plague. It is relatively less toxic.

It is prepared by condensing ASC with 2-aminothiazole and hydrolyzing the product in alkaline solution. The 2-aminothiazole required for the process is obtained by condensing thiourea with diethyl chloroacetal or with 1, 2-dichloro ethyl ether or 1,2-dichloro ethyl acetate.

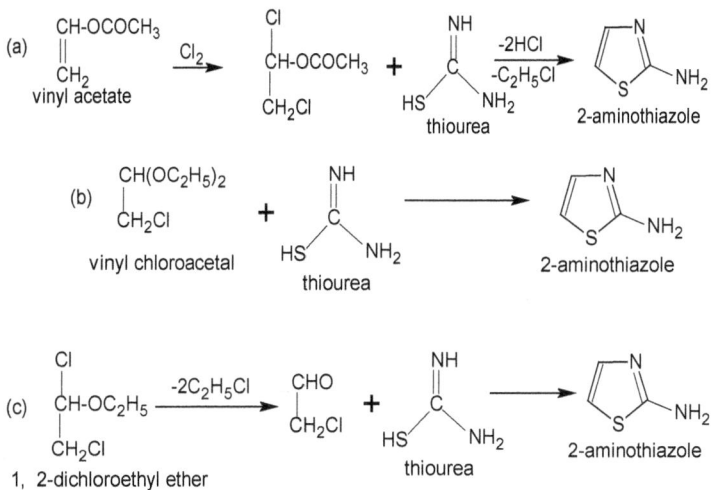

(a)
$$\begin{array}{c} CH\text{-}OCOCH_3 \\ \parallel \\ CH_2 \end{array}$$
vinyl acetate

$\xrightarrow{Cl_2}$

$$\begin{array}{c} Cl \\ | \\ CH\text{-}OCOCH_3 \\ | \\ CH_2Cl \end{array}$$

$+$

$$\begin{array}{c} NH \\ \parallel \\ C \\ HS \qquad NH_2 \end{array}$$
thiourea

$\xrightarrow[-C_2H_5Cl]{-2HCl}$

2-aminothiazole

(b)
$$\begin{array}{c} CH(OC_2H_5)_2 \\ | \\ CH_2Cl \end{array}$$
vinyl chloroacetal

$+$

$$\begin{array}{c} NH \\ \parallel \\ C \\ HS \qquad NH_2 \end{array}$$
thiourea

$\longrightarrow$

2-aminothiazole

(c)
$$\begin{array}{c} Cl \\ | \\ CH\text{-}OC_2H_5 \\ | \\ CH_2Cl \end{array}$$
1, 2-dichloroethyl ether

$\xrightarrow{-2C_2H_5Cl}$

$$\begin{array}{c} CHO \\ | \\ CH_2Cl \end{array}$$

$+$

$$\begin{array}{c} NH \\ \parallel \\ C \\ HS \qquad NH_2 \end{array}$$
thiourea

$\longrightarrow$

2-aminothiazole

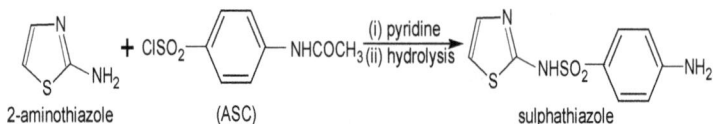

2-aminothiazole   (ASC)   sulphathiazole

Sulphathiazole is about 50 times more potent than sulphanilamide.

(iv) Succinyl sulphathiazole: It is prepared by heating Succinic anhydride with sulphathiazole.

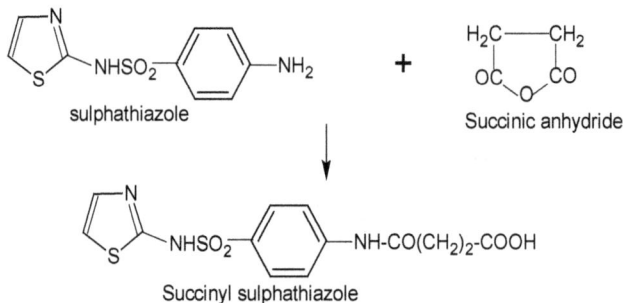

sulphathiazole   +   Succinic anhydride

Succinyl sulphathiazole

It is used in bacillary dysentery and cholera.

(v) Sulphadiazine or *sulphapyrimidine* (N^1-*2-pyrimidyl sulphanilamide):* Like sulphathiazole it is used against the *Coccus infection.* It is somewhat less toxic than sulphathiazole and hence is the most widely used sulpha drug.

It is prepared by condensing ASC with 2-aminopyrimidine in presence of pyridine and followed by the alkali hydrolysis. The 2-amino pyrimidine is obtained from the malic acid by the following method.

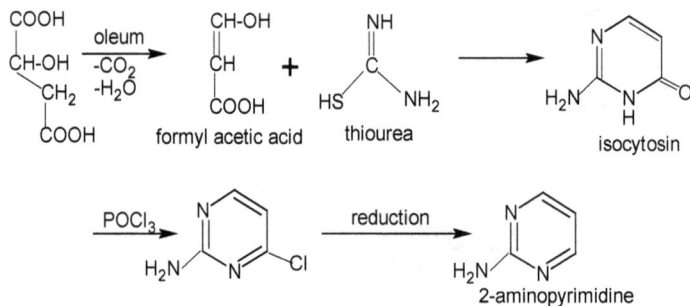

formyl acetic acid   thiourea   isocytosin

2-aminopyrimidine

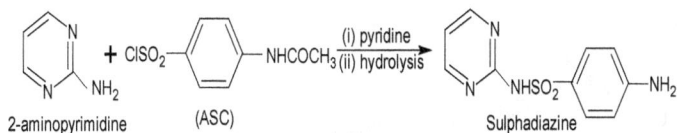

2-aminopyrimidine   (ASC)   Sulphadiazine

(vi) Sulphamerazine (N^1-2-(4-methyl) pyrimidyl sulphanilamide): It is prepared by condensing ASC with 2-amino-4-methylpyrimidine in presence of pyridine and followed by the alkali hydrolysis.

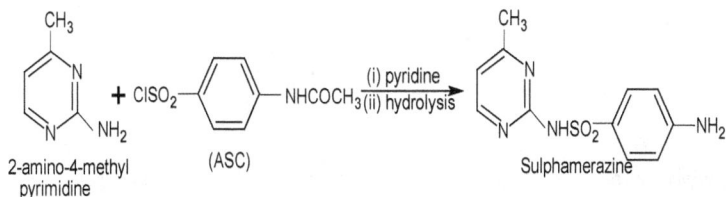

2-amino-4-methyl   (ASC)   Sulphamerazine
pyrimidine

Sulphamerazine is especially useful for combating pneumonia.

(vii) Sulphamezathine or sulphamethazine (N^1-2-(4, 6-dimethyl)-pyrimidyl-sulphanilamide): It is prepared by the condensing ASC with 2-amino -4,6-dimethylpyrimidine followed by the alkali hydrolysis.

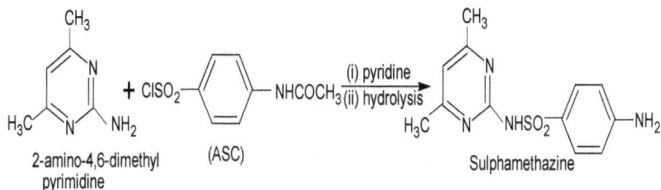

2-amino-4,6-dimethyl   (ASC)   Sulphamethazine
pyrimidine

The 2-amino-4, 6-dimethylpyrimidine is obtained by the condensation of acetylacetone with guanidine.

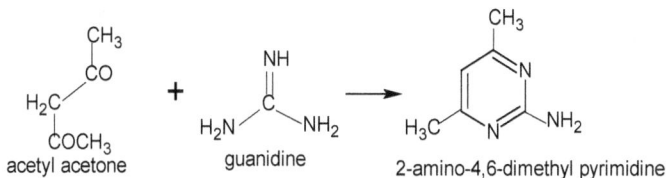

acetyl acetone   guanidine   2-amino-4,6-dimethyl pyrimidine

Sulphamezathine is used for general purposes. It has also been obtained by condensing sulphaguanidine with acetylacetone.

acetyl acetone + sulphaguanidine → Sulphamethazine

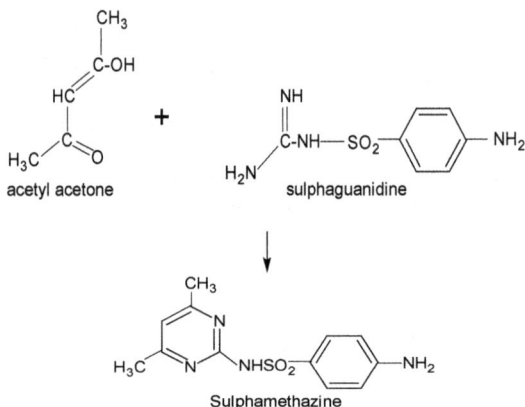

Sulphadimetine is an isomer of sulphamezathine and effective against both Gram-positive and Gram-negative bacteria whereas other sulphapyrimidine drug, *viz.* sulphadiazine, sulphamerazine and sulphamezathine, are effective only for Gram-positive bacteria. Due to this superiority and low toxic nature sulphadimetine has been recommended for urinary tract infections.

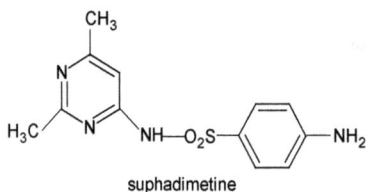

suphadimetine

(viii) *Sulphaguanidine:* It is obtained by the condensation of sulphanilamide with guanidine or dicyandiamide.

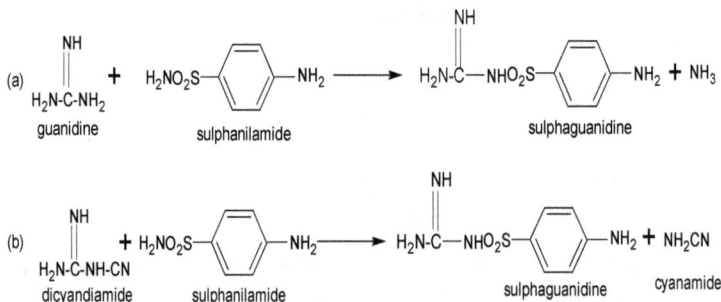

(a) guanidine + sulphanilamide → sulphaguanidine + NH_3

(b) dicyandiamide + sulphanilamide → sulphaguanidine + cyanamide

Since, sulphaguanidine is only slightly absorbed in the intestinal tract and possesses no toxic effects, it constitutes as one of the best drugs against bacillary dysentery.

(ix) Sulphacetamide: It is prepared by the careful and, of course, controlled hydrolysis of the diacetyl derivative of sulphanilamide.

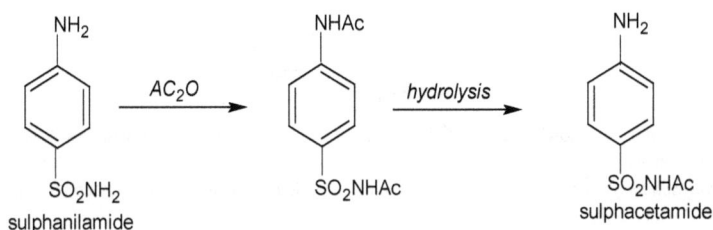

It is one of the highly soluble sulphanilamide drugs and used for urinary tract infections. But due to its toxic nature it is replaced by other highly soluble sulpha drugs, *viz,* sulphisoxazole.

(x) Sulphisoxazole or sulphafurazole: It is prepared by the condensation of ASC with 3:4-dimethyl-5-amino-isoxazole in the presence of pyridine followed by alkaline hydrolysis.

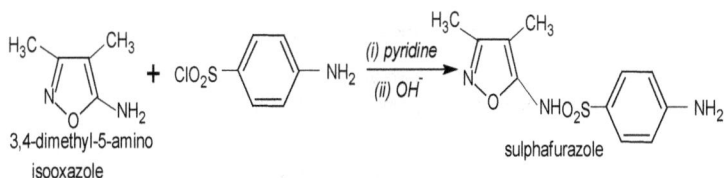

The 3,4-dimethyl-5-aminoisoxazole is prepared by condensing 3-cyanobutane-2-one with hydroxylamine.

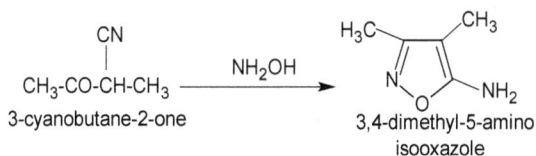

Sulphisoxasole is soluble over a wide range of pH and used for urinary tract infections.

(xi) Sulphaqunoxaline (N¹-2-quinoxalylsulphanilamide): It is prepared by the condensation of ASC with 2-amino-quinoxaline followed by hydrolysis.

sulphaquinoxaline

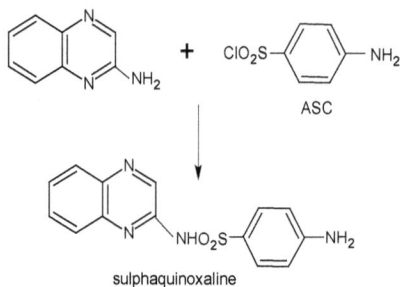

This drug is very slowly excreted and maintains a high blood concentration with low dosage for 24 hours.

(xii) Sulphapyrazine (N¹-2-pyrazinylsulphanilamide): It is prepared by the condensation of ASC with 2-aminopyrazine.

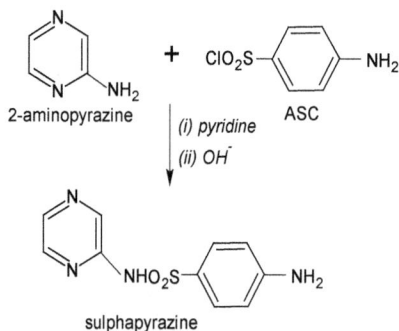

sulphapyrazine

The 2-aminopyrazine is obtained from quinoxaline by the following method.

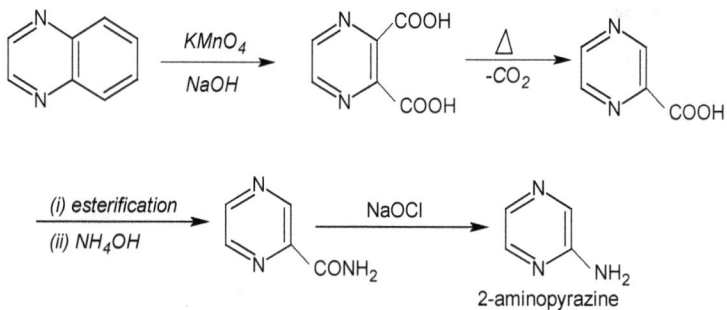

2-aminopyrazine

However, on the whole the yield is small. Sulphapyrazine is particularly useful against meningitis.

(xiii) Marfanil or sulphamylon: It is prepared from benzylamine by the following way.

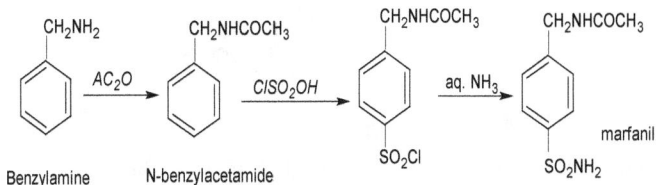

Benzylamine   N-benzylacetamide   marfanil

It must be noted that the replacement of benzene nucleus of sulphanilamide by thiophene ring leads to a compound extremely active against plague. It is nearly five times more powerful than sulphathizole, and also proportionately more toxic to the host. It is prepared from 2-nitrothiophene by the following method.

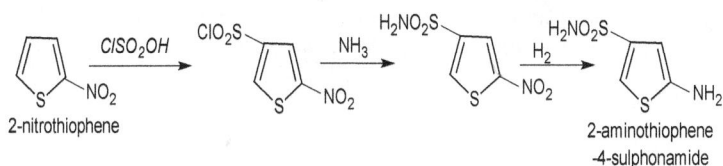

2-nitrothiophene   2-aminothiophene
-4-sulphonamide

Lastly, it must be noted that the compound, I, II, and III are although similar in structure to that of the active sulphonamides, they are not discussed here in detail.

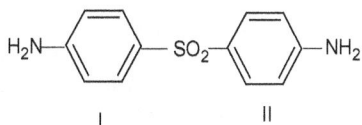

I promin, bis-glucosesulphonate

II diascne, bis-formaldehyde sulphoxylate

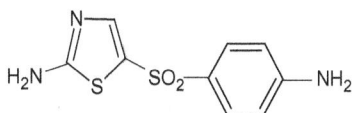

Promizole, III

Mechanism of action of sulpha drugs: The mode of action of sulpha drugs is fairly well understood. They do not kill bacteria directly at the concentrations in which they are used, *i.e.* they are not *bactericidal,* but on the other hand they prevent the growth and multiplication of bacteria *(bacteriostatic action).* The bacteriostatic nature of sulphonamides is due to their similarly in structure with that of *p*-aminobenzoic acid (PABA). An important vitamin require for the normal functioning of some of the vital process in bacteria.

sulphonamides

p- aminobenzoic acid

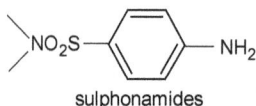

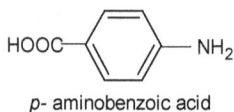

Due to similarity in structure with that of PABA the sulphonamides, when administered into the patient, are taken by the bacteria but they do not perform the necessary function of PABA during bacterial metabolism, and thus finally they inhibit the growth and reproduction of bacteria.

PABA is involved in the synthesis of folic acid, which is a member of vitamin B complex and an essential growth factor in some micro-organisms. Folic acid did not a single substances, the most important of them is pteroylglutamic acid.

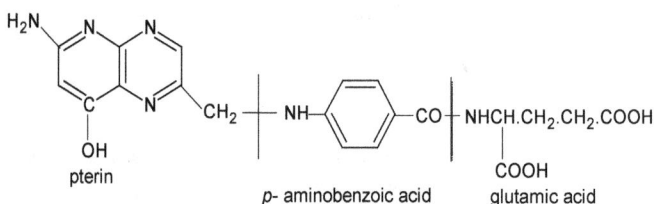

pterin

p- aminobenzoic acid

glutamic acid

6.7 Metallic therapeutics. Compound of mercury, arsenic, antimony, and bismuth have been used in medicine from time immemorial. Later on the presence of various other elements, *viz.* iron, copper, molybdenum, cobalt, etc, in some of the important natural products, *viz.* pigments from blood, fish and insects, and vitamin B_{12} led to the belief that the compounds of many metals play a biochemical role.

(a) Arsenic compound. Arsenious oxide and simple salts of arsenic have long been used as tonics and in the treatment of anaemia. The medicinal value of arsenic compounds was known as early as a 5th century B.C when the Greek physician Hippocrates recommended the use of an ointment containing arsenic tri sulphide for the treatment of ulcerative abscesses. But ignoring the history, let us discuss the more important and, of course, widely used arsenic compounds. Although some aliphatic arsenic compounds having medicinal value are also known, the most important and useful are the aromatic arsenic compounds.

1 Aliphatic arsenic compounds.

(i) Cacodylic acid. Bunsen in 1837 reported that cacodylic acid and cacodyl oxide are less toxic to animals then the equivalent amount of arsenious oxide. Cacodylic acid is obtained by the oxidation of cacodyl oxide with sodium hypochlorite solution.

Cacodyl oxide   Cacodylic acid

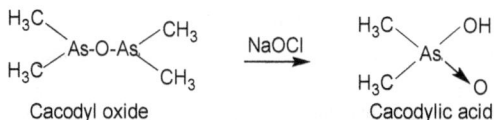

Salts of cacodylic acid are still prescribed for the treatment of certain anaemias.

(ii) Arrhenal: It is the neutral sodium salt of methyl arsonic acid, and prepared by dissolving arsenious oxide in excess of aqueous caustic soda followed by methylation with dimethyl sulphate.

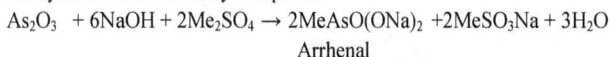

$$As_2O_3 + 6NaOH + 2Me_2SO_4 \rightarrow 2MeAsO(ONa)_2 + 2MeSO_3Na + 3H_2O$$

Arrhenal

(iii) Elarson: Elarson is a complex aliphatic arsenic compound prepared by heating bhenolic acid(docosyne-13, acid) with arsenic trichloride.

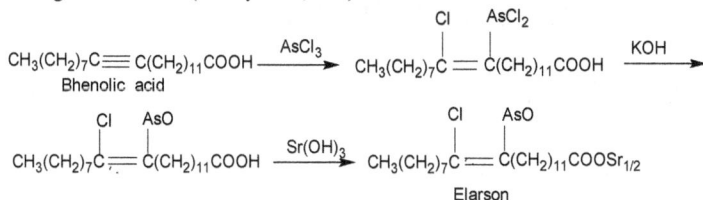

Bhenolic acid

Elarson

Elarson is used for the treatment of secondary anaemias.

2 Aromatic arsenicals.

(i) Atoxyl (Sodium salt of 4 - aminophenylarsonic acid or p - arsanilic acid) : Atoxyl is the key compound in the researches of aromatic arsenicals. Atoxyl may be prepared by the following methods, but none- of which gave satisfactory yield.

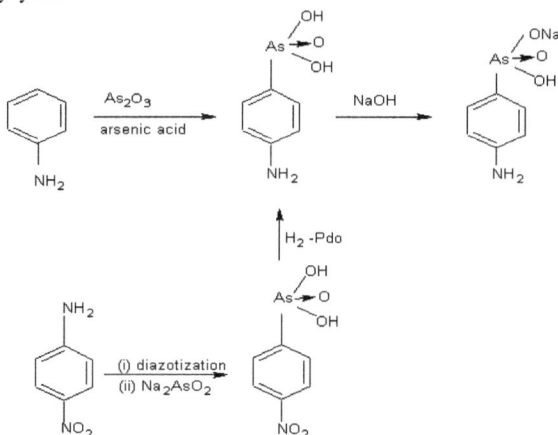

173

However, by using of diazonium borontriflouride the yield of 4-nitrophenylarsenic acid is increased to 8% which thus further increased the yield of arsenilic acid and hence atoxyl.

Aromatic arsenicals are particularly used for treating tripanosomiasis(sleeping sickness) and syphilis; but Atoxyl the parent compound of the various important aromatic arsenicals, does not itself exert any useful physiological action. On the other hand , it gives rise to other useful substances, some of them are mentioned below.

(ii) Trypersamide: It is prepared either by the action of chloroacetamide on Atoxyl or by the reaction of p- arsenilic acid with methyl chloroacetate followed by the treatment with ammonia.

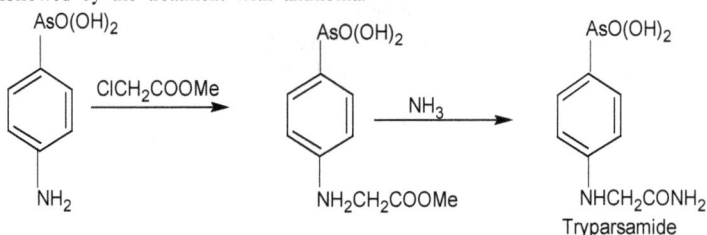

(iii) Arsacetin: It is the acetyl derivatives of Atoxyl. It is prepared by the acetylation of p- arsanilic acid followed by neutralization with soda.

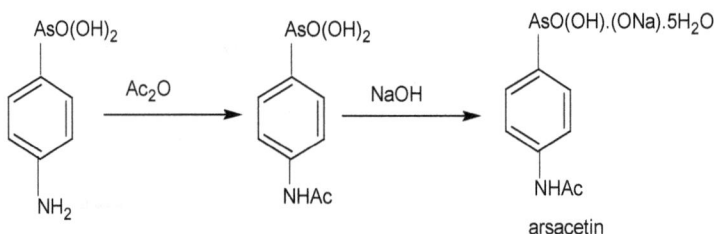

Arsacetin is less toxic to many animals than Atoxyl itself. Moreover, since it is more stable, its solution can be prepared very easily.

The methyl analogue of arsacetin is the orsudan which resembles arsacetin in its action with the additional advantage of being soluble in two and a half parts of water at body temperature.

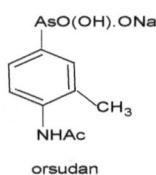

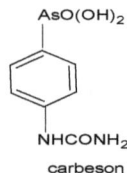

Other important derivative of arsanilic acid is carbeson.

Hydroxyphenyl arsonic acid derivatives: Hundred of substitution hydroxyphenyl arsonic acid have been prepared and tested for their activity but only a few of them which are more useful will be dealt below.

(iv) Stovarsol: The parent compound p –hydroxyphenyl arsonic acid is first synthesized by either of the two method , the acid is then converted into stovarsol by the following series of reactions.

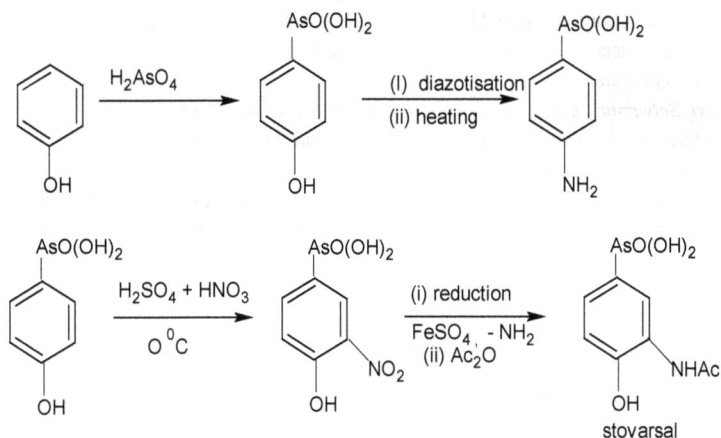

The nitro compound, It may also be converted into other active arsenical , according to following reaction.

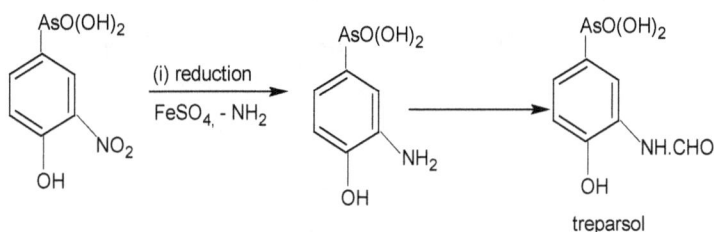

Orsamine, is an o-hydroxyphenyl arsonic acid derivative which was shown to be a strong trypanocide with a therapeutic index of twenty. It posses other important advantages are stability, solubility and rapid adsorption.

AsO(OH)ONa

OH

NHCOCH$_3$

Orsamine

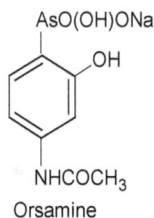

Arsenobenzene derivatives: It is the most important class of chemotherapeutic agents. They were first introduced by Ehrlich and have been widely successfully used for the treatment of syphilis. In these compounds the arsenic is present in the trivalent state.

(v) Salvarsan, arsphenamine, kharsivan, arsenobenzol or 606 (dihydroxydiaminoarseno benzene dihydrochloride): It is prepared by the following two processes.

(a) The 3-nitro-4-hydroxyphenyl arsonic acid is prepared as in stovarsol, which is then reduced with sodium hydrosulphite in the presence of caustic soda and magnesium chloride.

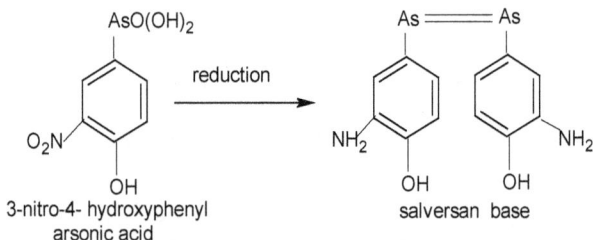

AsO(OH)$_2$

reduction

As $=$ As

O$_2$N

NH$_2$

NH$_2$

OH
3-nitro-4- hydroxyphenyl
arsonic acid

OH   OH
salversan base

(b) An alternative method for prepration of 3-nitro-4-hydroxy phenyl arsonic acid and hence salvarsan start from dimethyal aniline by the following method.

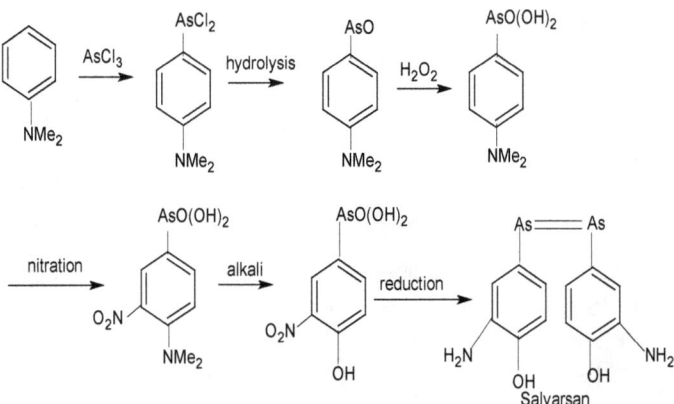

AsCl$_3$

NMe$_2$

AsCl$_2$

NMe$_2$

hydrolysis

AsO

NMe$_2$

H$_2$O$_2$

AsO(OH)$_2$

NMe$_2$

nitration

AsO(OH)$_2$

O$_2$N

NMe$_2$

alkali

AsO(OH)$_2$

O$_2$N

OH

reduction

As $=$ As

H$_2$N

NH$_2$

OH   OH
Salvarsan

In both the methods the first step the reduction of the arsonic acid may also be accomplished with zinc and acetic acid at 25-30 ^{0}C and then with hydrochloric acid sulphuric acid at 50-60 ^{0}C.

Salvarsan is administered either by intravenous or intramascular injection in the form of its sodium salt which is obtained by the addition of correct amount of alkali solution to its aqueous solution.

(vi) Neosalvarsan, neoaarsphnamine, neokharsivan of novarseaobenzol

To over come the difficulties of neutralizing the salavrsan at the time of injection neutral water soluble derivatives known as neosalvarsan was introduced. It is prepared by adding an aqueous solution of sodium formaldehyde sulphonylate to an aqueous solution of salvarsan.

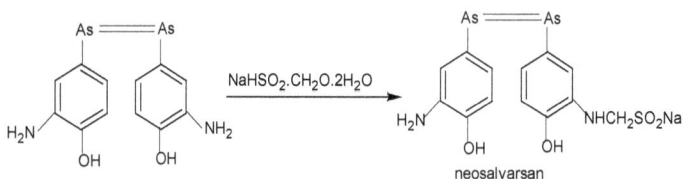

neosalvarsan

Neosalvarsan is more readily soluble in water than salvarsan to give neutral solution. Moreover, it is better tolerated by patients than the salvarsan itself. It has the same therapeutic properties as salvarsan and also administered either by intravenous or intramuscular injection.

(vii) Sulpharsphenamine or myosalversan

It is prepared by the action of excess of sodium formaldehyde bisulphite solution on salvarsan in base.

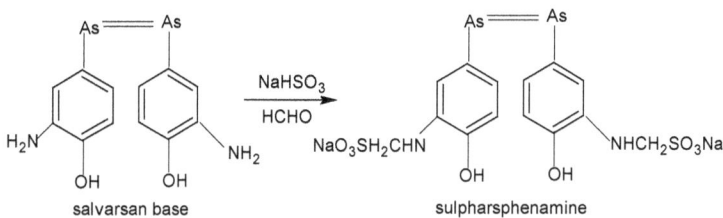

salvarsan base   sulpharsphenamine

(b) Antimony compounds

The organic antimonials are used in the treatment of trypanosomiasis, schistosomiasis and filariasis. All the medicinal antimonials can be classified under the two headings: (a) the salts of antimony compounds having -C-O-Sb or -C-S-Sb link and (b) orgenoantimony compounds having -C-Sb link. Only the important example of both of these classes will be discussed in the present book.

(i) **Tartar emetic** (potassium antimonyl tartrate) and analogous compounds: Tartar emetic is the most commonly used drug and has been known for several centuries. it is obtained by refluxing the aqueous solution of potassium hydrogen tartrate with freshly precipitate of antimony trioxide. Various structures were proposed for tartar emetic but the only given bellow accounts the chemical properties of the tartar emetic.

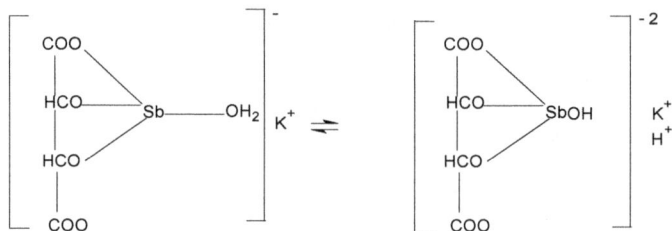

The analogous sodium antimonyl tartrate is equal in its efficiency to that of potassium antimonyl derivatives with the advantage of being less painful. The ethyl ester of antimonyl tartaric acid has some advantage over the above two alkali salts. Other alkali salt of antimony is potassium antimonyl tartrate (antiluetin).

(ii) **Stibosol:** It is obtained by the condensation of thiosalicylic acid with the cyclic ester of antimonous acid and pyrocatechol *I*.

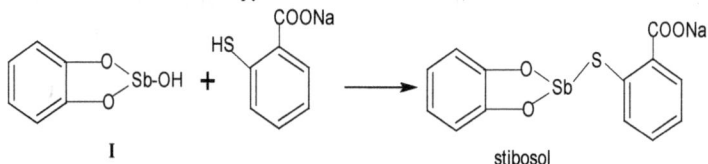

I   stibosol

It has been used in the treatment of heart worm infections in dog.

(iii) **Stibophen or fuadin:** It is obtained by the sulphonation of catechol followed by neutralization with sodium hydroxide containing antimony solution.

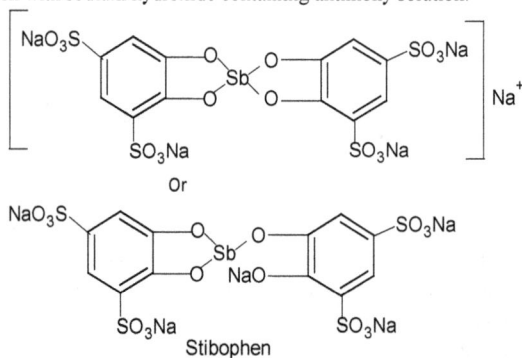

Or

Stibophen

Stibophen can be administered by intramuscular injection without any pain. The corresponding potassium salt of stibophen is known as *antimosam.*

(iv) Organoantimony compound: Most of the antimony compounds introduced into clinical practice are salts or derivatives of *stibanilic acid* (p-aminophenyl stibonic acid), e.g. the sodium salt (stibamine) has been successfully used in treating kala-azar fever.

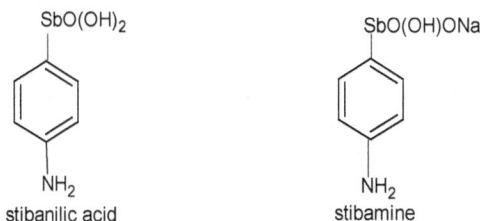

SbO(OH)$_2$

NH$_2$
stibanilic acid

SbO(OH)ONa

NH$_2$
stibamine

Among the substituted stibanilic acid the important compound are

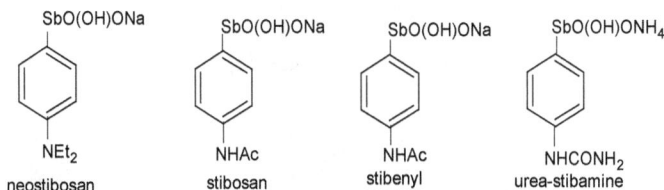

SbO(OH)ONa

NEt$_2$
neostibosan

SbO(OH)ONa

NHAc
stibosan

SbO(OH)ONa

NHAc
stibenyl

SbO(OH)ONH$_4$

NHCONH$_2$
urea-stibamine

(c) Bismuth compounds: Bismuth compounds, as chemotherapeutic, are of little importance because most of them are insoluble are injected in suspension. Although bismuth salts are used in various infection diseases, their main application is in syphilis. A large number of bismuth salts, *viz.,* oxychloride, oxyiodide, gallate, tartrate, malate, etc., have been prepared and used for one or other diseases. The compound bismuth butylthiolaurate (neo-cardyl) is of more interest. Their fatty oils solutions are used in the treatment of syphilis.

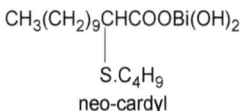

$$CH_3(CH_2)_9CHCOOBi(OH)_2$$

S.C$_4$H$_9$
neo-cardyl

The therapeutically most important bismuth compounds contain bismuth and arsenic atom in the same molecule and can be represented by the following general formula.

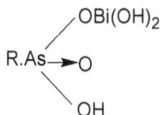

OBi(OH)$_2$

R.As→O

OH

The important examples of these groups are bismuth arsenilate, bismuth *m*-acetylamino *p*-hydroxyphenol arsonate (bistoval), and bismuth neosalvarsan (bimarsen).

(d) Mercury compound: The pharmaceutical uses of Mercurials were established long before the modern arsenicals and diuretics. The Mercurials used against syphilis belong to either of the following groups.

(a) Mercury or its simple salts mixed with fats, oils and antiseptics (*galenical mercurials*).

(b) Simple salts of mercury with organic acids.

(c) Half salts of mercury, Viz. phenyl mercuric acetate, Ph-Hg -OAc.

(d) Organo-metallic compounds of mercury, viz. diethyl mercury

$$.H_5C_2-Hg-C_2H_5$$

(e) Organic compounds in which mercury is attached to nitrigen atom, viz., mercury succinimide,

The important mercury compound used as antiseptics and diuretics are summarized below.

Name	Formula	Application
(i) Phenyl mercuriborate	$C_6H_5HgBO_2$	Surface sterilization antiseptic
(ii) Phenyl mercuriacetate	$C_6H_5.HgOAc$	surface sterilization antiseptic
(iii)4-Oxymercuri-o-toluic acid sodium salt (*afridol*)	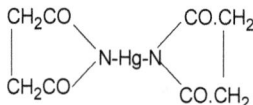	In sterilizing soap
(iv) Asterol	$C_{12}H_{10}O_8S_2Hg$ $4C_4H_4O_6(NH_4)_2.8H_2O$	General surgical antiseptic
(v) Mercurochrome	See 6.9	Antiseptic

(vi) Diethyl barbituric acid salt of Na-4-mercuri-2-chloro phenoxyacetic acid; novasurol (merbaphen)

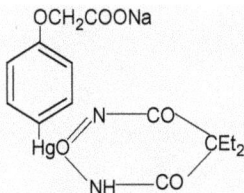

OCH$_2$COONa

HgO—N=CO—CEt$_2$
NH—CO

Diuretic

(e) Gold compounds: Gold compounds are used in medicine since the 13th century. For example, aurum potabile, a gold preparation, has been recommended by Roger Bacon and Paracelsus in leprosy. Chrestien in the 18th century introduced the double chloride of gold and sodium in syphilis and tuberculosis. Now a day a large number of complexes of gold have been used for the treatment of tuberculosis.

Ignoring the rarely used gold compound we will be discussing the most widely used one; among such compound the *krysolgan* group is the most important. The parent compound krysolgan (sodium salt of 4-amino-2auromercaptobenzoic acid) itself, is the best known anti-tuberculosis agent. It is prepared from acetylated anthranilic acid.

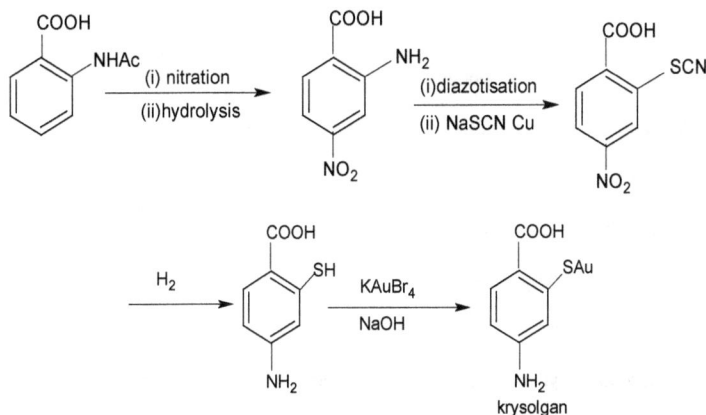

COOH NHAc → (i) nitration (ii)hydrolysis → COOH NH$_2$ NO$_2$ → (i)diazotisation (ii) NaSCN Cu → COOH SCN NO$_2$

→ H$_2$ → COOH SH NH$_2$ → KAuBr$_4$ NaOH → COOH SAu NH$_2$ krysolgan

Other gold compounds used in tuberculosis are solganol, triphal, and lopion.

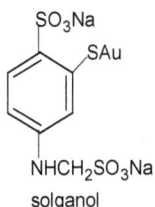

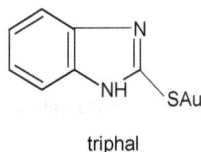

SO$_3$Na SAu NHCH$_2$SO$_3$Na solganol

triphal (benzimidazole) SAu

181

COONa

SAu

NHC

NCH$_2$CH=CH$_2$

lopion

CH$_2$SO$_3$Na

CHOH

CH$_3$SAu

allochrysine

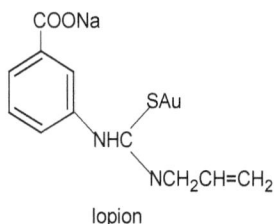

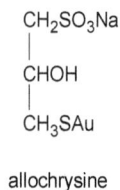

6.8 Anthelmintics: Anthelmintics are the compound used to kill or remove the parasitic worms, such as hook worm, liver fluke, tape worm, etc., from the infected host. It has been observed that nearly $^1/_3$ of the human race and also most of the economic animals suffer from helminth diseases (diseases caused by the above said parasitic worms). The parasitic worms are completely dependent upon the host for then permanent existence. The helminth infection is acquired either (i) by contact with infected animals, or (ii) by the infected animal or human excreta via, ground or water or (ii) by the ingestion of infected meal, or (iv) by means of certain mosquitoes.

The whole of anthelmintics discovered till now may be studied under the following headings.

1. Chlorinated hydrocarbons.

2. Phenols and related compounds.

3. Antimonials and arsenicals.

4. Piperazine derivatives.

5. Triphenylmethane and cyanine dyes.

6. Miscellaneous compounds.

7. Natural compounds.

(a) Chlorinated hydrocarbons: Among the various chlorinated hydrocarbons carbon tetrachloride, tetrachloroethylene-2, 2-dichlorobutane, butyl chloride and hexachloro ethane are the important anthelmintics.

Carbon tetrachloride, obtained by the interaction of carbon disulphide and sulphur monochloride, was the first anthelmintic used against hookworms.

$$CS_2 + 2S_2Cl_2 \longrightarrow CCl_4 + 6S$$

Although carbon tetrachloride is effective and inexpensive, its toxicity to liver cells makes it undesirable.

Tetrachloroethane is obtained by the action of alkali on pentachloroethane which is actually obtained as a byproduct in during manufacture of tetrachloroethane from acetylene and chlorine.

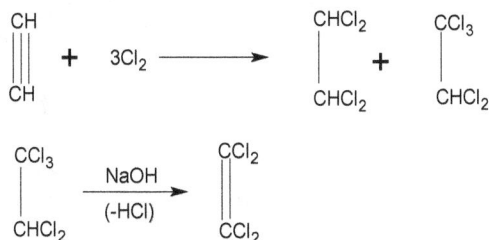

$$
\begin{array}{c} CH \\ \| \| \| \\ CH \end{array} + 3Cl_2 \longrightarrow \begin{array}{c} CHCl_2 \\ | \\ CHCl_2 \end{array} + \begin{array}{c} CCl_3 \\ | \\ CHCl_2 \end{array}
$$

$$
\begin{array}{c} CCl_3 \\ | \\ CHCl_2 \end{array} \xrightarrow[(-HCl)]{NaOH} \begin{array}{c} CCl_2 \\ \| \\ CCl_2 \end{array}
$$

It is the most important anthelmintic of this group. It is effective, inexpensive and tolerable even in hundreds of thousands of doses.

2, 2-Dichlorobutane is best prepared by the action of phosphorus pentachloride on methyl ethyl ketone.

$$
CH_3.CO.CH_2CH_3 \xrightarrow{PCl_5} CH_3.CCl_2.CH2.CH_3
$$

In activity it is equal to that of carbon tetrachloride and tetrachloroethylene, and has fewer side effect.

(b) phenol and related compounds (carbamates,-O.CONR₂): Thymol, 2-isopropyl-5-methyl phenol, which occurs in the oil of thyme, mint and other essential oils, and β-naphthol have been considered as remedies in eliminating hook-worms, but their toxic effects discarded them. Lamson and his coworkers, therefore, investigated a large number of phenols with a view to find more easily tolerable phenolic anthelmintics. Of course, their investigations contributed a large number of such phenolic compounds, out of which only the most important members are described here,

(i) 4-n-Hexylresorcinol (caprokal): It is synthesized from resorcinol in either of the following three ways.

OH
⬡ + ClCO(CH$_2$)$_4$CH$_3$ ⟶ OH
(resorcinol)   OH
         CO(CH$_2$)$_4$CH$_3$

Caproyl chloride

Reduction ↓

OH
⬡ + HO.(CH$_2$)$_5$CH$_3$ ⟶ OH
                    OH
              (CH$_2$)$_5$.CH$_3$

↑ H$_2$

OH
⬡ + CH$_2$=CH(CH$_2$)$_3$.CH$_3$ ⟶ OH
        OH   OH
                              CH=CH(CH$_2$)$_3$.CH$_3$

Hexene-1

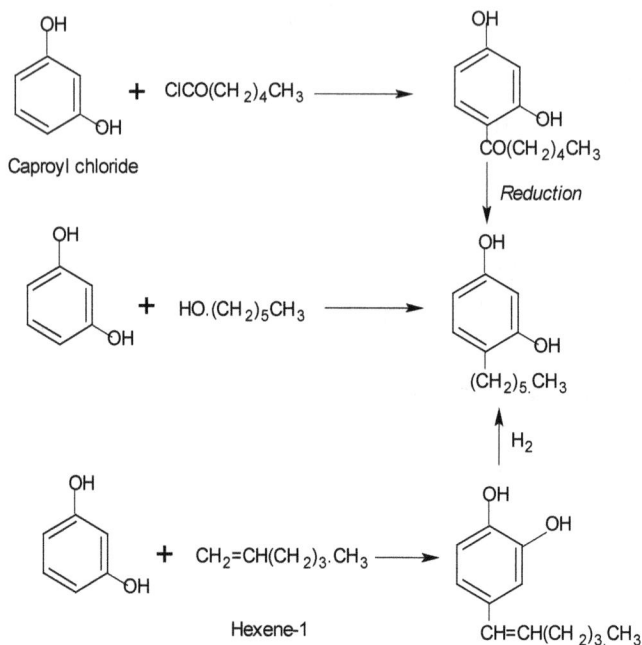

Hexylresorcinol, a well-known urinary antiseptic, is frequently used in the treatment for ascariasis, and tapeworm infections.

Other strongly active phenols with long alkyl chains are :

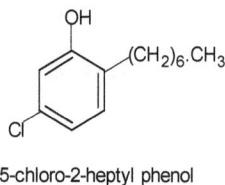

OH
⬡—(CH$_2$)$_5$.CH$_3$
Cl

5-chloro-2-hexyl phenol

OH
⬡—(CH$_2$)$_6$.CH$_3$
Cl

5-chloro-2-heptyl phenol

(ii) **Chloro-carvacrol:** It is also one of the strongest anthelmintics. It is prepared from p-cymene in the following way.

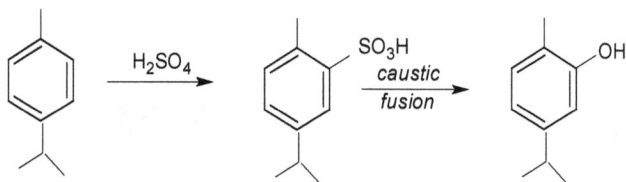

⬡ —H$_2$SO$_4$→ ⬡—SO$_3$H —caustic fusion→ ⬡—OH

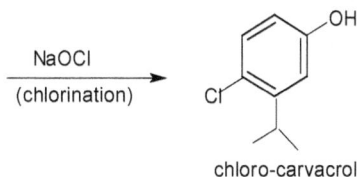

$$\xrightarrow[\text{(chlorination)}]{\text{NaOCl}}$$

chloro-carvacrol

(iii) Butyphen (p-ter-butyl phenol): It is prepared by the catalytic dehydration of a mixture of *tetra-butyl* alcohol and phenol.

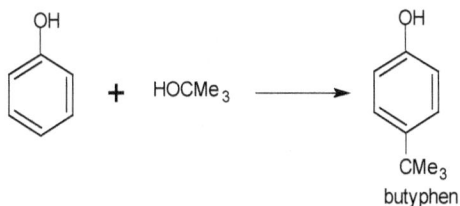

OH + HOCMe$_3$ ⟶

CMe$_3$
butyphen

It is particularly used for the remove of ascaris, hookworms, and whipworms particularly from dogs. It is a cheap less toxic anthelmintic.

Various halogenated diphenyl methanes, and diphenyl ethers were synthesized. The two important compounds in this series are found to be active against chicken tapeworm.

2,2-dihydroxy-3,3,5,5,6,6
tetrachlorodiphenylmethane

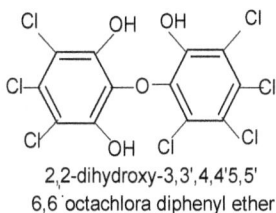

2,2-dihydroxy-3,3',4,4'5,5'
6,6 octachlora diphenyl ether

(iv) Egressin: It is *iso*-amylcarbamate derivative of thymol, It is obtained by the reaction of sodium thymol with chloroformic ester or phosgene (COCl$_2$) followed by the condensation of the product with iso-amylamine

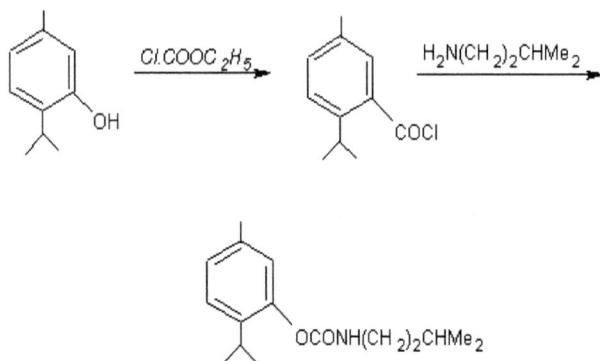

(v) Butolan or diphenan: It is the 4-carbamate of diphenyl methane. It is prepared by heating 4-hydroxydiphenyl methane with chloroformic ester or phosgene following by reaction with ammonia.

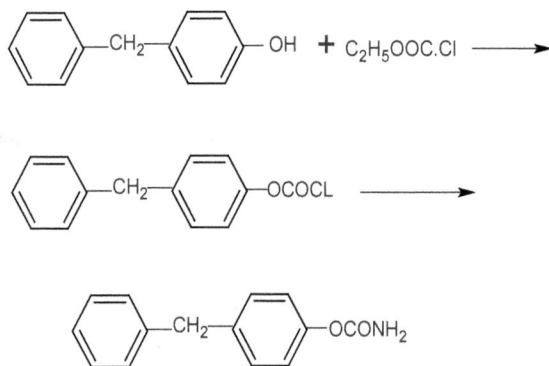

It is used for the expulsion of thread-worms.

(vi) Lubisn: It is the diethylcarbamate of resorcinol monobutyl ether. It is prepared from resorcinol monobutyl ether in the following way

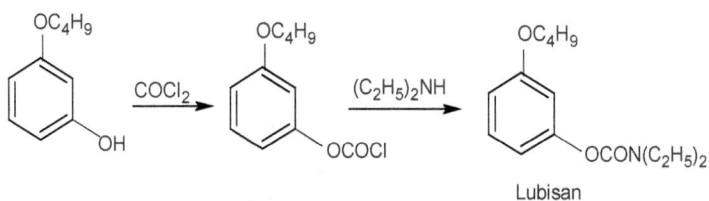

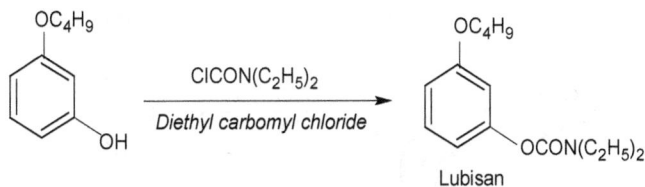

Lubisan

(c) **Antmonials and arsenicals:** certain compounds containing heavy metals, especially antimony any schistosomiasis diseases.

Filariasis is a group of diseases in man caused by insect-transmitted roundworm infection. Ethvlsitibamine, stiibophen tartaremetic urea stibamine and lithium antimony thiomalate are the drugs used for filariasis. Similarly schistosomisis represents a group of disease caused by human blood fluks antimony potassium tartrate and stibofen are the active drugs for schistosomisis for the details of these drugs pleased see 6.7(b).

Some copper compounds are also used as anthelmintics cupronate (a copper albumin complex) and gerlaverm (a form of copper silicate). However, now-a -day these metallic anthelmintics have been replaced by metal free compound like diethyl carbamazine, a piperazine derivative.

(d) **Piperazine derivative**: piperazine derivative with the following general formula are used as antiflarials.

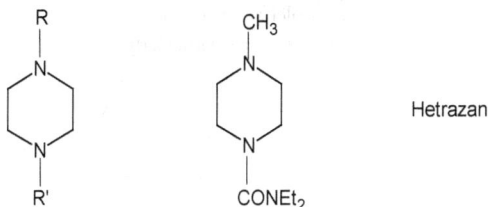

Hetrazan

The most important piperazine derivative is hetrazan (diethyl carbamazine). Hetrazan is synthesized from piperazine in the following manners.

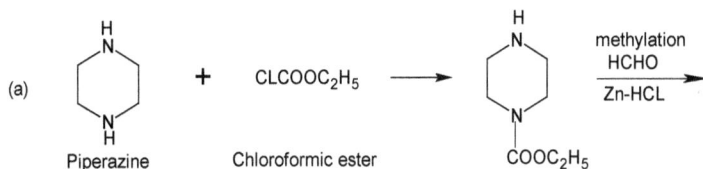

(a)

Piperazine   Chloroformic ester

Hetrazan

(b)

piperazine + ClCONEt$_2$ (Diethyl carbamyl chloride) $\longrightarrow$ $\xrightarrow{HCOOH+CH_2O}$

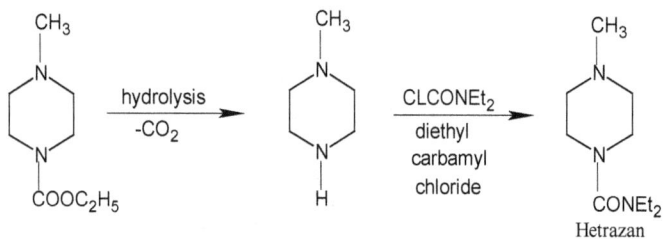

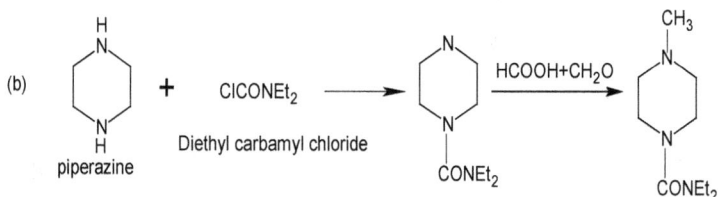

Diethyl carbamyl chloride is a quite   stable compound and prepare from diethylamine and phosgene.

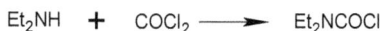

$$Et_2NH \quad + \quad COCl_2 \longrightarrow Et_2NCOCl$$

The citrate of diethyl carbamazine is also used as an anthelmintic and antiflarials under the name *banocide*. Many salt of piperazine adipate is used as an antiflarials agent.

(e)Triphenylmethane and cyanine dyes: The Gentian violet and its fully methylated derivative crystal violet are show to possess the anthelmintic activity in addition to antiseptic activity. They are particularly used in the treatment of skin-fluke *(strongyloides)* and the Chienese bile-fluke *(Clonorchis)* infection.

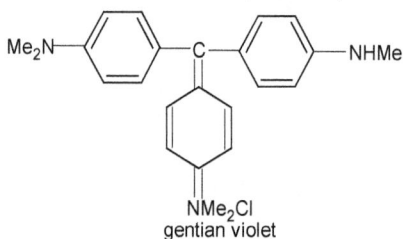

gentian violet

Certain cyanine dyes have also been found to be active as antifilariasis and anthelmintics, but so far they are not proved of practical use.

(f) Miscellaneous compound: Phenothiazine was firstly introduced as mosquito larvicide, but Harwood recognized its most important use as an anthelmintic. It is produced by melting together diphenylamine and sulphur.

diphenyl amine   phenothiazine

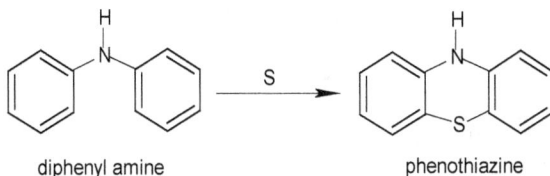

Because of its wide range of activity and inexpensiveness, phenothiazine is considered as one of the most important anthemintics for food producing animal is also active in human against pin and round worm infection but it is far from being a drug of choice.

Certain anti-malarials, *viz.*, quinacrine, azacrin and chloroquine, are also associated as anthelimintics. Mauss and his co-worker introduced a series of xanthones and thioxanthones as anti-schistosomiasis. The important ones are miracil A, B, C and D.

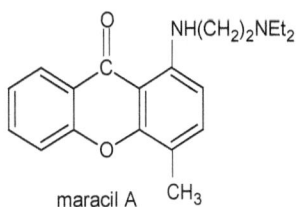

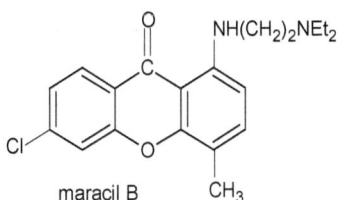

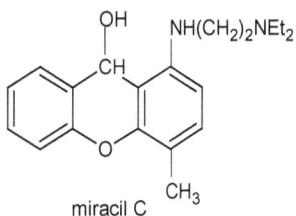

maracil A CH_3 maracil B CH_3

miracil C   miracil D CH_3

Out of the four miracils, miracil D, also known as nilodin or lucanthone, is the most important and prepared from p-chloroluene.

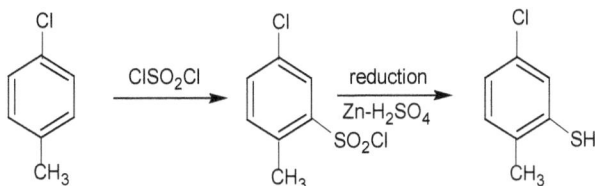

CI
COOK
+
HS
CH₃

→

CI
COOK
S
CH₃

Conc. H_2SO_4 →

O   CI

S
CH₃

$H_2N(CH_2)_2NEt_2$ →

O $NH(CH_2)_2NEt_2$

S
lucanthone   CH₃

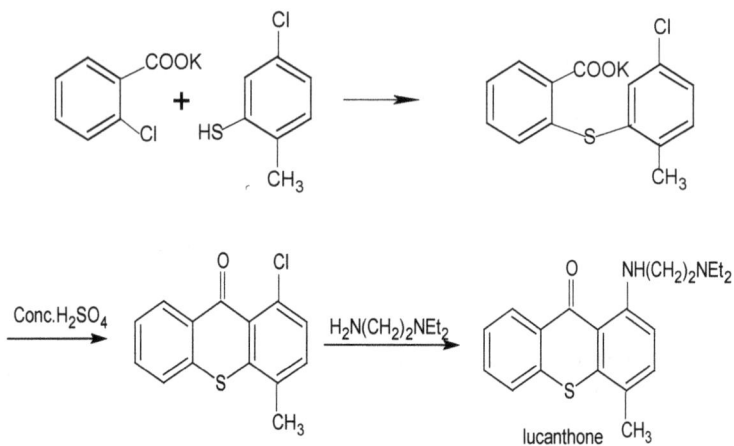

(g) Natural product: The anthelmintic properties of the various naturally occurring product have been known since times immemorial. In fact, a few of them are still used in medicine, the important one are as follow;

(i) Pelletierine: Pelletierine, an alkaloid from the pomegranatebark, is still used to some extent as an anthelmintic for tapeworms under the name punicine.

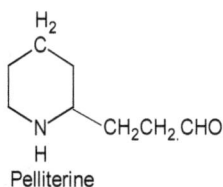

H_2
C

N
H
$CH_2CH_2 CHO$
Pelliterine

But it is highly toxic and has low therapeutic index.

(ii) Santonin: The leaves of Artemisia maritima have been used as a remedy for tapeworm infestation since time immemorial. The anthelmintic properties of leaves were found to be due to the presence of the compound, santonin. Santonin has also been used to expel the roundworm of swine, dogs, and cats. The therapeutic dose does not cause any side effect but the excess of the drugs may cause visual disturbance, headaches, nausea, and convulsions.

Later on, in 1930 the constitution of santonin was established followed by its synthesis in 1953.

methyl α–(3–ketocyclohexyl)propionate

streoisomeric acids

(i) seperation, m.p have 181 °C is taken
(ii) Br$_2$

-2HBr

santonin

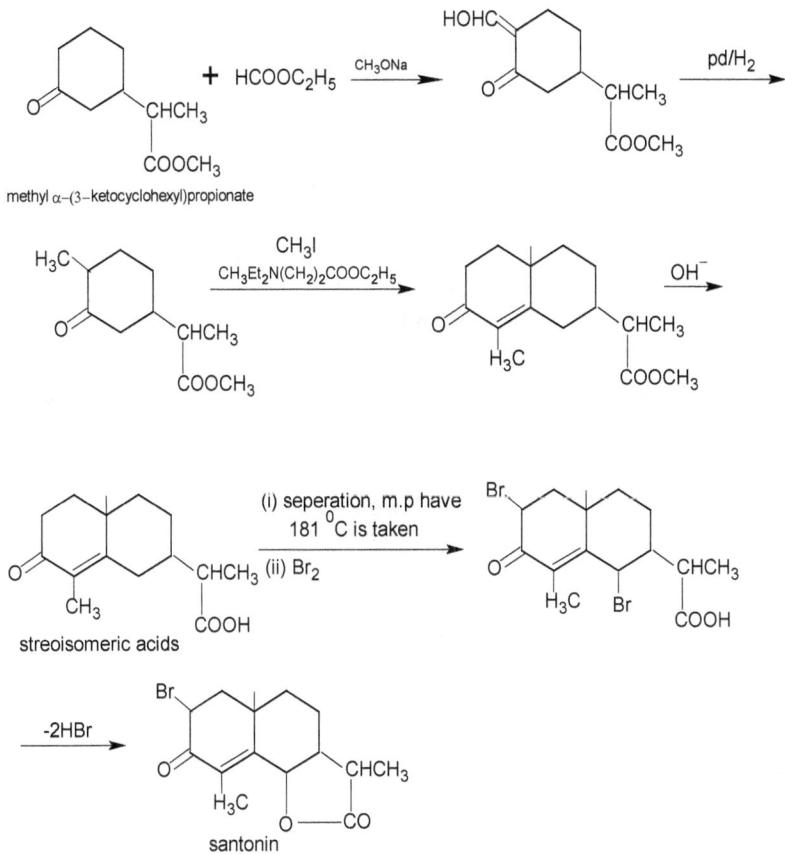

(iii) Ascardiol: - It is the principal constituent of the oil of wormseed (oil of chenopodium) obtained from the flowering plant of *chenopodium ambrosoides.* Ascardiol can be synthesized by the catalytic oxidation of terpinene which is obtained in plant from -pinene. In practice also -pinene and d-limonene are converted into ascaris-active preparation by treatment with hydrogen peroxide. Furthermore, Schenck and Ziegler reported the photosynthesis of Ascardiol from terpinene.

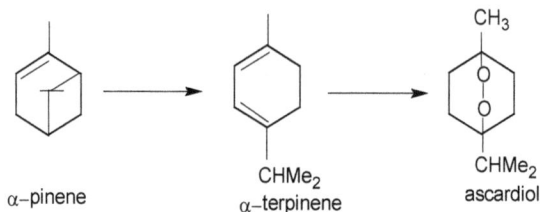

α-pinene → α-terpinene → ascardiol

Ascardiol has been used in treating dwarf tapeworms, ascardis in many animal, and dog hook worm infection. However the Ascardiol is found to be skin and mucous membrane irritant with some toxicity. Due to the above side-effect it is used in combination with tetrachloroethylene for treating hookworms and ascardis.

(iv) Aspidium (male fern, Filix mas) anthelmintics: The fern Aspidium filix-mas contain several anthelmintics substance; the importance ones are aspidinol(2-methylphloroglucinol propyl ketone mono methyl ether), filixic acid and albaspidin.

(a) Aspidinol: It was first isolated by Boehm and then synthesized by Karrer and Widmer which is further proved by Robertson and Sandrock work.

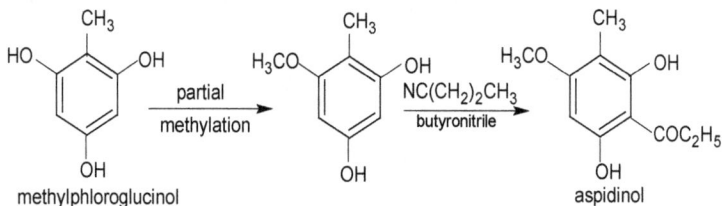

methylphloroglucinol → (partial methylation) → (NC(CH$_2$)$_2$CH$_3$ butyronitrile) → aspidinol

(v) Embelic acid (embelin): It is obtained by the extraction from *embelia ribes*. These anthel-mintics has a structure similar to filixic acid reminiscent and prepared from lauryl quinol.

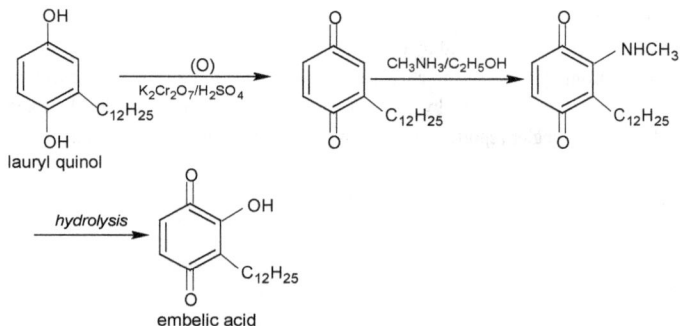

lauryl quinol → (O) K$_2$Cr$_2$O$_7$/H$_2$SO$_4$ → CH$_3$NH$_3$/C$_2$H$_5$OH → (hydrolysis) → embelic acid

(vi) Filicin: It is brownish powder obtained by drying the sap of laurifolia. Nothing is known about its structure but it has been established that it is an enzyme of the papain group. it digest ascardis and whipworms in vitro and vivo. It is almost nontoxic and is still in use as an intestinal anthelmintic.

(vii) Einetine: Emetine is the important *ipecacuanha* alkaloid It is especially active against liver fluke and lung fluke.

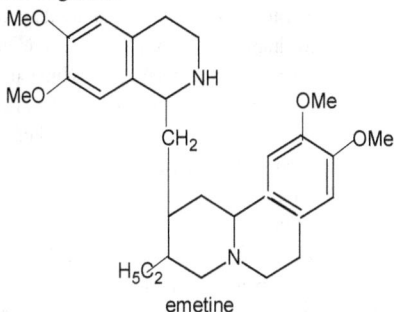

emetine

Paragonimiasis antibiotics, a parasitic infection which is very difficult to cure can actually be treated by means of emetine is also widely used for treating amoebic dysentery. Like other naturally anthelmintics the use of emetine is limited due to its toxicity.

(viii) Antibiotics: Several antibiotics, such as bacitracin, chlorotetracycline, oxytetracycline and henicillin have been found to be active against pinworm infection.

6.9 Antiseptics: Antiseptics are those chemical compounds which prevent the sepsis of wound, i.e. stops the action of micro organism either by inhibiting their reproduction or causing their death. On the other hand, the term disinfectant is used for the times it is used not correctly, in certain combined terms such as skin disinfectant. The antiseptic power of a drug is generally measured in terms of *phenol coefficient* which is defined as the dilution at which it will kill bacteria in a given time (generally 10 minutes) compared to the dilution required for phenol.

$$\text{Phenol coefficient} = \frac{germicidal\ dilution\ of\ the\ compound}{germicidal\ dilution\ of\ phenol}$$

The commonly used antiseptics may be studied under the following heading
1. Halogens and halogen compound.
2. Aromatic acids and esters.
3. Phenol and related compound.
4. Aldeheyds and compound.
5. Synthetic dyestuffs.

(a) The Halogen and halogen compounds

Chlorine compound. The germicidal properties of chlorine and hypochlorous acid are know since a long time and even now the chlorine is frequently used as disinfectant for drinking water. Since bleaching powder is a cheap compound and easily gives availble chlorine and the compound whose antisepics properties were due to available chlorine were soon replaced by hypochlorites.

Water soluble sodium hypochlorite NaOCl was also not used as an antiseptic because and irritant properties. The two importance inorganic antiseptic are Dakin s solution and eusol. Dakin solution is prepared by treating bleaching powder with sodium carbonate solution follwed by the addition of boric acid to the filterate. The antiseptic property of Dakin s solution is due to the presence of sodium hypochlorite (0.45-0.5%). it has been largely and successfully used for irrigation of wounds, especially in world war I. Eusol known as eupad in the powder form is obtained by treating bleaching powder with boric acid.

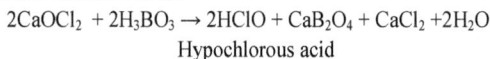

$$2CaOCl_2 + 2H_3BO_3 \rightarrow 2HClO + CaB_2O_4 + CaCl_2 + 2H_2O$$

Hypochlorous acid

The antiseptic property of eusol solution is due to the presence of free hypochlorous acid, it has been used for dressing of wounds and in the treatment of septic condition.

The importance of hypochlorous acid led tothe discovery of organic compounds capable of giving free hypochlorous acid. The important organic compound, exhibiting thus property, are chloramines were first discovered and prepared by chattaway by the action of hypochlorite solution upon organic compound containing free imino (-NH-) or amino (-NH$_2$) groups

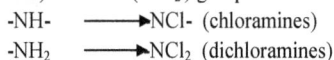

-NH- $\longrightarrow$ NCl- (chloramines)

-NH$_2$ $\longrightarrow$ NCl$_2$ (dichloramines)

Chloramines were introduced by Dakin as substitutes for sodium hypochlorite. Chloramines resemble hypochlorite in its bacteriological action; they act by the release of hypochlorous acid.

$$RNHCl + H_2O \longrightarrow RNH_2 + HOCl$$

But therapeutically they possess some advantages over the latter they are less irritant and are stable water soluble solids capable of producing therapeutic solution of a definite strength, whereas sodium hypochlorite cannot be kept in the solid state and always yields solutions of uncertain strength. Moreover since the active hypochlorous acid is released slowly the chloramines possess an additional advantage of prolonged bacteriostatic action.

Chloramine itself, NH$_2$Cl, is not a powerful antiseptic, although it is frequently used in the sterilization of water supplies. The important organic chloramines belong to the aromatic series which can only be applied locally since they are strongly toxic by injection. The important aromatic chloramines used as antiseptic are given below.

SO_2NClNa SO_2NClNa SO_2NCl_2 SO_2NCl_2

(benzene ring)   (benzene ring .3H$_2$O, CH$_3$)   (benzene ring, CH$_3$)   (benzene ring, COOH)

chloramine-B   chloramine-T   dichloramine-T   halazone

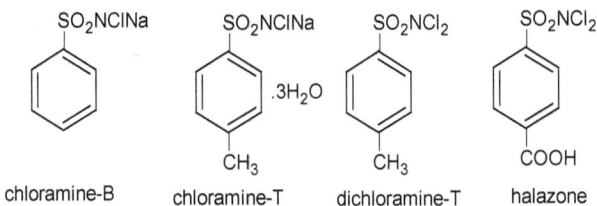

(i) chloramine-T or chlorozone (sodium p-toluenesulphon chloramine): It is prepared from p-toluenesulphonyl chloride. It is also obtained as a byproduct in the manufacture of saccharin.

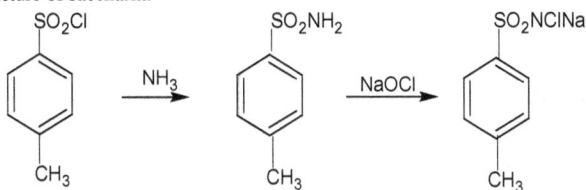

SO_2Cl (benzene ring, CH$_3$) $\xrightarrow{NH_3}$ SO_2NH_2 (benzene ring, CH$_3$) $\xrightarrow{NaOCl}$ SO_2NClNa (benzene ring, CH$_3$)

It is a stable crystalline water soluble compound, and has been widely used during world war I far the treatment of infected wounds. Its freedom from irritant properties has rendered it particularly useful for hyginic purposes such as mouth washes, douches etc. It is also used as a disinfectant lotion in some infection diseases such as scarlet fever, measles etc.

(ii) Dichloramine-T (p-toluenesulphondichloramide): It is obtained by dissolving p-toluene sulphonamide in bleaching powder solution or by the action of hypochlorous acid upon chloramine-T (prepared above).

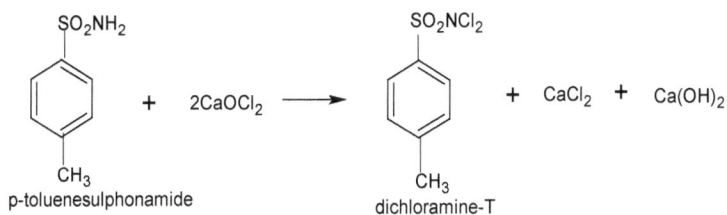

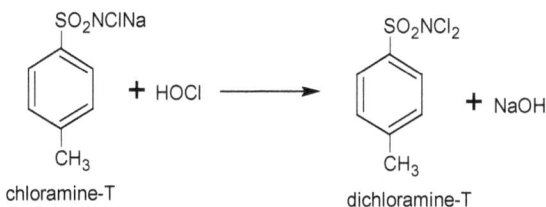

SO_2NH_2 (benzene ring, CH$_3$) p-toluenesulphonamide $+$ $2CaOCl_2$ $\longrightarrow$ SO_2NCl_2 (benzene ring, CH$_3$) dichloramine-T $+$ $CaCl_2$ $+$ $Ca(OH)_2$

SO_2NClNa (benzene ring, CH$_3$) chloramine-T $+$ HOCl $\longrightarrow$ SO_2NCl_2 (benzene ring, CH$_3$) dichloramine-T $+$ NaOH

It is insoluble in water but soluble in a number of organic solvents, such chlorinated paraffin wax and chlorinated eucalyptol in which its solution is used for treatment of wound, Its oil solution has also been used as a nasophyngeal sterilizer in meaningities.

(iii) Halazone (p- sulphondichloramine- benzoic acid): It is prepared from p-toluenesulphon- namide by the following method.

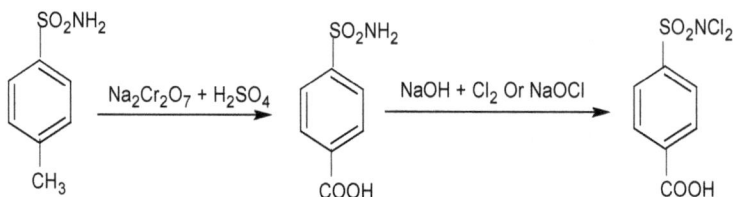

Its soluble sodium salts is used for sterilization of drinking water.

(iv)other chloramines: Its must be noted that antiseptics activity is not only limited to aromatic chloramines other than aromatic chloramines have also been investigated e.g., succinchlorimide (having 25% available chlorine) has also been successfully used for sterilization of drinking water similarly 1,3-dichlo-5,5-dimethydantionby hypochlorous acid, has been found to be sanilizing agent.

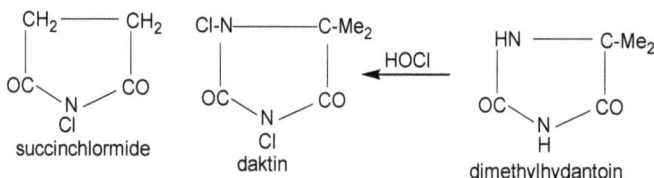

Chloroazodin (azochloramide): It is a modern antiseptic and is claimed to be relatively nontoxic to tissue. It is obtained by dissolving guanidine as the nitrate in glacial acetic acid sodium acetate solution and adding NaOCl at the temperature below $0\ ^0C$.

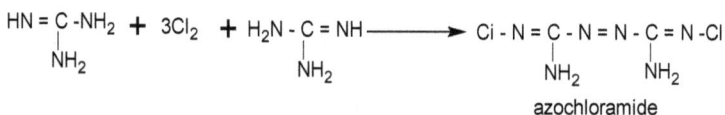

Bromine compound: Bromine compound are not important as, however, two compound have been used as antiseptics viz, *providoform (sodium salt of tribrom-B-naphthol)* and *dibromin (dibromonalon urea).*

dibromin

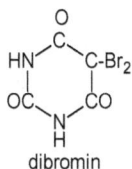

Iodine compound: Iodine is widely used as tincture of iodine as an emergency antiseptic. Since, it is undesirable as a wound antiseptic; attempts were made to find more suitable iodine antiseptics. By the antiseptic properties of the iodine compound it was concluded that medicinal value is due to the liberation of free iodine, when they come in contact with fat which purposes iodine antiseptics are described below.

(i) Iodoform: Iodoform is the most common antiseptic of the iodine compounds. It is prepared by the action of free iodine on a warm mixture of acetone or alcohol and sodium hydroxide or sodium carbonate solution.

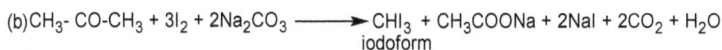

(a) $C_2H_5OH + 4I_2 + 3Na_2CO_3 \longrightarrow CHI_3 + HCOONa + 2NaI + 2CO_2 + 2H_2O$
iodoform

(b) $CH_3\text{-}CO\text{-}CH_3 + 3I_2 + 2Na_2CO_3 \longrightarrow CHI_3 + CH_3COONa + 2NaI + 2CO_2 + H_2O$
iodoform

Iodoform is prepared by electrolyzing potassium iodide solution in a current of carbon dioxide.

Iodoform was extensively used as an antiseptic in dressing for wound but now-a-days replaced by other antiseptics due to its some drawbacks viz objectionable odour, irritant action on skin to cause a kind of eczema and toxicity.

(ii) Tetraiodopyrrole (Iodol): Iodol resemble iodoform in its antiseptic action with the various advantage, viz. it is odourless non-irritant and insoluble. It is obtained by the iodination of alkaline solution of pyrrole.

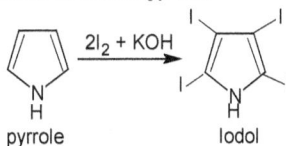

pyrrole   Iodol

(iii) Aristol (dithymol-di-iodide): Aristol is a useful antiseptic with the following structure.

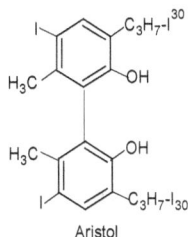

Aristol

However it suffers from the drawback of being rather unstable and expensive

(iv) *Vioform:* It is a non-poisonous iodoform substitute. It is prepared by the iodination of 8-hydroxyquinoline with one mole of sodium iodide and excess of sodium hypochlorite.

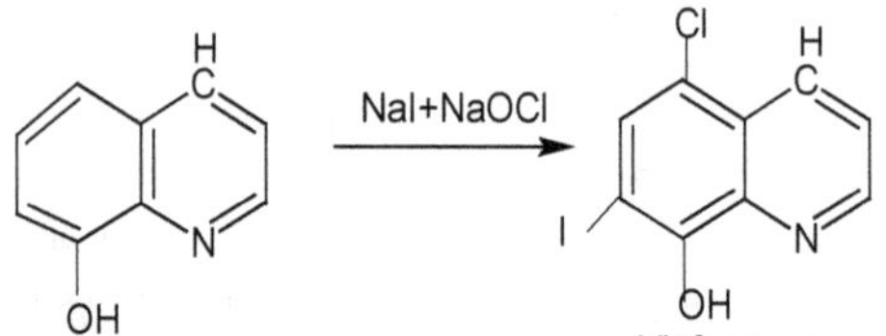

Vioform

(b) Phenol and its derivatives

Phenol was introduced as an antiseptic by Lister in 1865, but its irritant action led to the discovery of the more effective and less toxic alky and chloros substituted phenol. Some of the important members of these two substituted phenolic compound are given below.

Alkylated phenol: The introduction of alkyl group in the phenol makes the phenol more effective and lesser toxic. For example cresols (phenol coefficient 2-5) are more active than phenol. However, since they are insoluble in water their emulsions in soap (Lysol) are used as antiseptics.

(i) *Carvacrol:* It is a naturally occurring phenol and synthesized from p-cymene by the following method.

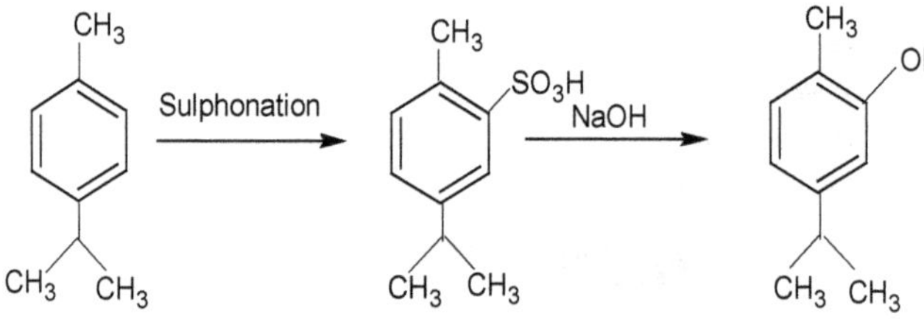

p-cymene   p-cymene sulphonic acid   carvacrol

(ii) *Thymol:* It is also naturally occurring antiseptic (from oil of thyme). It is synthesized from m-cresol in the following way.

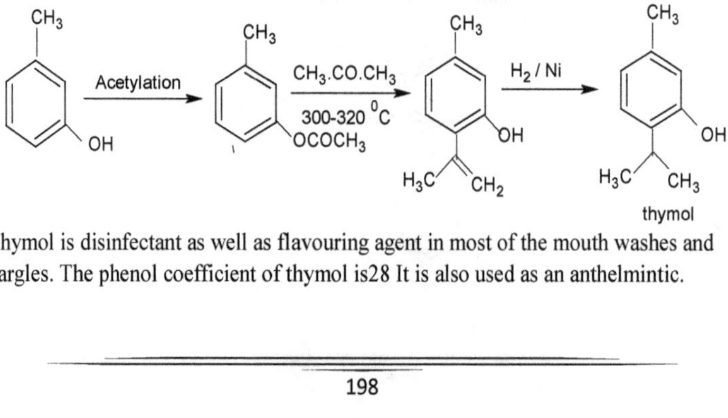

thymol

Thymol is disinfectant as well as flavouring agent in most of the mouth washes and gargles. The phenol coefficient of thymol is28 It is also used as an anthelmintic.

(iii) n-Hexylresorcinol: Although n-hexylresocinol has some irritant action on the skin it is used as general antiseptic. It is prepared by the condensation of caproic acid chloride and resorcinol in the presence of zinc chloride followed by Clemmensen reduction.

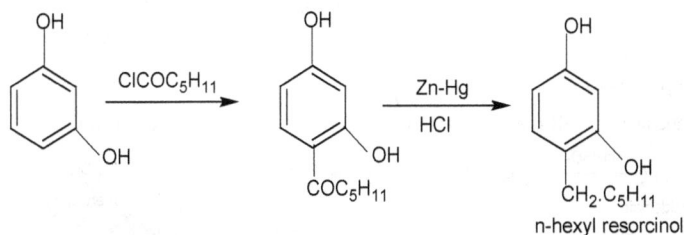

n-hexyl resorcinol

It has phenol coefficient 147. It is also as an anthelmintic.

Chlorinated phenol: The antiseptic property of phenol is greatly enhanced by the presence of halogen atoms. The two medicinally important chlorinated phenols are chloros - m-cresol, used in sterilizing soap and chloros -m-xylenol.

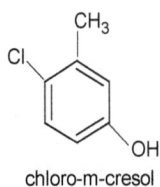

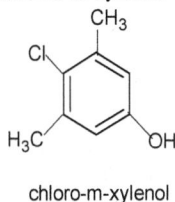

chloro-m-cresol

chloro-m-xylenol

Chloro-m-xylenol is an important antiseptic of widespread use. Known and commonly used antiseptic *dettol*.

Among hydroxyl derivatives of polycyclic system the bis-(3.4.6-trichloro -2-hydroxyphenyl) methane (hexachlorophene, GII) is the most important.

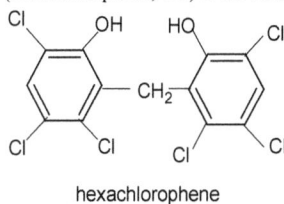

hexachlorophene

The chlorinated phenols are prepared by direct action of chlorine on the aqueous solution of the corresponding sodium phenates.

(c) Aromatic acid and their derivatives: Ester of p-hydroxy and o-hydroxy benzoic acids are powerful antiseptic properties. The important examples are;

ethyl parasept — OH ... COOC$_2$H$_5$

phenyl salicylate (salol) — OH ... COOC$_6$H$_5$

betol — OH ... COO ...

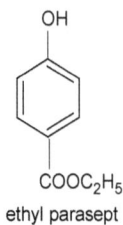

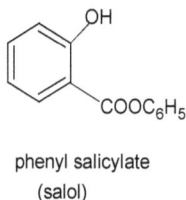

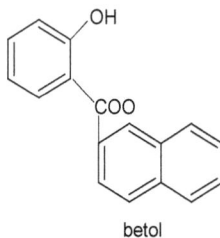

Mandelic acid itself is a valuable urinary antiseptic. It is synthesized as below.

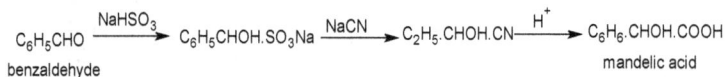

$$C_6H_5CHO \xrightarrow{NaHSO_3} C_6H_5CHOH.SO_3Na \xrightarrow{NaCN} C_2H_5.CHOH.CN \xrightarrow{H^+} C_6H_6.CHOH.COOH$$

benzaldehyde — mandelic acid

(d) Synthetic dyestuffs: Certain dyes are important and widely used as an antiseptic. The commonly used dyes for antiseptic purposes are grouped under the following heading.

(i) Triphenylmethane dyes: The important members of this group of synthetic dyes used as antiseptics are methyl violet, crystal violet and malachite green.

(ii) Acridine dyes: Proflavine and acriflavine are the common wound antiseptic particularly acriflavine.

(iii) Thiazine dyes: Methylene blue and phenothiazine are the examples of this group of dyes used as an antiseptic. Methyene blue has occupied an historical importance in the development o f dye chemotherapy .It was the first synthetic dyestuff used as a chemotherapeutic agent. It has been used for a long time as an intestinal and urinary antiseptic. It is preparing by the treating hydrogensulphide with the mixture of p-dimethylamino aniline and dimethyl aniline in presence of mild oxidizing agent, viz., FeCl$_3$.

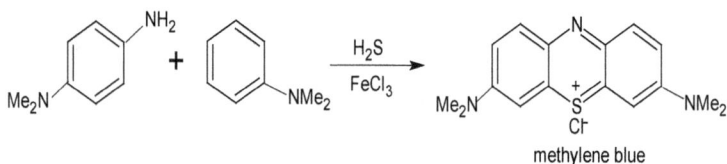

methylene blue

The two components may also be converted into methylene blue in the following way.

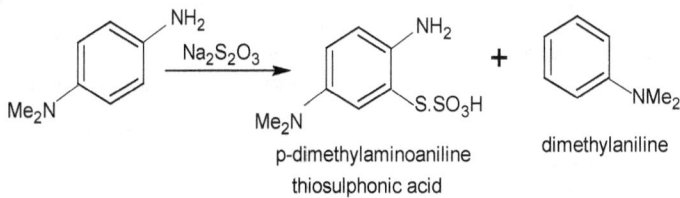

p-dimethylaminoaniline
thiosulphonic acid

dimethylaniline

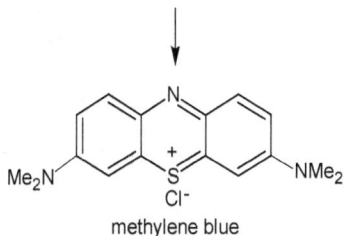

methylene blue

Phenothiazine is also a valuable urinary antiseptic. It is a prepared by heating diphenylamine and sulphur in the presence of a catalyst.

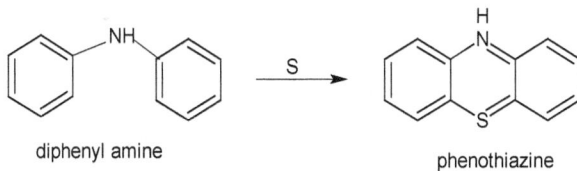

diphenyl amine

phenothiazine

(e) Metal and their compounds: The most important compounds among this group of antiseptics are mercury compound . an organometallic compound is the most important member of this group. It is obtained by the mercuration of dibromofluorescein with mercuric acetate in an alkaline medium.

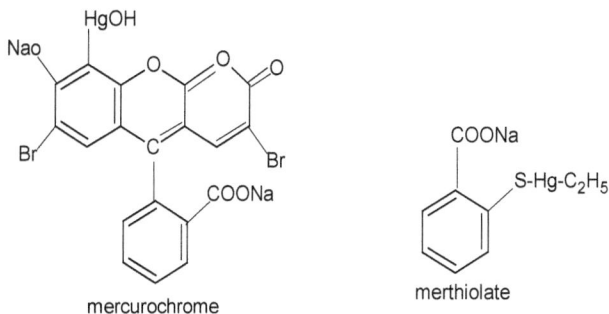

mercurochrome

merthiolate

Another one important antiseptic of mercury is Merthiolate. Both of these antiseptic are sold in tincture solution.

Colloidal suspensions of silver and its salts, particularly silver chloride and silver iodide are used as antiseptics on mucous membranes. Similarly, certain zinc compounds, *viz.* zinc sulphate, zinc oxide zinc soaps have been used as valuable moderate antiseptics.

(f) Formaldehyde and its derivatives: Formaldehyde has long been used as a room disinfectant. The disinfectant property of formaldehyde is due to its volatility. But on the other hand , its corrosive and toxic action on the other hand, its corrosive and toxic action on skin hindered its direct use as a strong antiseptic in medicine. In medicine formaldehyde is used as its compounds which slowly split off HCHO under the influence of secretions of the organism.

Formaldehyde react with a variety of natural product, viz. gelatin[8], casein, nucleic acid and carbohydrate. The compounds of HCHO and carbohydrate are of immense importance and Classen found that it react with starch , dextrin , lactose ,etc. to from insoluble odourless, non-irritant compounds which slowly split off formaldehyde and thus the compounds function as antiseptics without being poisonous . The soluble condensation product of HCHO and lactose liberates HCHO when dissolved in the mouth, and hence it is used in the form of tablets , under the name from mint, to check and prevent the septic conditions in throat and mouth. The well known product of HCHO and ammonia, the hexamine , has also been used as an important urinary antiseptics under the name urotropine, cystamine and cystogen. Urotropine is also used in laryngitis, pharyngitis, etc.

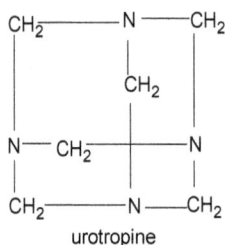

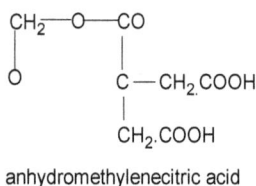

urotropine

anhydromethylenecitric acid

A large number of derivatives and additive compounds of urotropine and hexamine camphorate are the examples of urotropine derivatives used as urinary antiseptics . Similarly, the additive compounds of urotropine with sodium acetate, sodium citrate, and sodium benzoate, respectively known as cystopurin, formurol and cystazol, are also used as urinary antiseptics.

6.10 Anaesthetics: The tetra anaesthetics is derived from the Greek word, anesthesia, meaning, insensibility and hence anaesthetics may be defined as those drug which bring about the insensibility to the vital function of all types of cell, especially those of nervous system. The effect produced by these drugs is reversible that means the affected organs return to a normal state when the concentration of the drug has been decreased; and hence we can say that the anaesthetics bring about the temporary insensibility to pain or feeling in the body or particular organs required for the surgical operations. The anaesthetics may be classified into two groups on the basis of their applications .

(i) General or central anaesthetics : These anaesthetics depress the central nervous system to such an extent that all sensitivity to pain or feeling is lost , i.e. they cause unconsciousness all over the body

*(ii)*Local anaesthetics : Instead of affecting the whole body these anaesthetics make a particular organs insensible to pain or feeling , i.e. the individual become unconscious only to the affected part of the body and not to the whole of the body.

(a) General anaesthetics: the general anaesthetics may be subdivided into volatile or gaseous and non-volatile or fixed anaesthetics. The former are administered by the respiratory route and the latter by injection or by alimentary canal.

Volatile general anaesthetics

(i) Ether: Ether was the first compounds to be used as anaesthetics by an American Dr. Crawford Long, but his results were not published In 1846 Morton successfully demonstrated the use of ether as an anaesthetics in surgical practice and hence its name is generally associated with the use of ether as an anaesthetics . But it was Sir James Simpson who popularized the use of ether in surgical operations. Ether is best prepared by using Williamson's synthesis.

$$C_2H_5Br \;+\; NaOC_2H_5 \longrightarrow C_2H_5OC_2H_5 \;+\; NaBr$$

But due to its objectionable properties. like the very easy formation of explosive and dangerous peroxide on exposure to light, and high inflammability. it was not used for long time as an anaesthetics.

(ii) Ethylene: Ethylene was first introduced as general anaesthetics, in 1923. It causes a rapid and pleasant induction of unconsciousness with quick recovery. But since it forms an explosive mixture with air, ethylene also did not a good general anaesthetics.

*(iii) Divinyl ether(vinethene):*The divinyl ether possesses the structural feature of both of the above compounds, i.e., ether and ethylene. It was introduced by Leake

In1930.It is prepared from ethylene from ethylene chlorohydrins in the following manner.

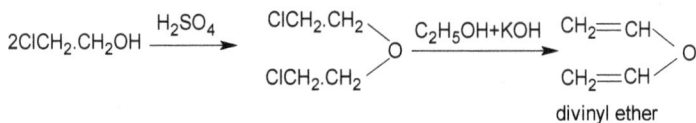

$$2ClCH_2.CH_2OH \xrightarrow{H_2SO_4} \begin{matrix} ClCH_2.CH_2 \\ \\ ClCH_2.CH_2 \end{matrix} O \xrightarrow{C_2H_5OH+KOH} \begin{matrix} CH_2\!=\!CH \\ \\ CH_2\!=\!CH \end{matrix} O$$

divinyl ether

Divinyl ether has a few advantage over ether and ethylene , e.g., it is nearly seven times more potent then ether with a rapid recovery. On the other hand, it has also some serious disadvantage, e.g., high volatility and instability. Due to the latter property it undergoes easy oxidation and hence it require some antioxidant, like α-phenylamino naphthalene, for its stability. Moreover, its prolonged inhalation may also damage the liver.

(iv) Methyl n-propyl ether (methopryl): It was introduce in 1964, and is found to be more powerful and less irritating than the ethyl ether. Moreover, it has also objectionable after effects than the ether.

(v) Cyclopropane: Among the hydrocarbon general anaesthetics, viz., acetylene, ethylene, and propylene,[8]cyclopropane is undoubtedly the most useful and powerful. Its preparation is, however, costly. It is prepared by the action of metals, generally either sodium, magnesium, on 1,3 -dichloropropane obtained by the careful chlorination of propane.

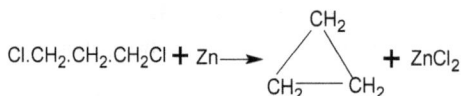

$$Cl.CH_2.CH_2.CH_2Cl + Zn \longrightarrow \begin{matrix} CH_2 \\ \diagup \;\; \diagdown \\ CH_2 \!-\! CH_2 \end{matrix} + ZnCl_2$$

Cyclopropane is the most potent among the anaesthetics gasses , and its disadvantage are that it is explosive , expensive, and flammable. But in combination with avertin it is nearly the best general anaesthetics .

(vi) Nitrous oxide: Although nitrous oxide was the first anaesthetics to be known, yet it was used after the ether . It is prepared by heating ammonium nitrate up to 200 ^{0}C.

$$NH_4NO_3 \rightleftharpoons N_2O + 2H_2O$$

Now a days it is used only in dentistry and in a few minor surgical operations.

(vii)Chloroform: Chloroform was among the most widely used general anaesthetics. It is prepared from bleaching powder and ethyl alcohol as follows;

$$CaOCl_2 + H_2O \longrightarrow Ca(OH)_2 + Cl_2$$

$$C_2H_5OH + Cl_2 \longrightarrow CH_3CHO + 2HCl$$

$$CH_3CHO + Cl_2 \longrightarrow CCl_3.CHO + 3HCl$$

$$2CCl_3.CHO + Ca(OH)_2 \longrightarrow CHCl_3 + (HCOO)_2Ca$$

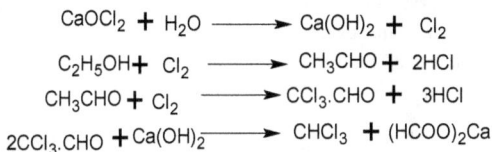

It is got a few advantage as anaesthetics , viz., it is noninflammable and required only in small quantity to produced surgical anaesthesia . but its serious and, of course, dangerous disadvantage is its rapid decomposition by air , light or moisture into phosgene $(COCl_2)$,a poisonous gas

(viii) Fluothane (2-bromo-2-chloro-1,1,1,-trifluoroethane,CF_3,CHBrCl): Fluothane has recently been introduce in the hospitals as a general volatile anaesthetics. It is considered to be the best general anaesthetics yet produced . It is nearly four times more active than ether. Moreover, it is non-toxic, nonexplosive and non-inflammable.

(ix) Trilene(trichloroethylene: Trilene is an important anaesthetics mainly used to induce an analgesics state in the relief of pain in migraine, angina pectoris and child birth. Commercially, it may be synthesized by the controlled addition of chlorine to acetylene diluted with carbon dioxide and cooled at-60 ^0C , but generally it is prepared by the alkaline decomposition of tetrachloroethane.

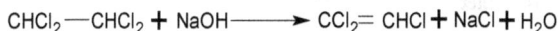

$$CHCl_2 - CHCl_2 + NaOH \longrightarrow CCl_2 = CHCl + NaCl + H_2O$$

(x)Viadril: Rudel et al (1995) introduced a steroid horomones Viadril, as an anaesthetic. Viadril is prepared by the palladium reduction of 11-deoxycorticosterone followed by treatment with succinic anhydride , formation of a sodium salt.

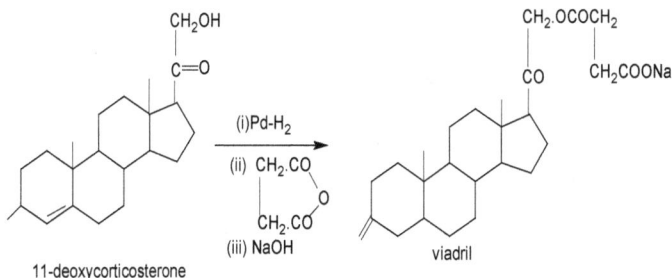

11-deoxycorticosterone

viadril

Its aqueous solution is used by intravenous injection to introduce complete anaesthesia.

Non-volatile general anaesthetics: Among them the most important is avertin which is a solution of tribromoethanol in *tert*-amylalcohol . The tribromoethanol is prepared by the reaction of bromal by aluminium isopropoxide.

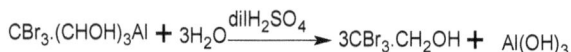

$$3CBr_3CHO \quad CBr.(CHOH)_3 \longrightarrow 3CBr.(CHOH)_3Al \quad 3Me_2CO$$

$$CBr_3.(CHOH)_3Al + 3H_2O \xrightarrow{dilH_2SO_4} 3CBr_3.CH_2OH + Al(OH)_3$$

Avertin is administered by the rectal route. It is generally used to produce a short sleep and then the full surgical anaesthetics is completed by one of the volatile anaesthetics described above.

Another non-volatile general anaesthetics used only on laboratory animals is chloralose, obtained by the condensation of glucose and chloral. It is must by noted that the central nervous system is depressed by the use of many therapeutics which , in general, are known as narcotics. The narcotics may be classified into various classes according to the degree of their physiological action an analgesics, e.g., an analgesic lower sensitivity to pain without causing unconsciousness , a hypnotic lead to an unconsciousness state akin to sleep, usually without anaesthesia , and an anaesthetic produced insensibility to pain either throughout the whole system or in certain localized area. Furthermore , it is important to note that the same drug can induced either analgesia, unconsciousness or anaesthesia depending upon the dosage

(b) Local anaesthetic

Cocaine, alkaloid from the cocaleaves was the first to be used as a local anaesthetics. But due to its high toxicity and habit-forming nature, attempts were made to synthesize more effective and less toxic substitutes for cocaine.

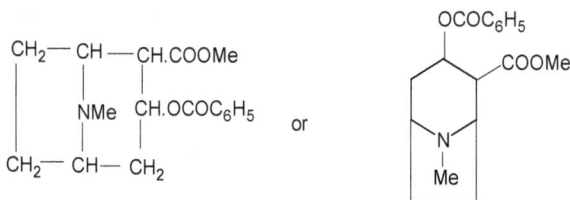

cocain

The important synthetics substitutes for cocaine may be studied under the following headings.

1. Piperidine or tropane derivatives: The two important example of this group are as follows;

(i) α-Eucaine: It is prepared by condensation of three molecules of acetone with one molecule of ammonia to give triacetoneamine, I. The N-methyl derivative of triacetoneamine II is treated with HCN to yield cyanohydrin which on hydrolysis followed by o-benzylation and N-methylation gives α-eucaine

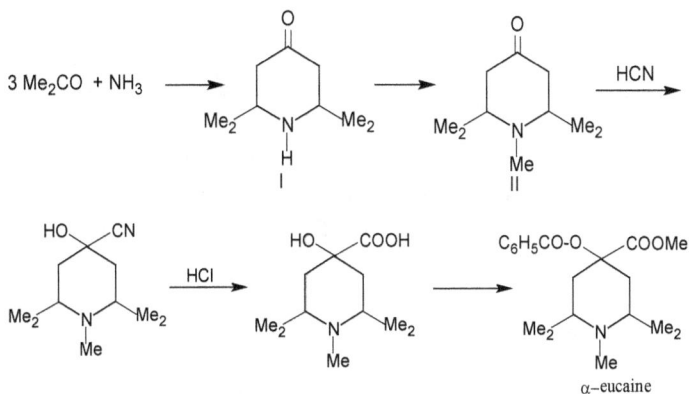

It is a cheap and less toxic substitute of cocaine. Moreover, it is more stable to boiling water than cocaine; hence its solution can be sterilized by boiling. However, it has the drawback of being somewhat painful and irritant when injected, so it is rapidly replaced by β-eucaine.

(ii) β-eucaine (benzamine): It is prepared by the interaction of diacetonamine and acetaldehyde (or better by boiling the acid oxalate of diacetonamine with diethylacetal) to give vinyl diacetonamine which is reduced and then benzoylated to β-eucaine.

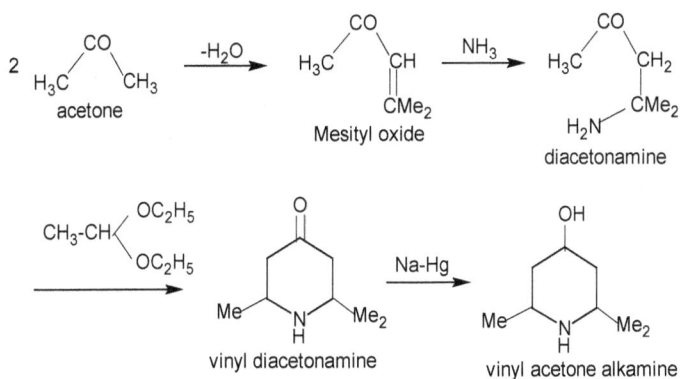

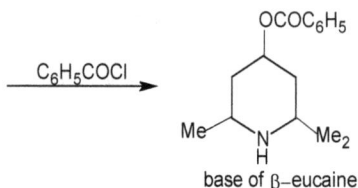

base of β–eucaine

The hydrochloride of the above compound is the proper β-eucaine. It is less toxic and easily soluble in water than α-eucaine. It is soluble in water and hence can be easily sterilized. In anaesthetic properties it is equal to cocaine and widely used in many branches of surgery.

(iii) Euphthalmin: In its synthesis the alcoholic group of vinyl-diacetone alkamine (prepared above) is esterifies by means of mandelic acid in place of benzoyl chloride.

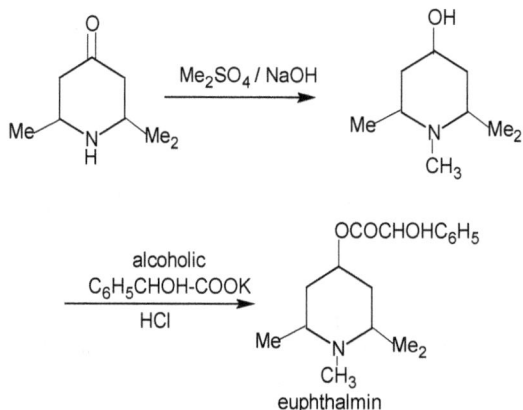

euphthalmin

2. Aminobenzoic acid derivative: Some of the m-and p-amino benzoic acid derivative were found to be useful local anaesthetics. The p-amino benzoic acid derivatives are more important anaesthetic than the m-amino benzoic acid derivatives. In the foregoing example the first belongs to the latter group whereas the other rest belongs to the former group.

(i) Orthocaine or new orthoform: It is obtained from p-hydroxy benzoic acid in the following two methods.

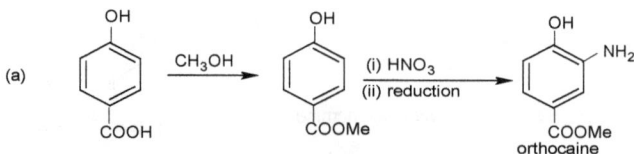

orthocaine

(b)

OH / COOMe benzene + N₂Cl / SO₃H benzene → OH / N=N-benzene-SO₃H / COOMe

reduction →

OH / NH₂ / COOMe
orthocaine

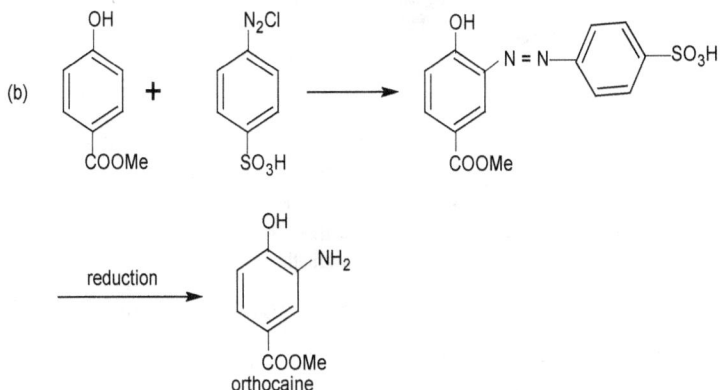

It is used as a dusting power for painful wounds.

(ii) Benzocaine or anaesthesin (ethyl p-aminobenzoate): It is prepared by from p-nitrotoluene or p-acetyl toluide in the following way.

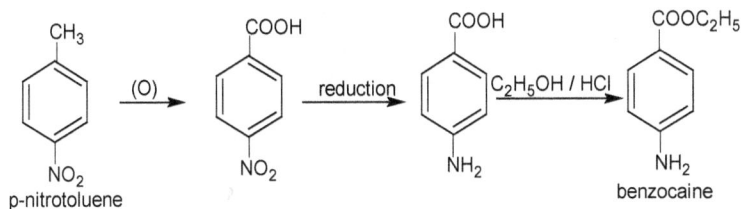

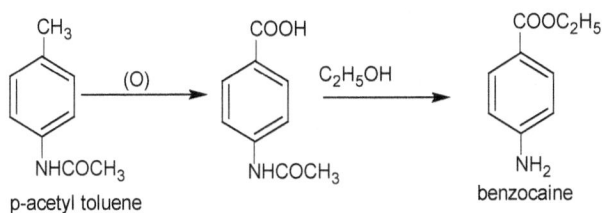

CH₃ / NO₂ (p-nitrotoluene) $\xrightarrow{(O)}$ COOH / NO₂ $\xrightarrow{reduction}$ COOH / NH₂ $\xrightarrow{C_2H_5OH \,/\, HCl}$ COOC₂H₅ / NH₂ (benzocaine)

CH₃ / NHCOCH₃ (p-acetyl toluene) $\xrightarrow{(O)}$ COOH / NHCOCH₃ $\xrightarrow{C_2H_5OH}$ COOC₂H₅ / NH₂ (benzocaine)

It is itself insoluble in water, but its salt with *p*-phenolsulphonic acid is soluble in water, and hence this salt is used for hypodermic injection under the name of *subcutin*. The two other similar local anaesthetics are *cycloform* and *butesin*.

COOC$_4$H$_9$-iso

cycloform

COOC$_4$H$_9$-n

butesin

NH$_2$

NH$_2$

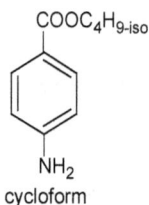

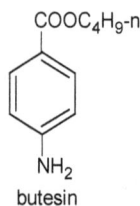

(iii) Procain: Ethocaine hydrochloride, novocaine or scurocaine. It is the most important local anaesthetic. It is a non irritant, least toxic and powerful local anaesthetic. It is prepared by the following two methods.

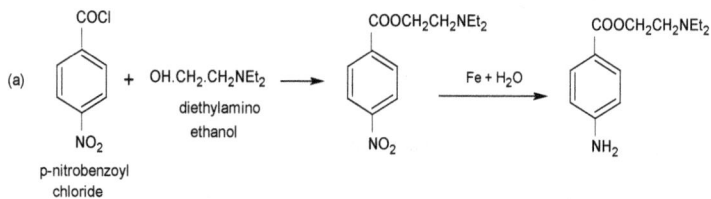

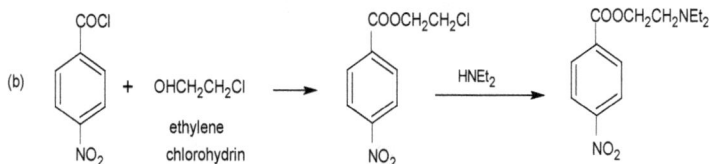

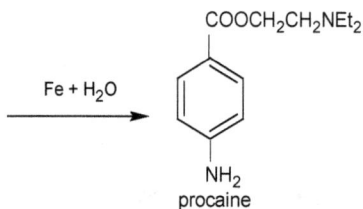

(a)

COCl

NO$_2$

p-nitrobenzoyl chloride

+ OH.CH$_2$.CH$_2$NEt$_2$

diethylamino ethanol

⟶

COOCH$_2$CH$_2$NEt$_2$

NO$_2$

Fe + H$_2$O ⟶

COOCH$_2$CH$_2$NEt$_2$

NH$_2$

(b)

COCl

NO$_2$

+ OHCH$_2$CH$_2$Cl

ethylene chlorohydrin

⟶

COOCH$_2$CH$_2$Cl

NO$_2$

HNEt$_2$ ⟶

COOCH$_2$CH$_2$NEt$_2$

NO$_2$

Fe + H$_2$O ⟶

COOCH$_2$CH$_2$NEt$_2$

NH$_2$
procaine

(iv) Tutocaine: It is also a p-aminobenzoic acid derivative, prepared according to the following set of reaction.

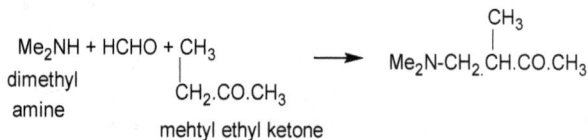

Me$_2$NH + HCHO + CH$_3$

dimethyl amine

CH$_2$.CO.CH$_3$

mehtyl ethyl ketone

⟶

CH$_3$
|
Me$_2$N-CH$_2$.CH.CO.CH$_3$

$$\xrightarrow{\textit{Na-alcohol}} \quad \overset{\underset{\displaystyle |}{CH_3}}{Me_2N\text{-}CH_2.\,\overset{}{CH}OH.CH_3} \quad \xrightarrow{\quad H_2N\text{—}\bigcirc\text{—}COCl \quad}$$

$$H_2N\text{—}\bigcirc\text{—}COO.\overset{\underset{\displaystyle |}{CH_3}}{CH}\text{-}\overset{\underset{\displaystyle |}{CH_3}}{CH}.CH_2NMe_2$$

totocaine

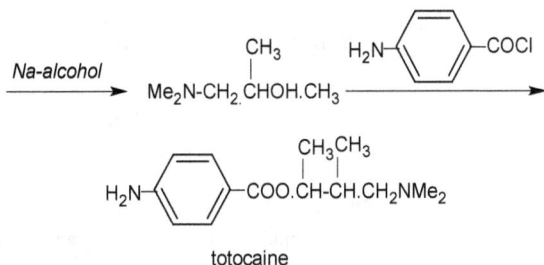

3. Stovaine and Alypine: These anaesthetics are b characterized by the presence of the grouping

$$\overset{\underset{\displaystyle |}{R}}{-N\text{-}C\text{-}C\text{-}O\text{-}CO\text{-}R'}$$

(i) Stovaine *(amylocaine hydrochloride):* It is prepared in the following manner.

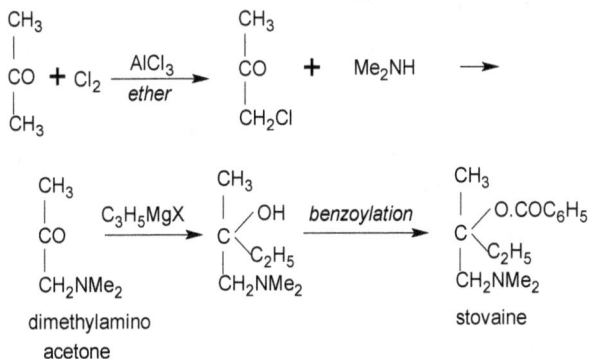

$$\begin{array}{c} CH_3 \\ | \\ CO \\ | \\ CH_3 \end{array} + Cl_2 \xrightarrow[\textit{ether}]{AlCl_3} \begin{array}{c} CH_3 \\ | \\ CO \\ | \\ CH_2Cl \end{array} + \quad Me_2NH \quad \longrightarrow$$

$$\begin{array}{c} CH_3 \\ | \\ CO \\ | \\ CH_2NMe_2 \end{array} \xrightarrow{C_3H_5MgX} \underset{\substack{\\ CH_2NMe_2}}{\overset{CH_3}{\underset{|}{C}}}\!\!\big\langle\substack{OH \\ C_2H_5} \xrightarrow{\textit{benzoylation}} \underset{\substack{\\ CH_2NMe_2}}{\overset{CH_3}{\underset{|}{C}}}\!\!\big\langle\substack{O.COC_6H_5 \\ C_2H_5}$$

dimethylamino acetone   stovaine

It is widely used as a spinal anaesthesia.

(ii) Alypin *(amydricaine hydrochloride):* It is the dimethylamino derivative of the stovaine. It is prepared in following manner.

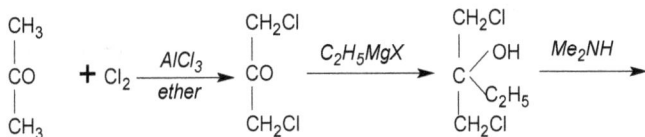

$$\begin{array}{c} CH_3 \\ | \\ CO \\ | \\ CH_3 \end{array} + Cl_2 \xrightarrow[\textit{ether}]{AlCl_3} \begin{array}{c} CH_2Cl \\ | \\ CO \\ | \\ CH_2Cl \end{array} \xrightarrow{C_2H_5MgX} \underset{\substack{\\ CH_2Cl}}{\overset{CH_2Cl}{\underset{|}{C}}}\!\!\big\langle\substack{OH \\ C_2H_5} \xrightarrow{Me_2NH}$$

$$CH_2NMe_2$$

(structure) C with OH, C_2H_5, CH_2NMe_2 $\xrightarrow{C_6H_5COCl}$ (structure) C with $OCOC_6H_5$, C_2H_5, CH_2NMe_2, CH_2NMe_2

alypin

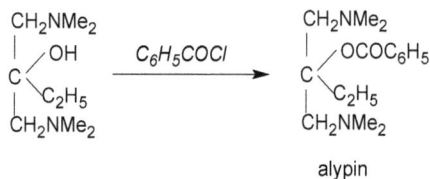

Alypin is a useful anaesthetic of the order of cocaine without most of its drawbacks; it produces rapid anaesthesia free from injurious effects on the heart and respiratory system.

4. Amide and Quinoline anaesthetics. The use of acetanilide and methyl acetanilide (*exalgin*) as antipyretic and analgesic drugs gave rise to the bee life that when the -$COCH_3$ group of these simple amides is elaborated to the -$COCH_2NR_2$, compounds of local anaesthetic properties are obtained. Some of the important examples are given below.

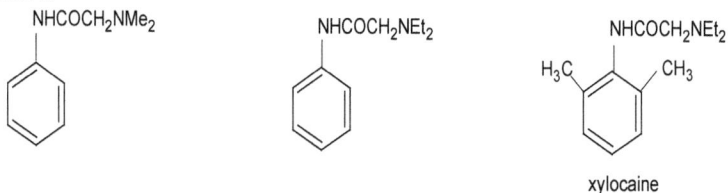

$NHCOCH_2NMe_2$ (benzene ring) $NHCOCH_2NEt_2$ (benzene ring) $NHCOCH_2NEt_2$, H_3C, CH_3 (benzene ring)

xylocaine

(i) Xylocaine. The most important example of this group is **xylocaine** (lidocaine), prepared in the following way.

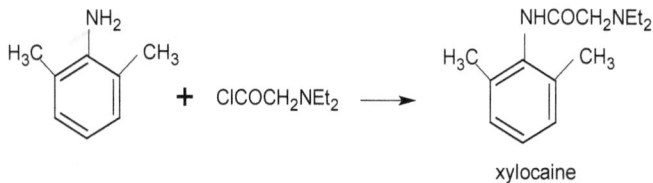

NH_2, H_3C, CH_3 (benzene ring) $+$ $ClCOCH_2NEt_2$ $\longrightarrow$ $NHCOCH_2NEt_2$, H_3C, CH_3 (benzene ring)

xylocaine

Xylocaine is used for dermal and surface anaesthesia. Since the side reactions are very few the xylocaine is regarded as one of the most satisfactory local anaesthetics and approaches to an ideal anaesthetic.

(ii) Nupercaine: However, amides of quinoline derivatives are more important and common local anaesthetics, the oldest and most important example is *nupercaine* (percaine, cinchocaine or dibucaine hydrochloride). Nupercaine is synthesized from isatin in the following manner.

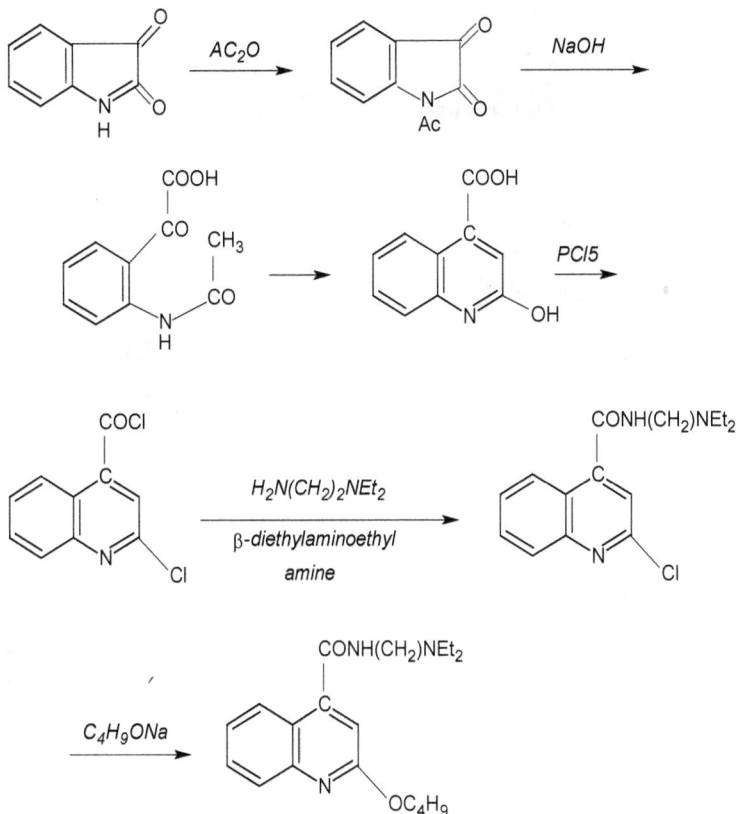

Nupercaine is used as a mucous surface anaesthetic. It is nearly 120 times more active than cocaine, and 60 times more active than procaine. Moreover, its action is of long duration.

5. Isoquinoline group: The only important example is dimethisoquin (quotane). It is prepared from phthalic anhydride in the following manner.

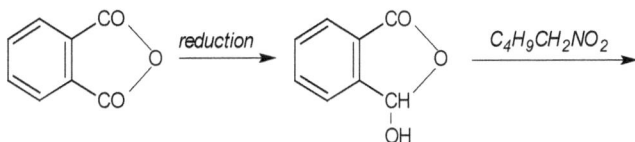

dimethisoquin

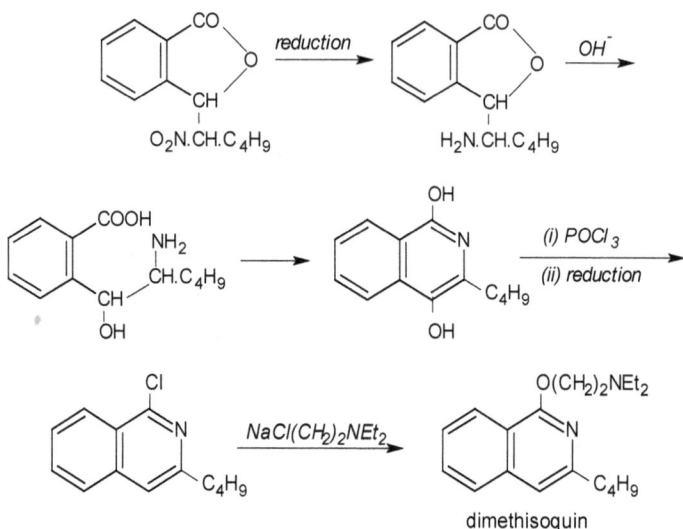

Dimethisoquin is 1000 times more active than cocaine, 100 times more active than procaine, and 10 times more active than dibucaine. It is used painful and irritating skin conditions.

6.11 Antipyretics and analgesics: Antipyretics are defined as those substances which reduce body temperature in fever, whereas analgesics are the compounds which relieve pain. The antipyretic compounds studied and synthesized in the earlier days were also found to possess the analgesic properties and hence these two types of drugs are generally studied together, but now-a-days some compounds are known which possess only the antipyretic property without any analgesic fall into the following groups.

(a) Tetrahydroquinoline derivatives: Quinine was known since a long time earlier as a valuable antipyretic with the additional action against malarial fever. Since the quinine molecule was found to possess quinoline nucleus, which itself also has some antipyretic activity, attempts were made to synthesize the quinoline derivatives having antipyretic properties. 6-Methoxyquinoline and its 1, 2, 3, 4-tetrahydro derivative *(thalline)* were the synthetic quinoline derivatives having some antipyretic action.

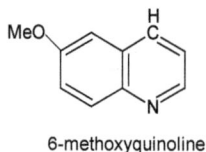

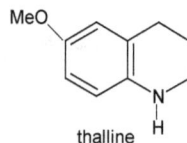

6-methoxyquinoline

thalline

Thalline has a strong antipyretic action but its toxic effects on the blood and kidneys restricted its use as drugs. Other synthetic quinoline derivatives which could not gain importance in chemotherapy, due to their toxicity towards the red blood corpuseles, are *kairine, kairoline A* and *kairoline B*.

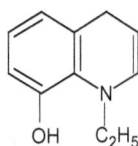

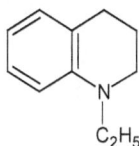

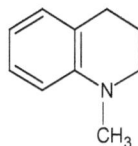

| kairine | kairoline A | kairoline B |

(b) Pyrazolone derivatives

(i) Antipyrine (phenazone or sedatine): Antipyrine is the outcome of the synthesis of quinoline derivatives. Knorr (1884) while synthesizing the quinoline derivatives (having antipyretic property). From phenyl hydrazine and acetoacetic ester actually obtained a pyrazolone derivative, the *antipyrine*. Antipyrine may now be synthesized by the condensation of phenyl hydrazine with acetoacetic ester or with diketen, followed by methylation which is best carried out by heating the product pyrazolone in anisole under pressure with methyl chloride or bromide.

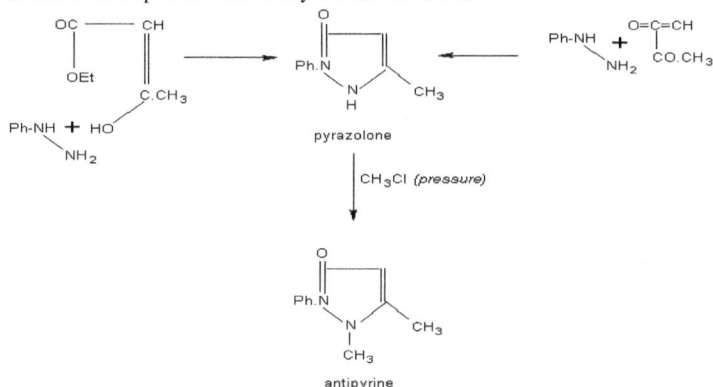

Antipyrine was used as an antipyretic against influenza, but it has action on the heart.

(ii)Aminopyrine, amidopyrine or pyramidon: It is the dimethylamino compound of antipyrine. It is prepared from antipyrine in the following manner.

aminopyrine, III

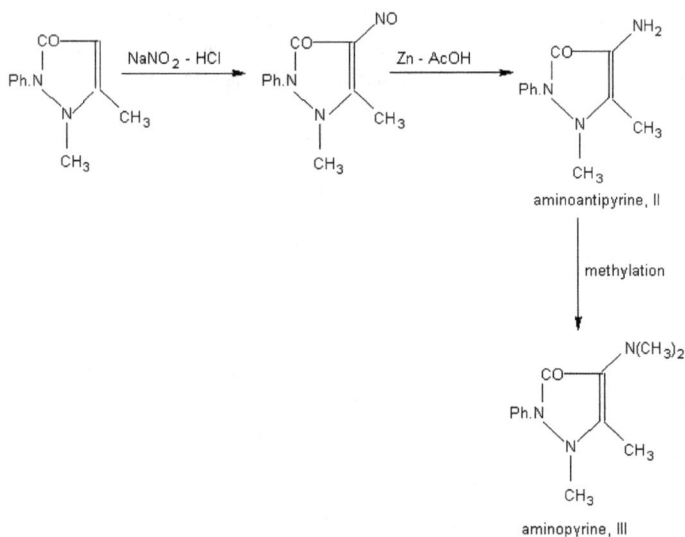

The amino antipyrine, II was methylated directly either by methyl iodide or dimethyl sulphate. On the other hand, it may also be methylated by dichloroacetic acid or dimethyl ether in the following way.

$$-NH_2 \xrightarrow{2\ ClCH_3COOH} -N(CH_2COOH)_2 \xrightarrow{-2CO_2} -N(CH_3)_2$$

$$-NH_2 + O(CH_3)_2 \xrightarrow[\text{high pressure}]{\text{catalyst}} -N(CH_3)_2 + H_2O$$

The nitroso antipyrine, I may also be directly converted into aminopyrine, III by the following two methods.

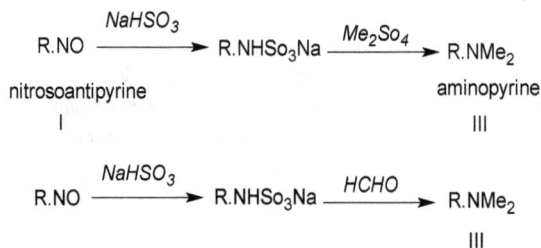

$$R.NO \xrightarrow{NaHSO_3} R.NHSo_3Na \xrightarrow{Me_2So_4} R.NMe_2$$

nitrosoantipyrine   aminopyrine

I   III

$$R.NO \xrightarrow{NaHSO_3} R.NHSo_3Na \xrightarrow{HCHO} R.NMe_2$$

III

Most of the commercial aminopyrine, III is prepared by the above two methods. Other methods, but of only theoretical interest, starting from antipyrine are as follows.

Aminopyrine has a very strong antipyretic and analgesic action without any injurious effect on the heart. It is nearly three times more powerful than antipyrine.

(iii) Melubrin and Novalgin: Novalgin the most widely used analgesic is the N-methyl derivative of melubrin. The latter is obtained by the interaction of amino antipyrine and sodium formaldehyde bisulphate, and *novalgin* is prepared by methylation of melubrin with dimethyl sulphate.

Melubrin is approximately equal in activity to aminopyrine, whereas novalgin has twice the activity of melubrin.

(iv) Butazolidine or igrapyrin (4-n-butyl-1, 2-diphenyl-3, 5-diketopyrazolidine): It is obtained by the condensation of *n*-butylmalonic ester and hydrazobenzene.

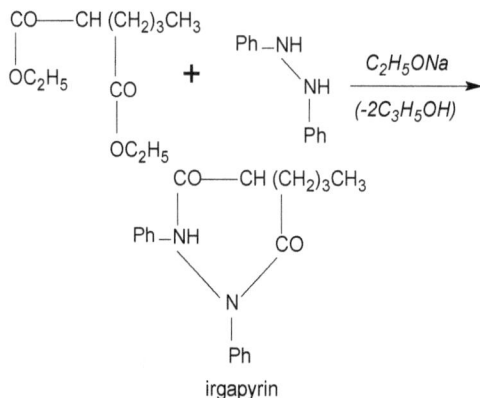

$$CO—CH(CH_2)_3CH_3$$

$$OC_2H_5 \quad CO \quad + \quad \begin{array}{c} Ph—NH \\ \diagdown \\ NH \\ | \\ Ph \end{array} \quad \xrightarrow[(-2C_3H_5OH)]{C_2H_5ONa}$$

$$OC_2H_5$$

$$CO—CH(CH_2)_3CH_3$$

$$Ph—NH \qquad CO$$

$$\diagdown \qquad \diagup$$

$$N$$

$$|$$

$$Ph$$

irgapyrin

It is an effective drug for gout, rheumatic fever, arthritis, etc. Its sodium derivatives are sufficiently stable, soluble and non-irritant to be used by intramuscular injection also in the treatment of rheumatism.

(c)Aniline derivatives: Aniline and its salts possess a strong antipyretic action, but their easy absorption and action on red blood corpuseles give rise to serious symptoms. Hence attempts were made to find out the aniline derivatives having antipyretic action without such serious after-effects; some of the important derivatives are mentioned below.

(i) Acetanilide or antifebrin: Acetanilide is prepared by the acetylation of aniline with glacial acetic acid in a special type of vessel at a temperature of 120-125 ^{0}C in the presence of $ZnCl_2$ as a condensing agent. Acetanilide is separated out by pouring the reaction mixture in cold water.

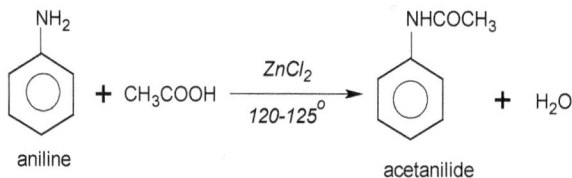

$$NH_2 \qquad\qquad\qquad NHCOCH_3$$

aniline + CH_3COOH $\xrightarrow[120\text{-}125^0]{ZnCl_2}$ acetanilide + H_2O

Acetanilide is the cheapest antipyretic. Since its antipyretic action is owing to the slow liberation of free aniline, it is not free from harmful effects of aniline and after some time symptoms of aniline poisoning may be observed. Due to its cheapness, acetanilide is used to adulterate other expensive drugs, such as phenacetin. Many others compounds analogous to acetanilide *viz.* benzanilide, salicylanilide, etc., were synthesized and tested for their antipyretic properties but none of them was found to be free from the after-effects.

(ii) Exalgin: Exalgin, the N-methyl- acetanilide, $C_6H_5N(CH_3)COCH_3$, is another important example of aniline derivative having antipyretic action with toxic effects.

(iii) Euphorin: The most important derivative of aniline is euphorin (phenylurethan), $C_6H_5NHCOOC_2H_5$. It is obtained by the interaction of aniline and chloroformic ester.

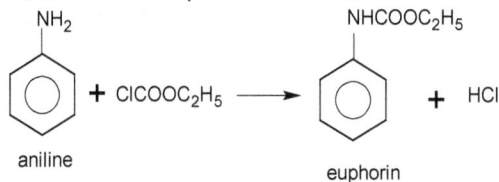

It is very much less toxic than aniline, and has a strong analgesic action. As far as antipyretic action is concerned it is of little importance.

(d) Derivatives of p-aminophenol: The oxidation of acetanilide in the body to *p*-aminophenol led to the introduction of *p*-aminophenol derivatives as antipyretics. Although, *p*-aminophenol, itself, also is far less toxic than aniline, it is not from the action on red blood corpuseles to be harmful in moderate doses. But on the other hand, the substituted *p*-amino phenols constitute the most important and widely used antipyretics of to-day. The important substituted *p*-amino phenols used as antipyretics are phenacetin, methacetin and portonal.

(i) Phenacetin: First of all phenacetin was prepared from *p*-nitrophenol a byproduct in the manufacture of *o*-nitrophenol, a dyestuff intermediate.

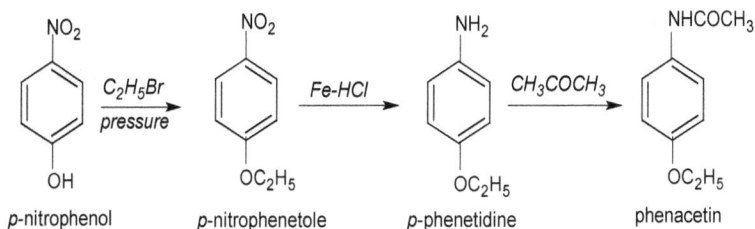

Phenetidine and hence phenacetin may also be prepared by the following two methods.

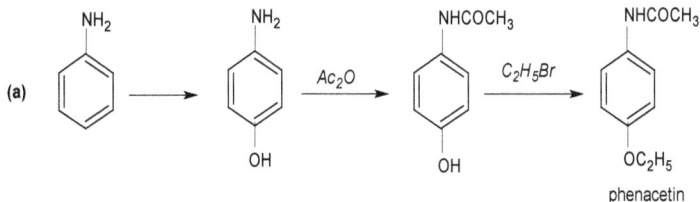

219

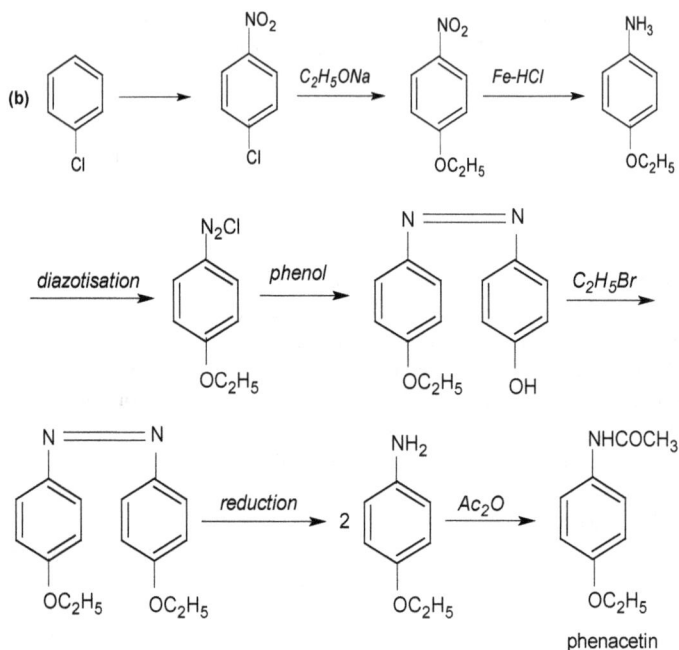

(b)

Phenacetin is most important antipyretic as well as analgesic. It is the least toxic.

The other two antipyretics of this group are *methacetin* and *pertonal*. Methacetin is more powerful antipyretic than phenacetin,

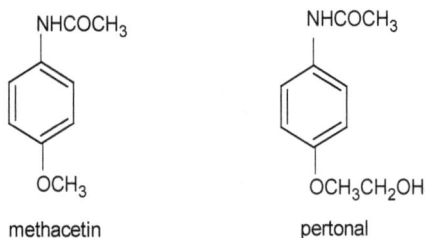

methacetin

pertonal

But its toxicity is greater. Pertonal is a moderate antipyretic with very low toxicity.

(e) Quinoline derivatives: The tetrahydroquinoline derivatives have already been described in the beginning of the topic. The only important quinoline derivative having antipyretic and analgesic properties is cinchophen, atophan or phenylcinchoninic acid.

(i) Cinchophen. It may be prepared by the following methods.

(a) p-amino benzaldehyde cynohydrin → cinchophen

(b)

sehif's base + enolic form of pyruvic acid → cinchophen

(c) isatin + Acetophenone, excess Liq. NH₃ → cinchophen

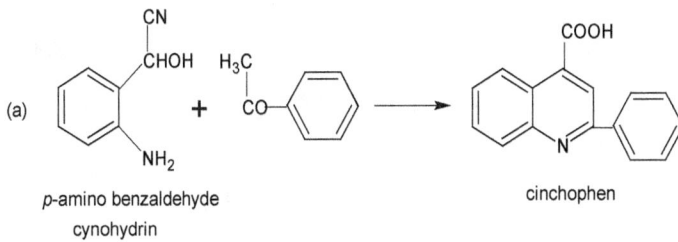

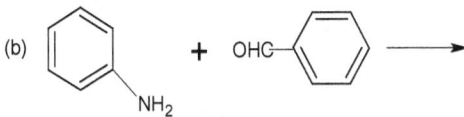

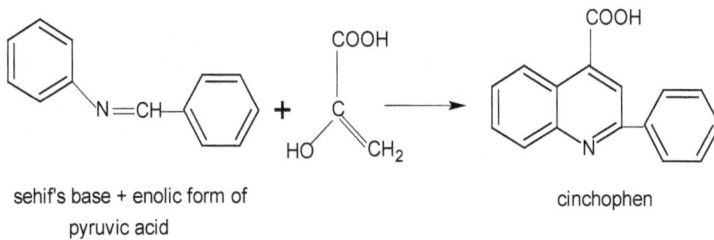

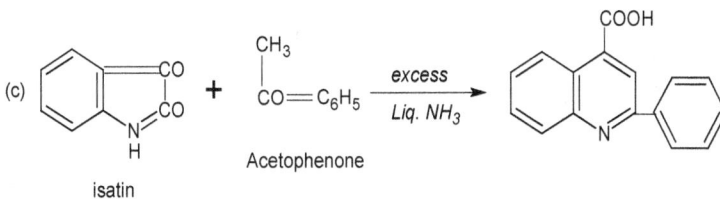

On commercial scale cinchophen is prepared by the method (c). Cinchophen is used as an antipyretic and analgesic for the treatment of gout and rheumatism conditions. But its use causes liver poisoning leading to acute jaundice.

(f) **Salicylic acid and its derivatives:** Salicin was the first compound of this series to have a medicinal value. It was obtained from the bark of willows (Leroux, 1823) and used as a substitute for quinine as a febrifuge. It is still obtained industrially from the natural source which has satisfies the current of course very little, demand. It required in large quantities it may readily be synthesized by the interaction of acetobromoglucose and salicyladehyde by hydrolysis and reduction.

$$\underset{\text{CHO}}{\text{C}_6H_4}\text{-OH} + Br.C_6H_7O_5(COCH_3)_4 \longrightarrow \underset{\text{CHO}}{\text{C}_6H_4}\text{-O.C}_6H_7O_5(COCH_3)_4$$

$$\downarrow H_2O$$

sallein $\underset{CH_2OH}{}$ .O.C$_6$H$_{11}$O$_6$ $\xleftarrow{\ Na\text{-}Hg/H_2O\ }$ beliein $\underset{CHO}{}$ .O.C$_6$H$_{11}$O$_6$

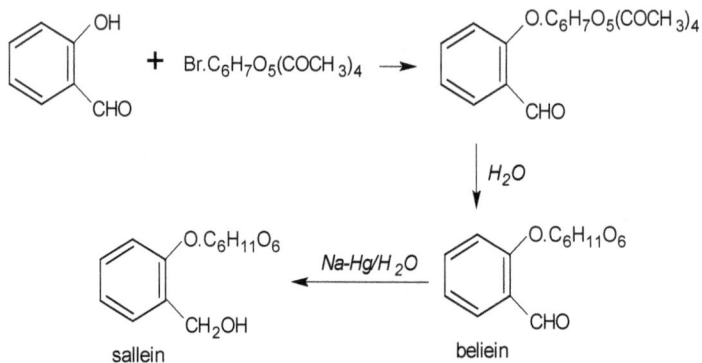

After sometime the structure of salicylic acid was established by Hofmann, and then Gerland in 1853 obtained the salicylic acid by the action of nitrous acid on anthranilic acid. But now a days most of the commercial salicylic acid is obtained by the Kolhe's synthesis in which the dry sodium phenate is heated with dry carbon dioxide under pressure at 140° to give sodium phenyl carbonate. The latter compound rearranges to the sodium salt of phenol-*o*-carboxylic acid which on hydrolysis by means of dilute hydrochloride acid gives salicylic acid.

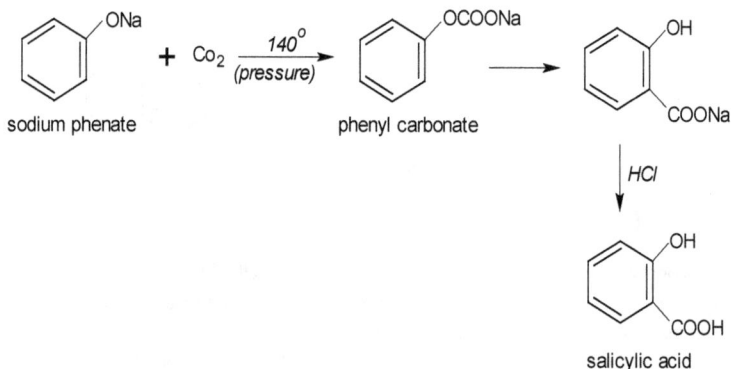

sodium phenate + CO_2 $\xrightarrow[\text{(pressure)}]{140^0}$ phenyl carbonate (OCOONa) $\longrightarrow$ (OH, COONa)

$$\downarrow HCl$$

salicylic acid (OH, COOH)

Salicylic acid has the property for controlling rheumatic pain and temperature but due to its bad effect of producing gastric disturbances it was not used medicinally. A large number of salicylic acid derivatives were prepared and tested for their antipyretic and analgesic action, but out of them only two gained more importance, *viz.* aspirin and salol.

(i) Aspirin (acetyl salicylic acid): Aspirin is obtained by the acetylation of salicylic acid the acetylation can be affected either by a mixture of acetic acid and acetic anhydride in the presence of sulphuric acid or by acetic anhydride also in the presence of pyridine.

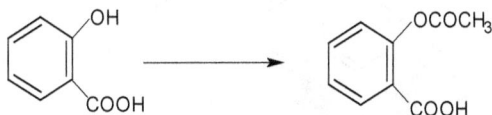

The acetylation by diketen $(CH_2=CO)_2$ has not proved industrially economic.

Aspirin has the property of salicylic acid against rheumatic fever, with the additional use against headaches and colds. Various soluble salts of aspirin have been prepared and examined, *viz.* calcium salt *(soluble aspirin or kalmopyrin).* Sodium salt *(tylmartin),* lithium salt *(hydroprin),* magnesium salt *(novacyl),* and urea salt *(diafar),* but none of them have attained widespread use as aspirin.

(ii) Salol (phenyl salicylate): Salol was introduced in medicine by Nenoki in 1886. It is prepared by the following two methods.

(a) By the heating salicylic acid at 160-240°C in the absence of air and distilling the water formed.

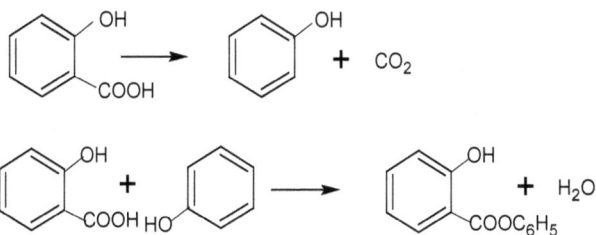

(b) By healing salicylic acid with phenol in presence of an acid chloride such as phosphorus oxychloride.

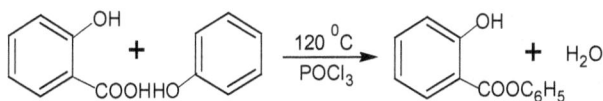

Salol is used as an antipyretic as well as internal and external antiseptic. In the system it pass unhydrolyzed in the stomach but hydrolyzed slowly in the intestine into two active salicylic acid and phenol; the hydrolysis takes place so gradually that the phenol exerts its antiseptic effect without any toxic effect administration of compound on the above basis is known as either salol principle or Nencki principle. All the drugs used on the salol principle may be classified under two heading; *viz,* true salol and partial salols.

True salols are those in both the compounds of the ester, i.e. acid as well as phenol or alcohol is active. The most important example of true salol is itself followed by β-naphthyl salicylate or naphthyl salol.

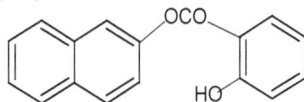

betol

Partial salols: Those ester in which either the acids or the hydroxylic fragment is active are known as partial salols. The best known example of the partial salols is methyl salicylate in which salicylic acid is the active moiety.

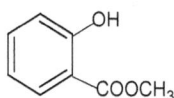

methyl salicylate

(g) Morphine and related compounds: Morphine is the most important alkaloid of the opium. Morphine and its methyl analogue, codeine are used as analgesics.

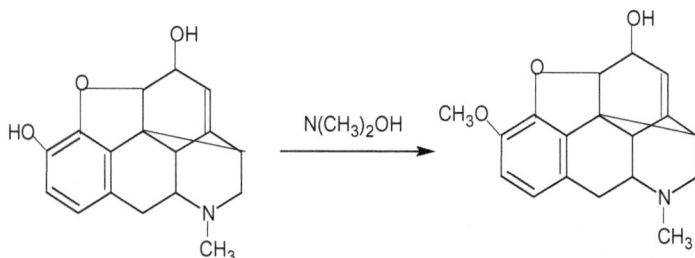

Among the various important morphine and codeine derivatives having the advantage over the compound the following compounds require attention.

(i) Codesine methyl bromide is less toxic than codeine.

(ii) Ethyl morphine is a stronger analogous with the prolonged action and is generally recommend against coughing. The hydrochloride of ethyl morphine is used therapy under the name of dionin.

(iii) Benzyl derivatives of morphine also resemble with the codeine in the action and hence, mering introduced its hydrochloride under the name of peronine.

6.12 Sedatives and hypnotics: sedatives and hypnotics both depress the central nervous system. A sedative functions by calming the nervous system and thus induces the natural sleep. On the other hand, a hypnotic indicates such a deep sleep that the patient cannot easily be wakened until effects of the drug wear. Hence, there is no clear but difference between a sedative and a hypnotic and they differ only in their degree of action moreover in some cases a small dose of the given drug produces sedation, where as a larger dose of the same compound induces hypnotic sleep. Furthermore, it must be noted that in some cases a compound exerts only one effect, e.g. simple bromides are good sedative and posses little or do hypnotic action, similarly certain powerful hypnotics. The commonly used sedatives or hypnotics fall in one or other of the following groups.

1. Alcohols.

2. Aldehydes, ketones and sulphones.

 3. Tranquilisers.

(a) Alcohols: Ethyl alcohol the most important member of the series, was used as a sedative and hypnotic for a long time but it continuous use leads to alcoholic and is replaced by ether. Among the halogenated alcohols trichloroethanol, tribromoethanol and chloretone have been used widely. The first two are prepared by the reduction of the chloral or bromal with aluminium isopropoxide and iso alcohol.

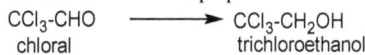

$$CCl_3\text{-CHO} \longrightarrow CCl_3\text{-CH}_2\text{OH}$$
chloral trichloroethanol

$$CCl_3\text{-CHO} \longrightarrow CBr_3\text{-CH}_2\text{OH}$$
chloral tribromoethanol

On other hand, chloretone and analogues bromine compound brometone are prepared by condensation of acetone with chloroform or bromoform.

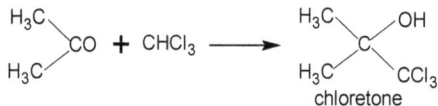

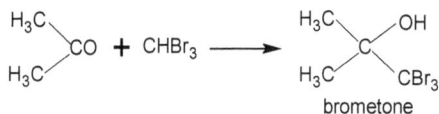

$$\underset{H_3C}{\overset{H_3C}{>}}CO + CHCl_3 \longrightarrow \underset{H_3C}{\overset{H_3C}{>}}C\underset{CCl_3}{\overset{OH}{<}}$$
chloretone

$$\underset{H_3C}{\overset{H_3C}{>}}CO + CHBr_3 \longrightarrow \underset{H_3C}{\overset{H_3C}{>}}C\underset{CBr_3}{\overset{OH}{<}}$$
brometone

(b) Aldehyeds, ketones and sulphones: Aliphatic aldehydes produce a powerful depression in the central nervous system, e.g., paraldehyde, is valuable hypnotics. Paraldehyde is often used in mental conditions, but it suffers from some disadvantages such as unpleasant taste, pungent odour and effect on mucous membranes.

Halogenation of aldehyde generally increases the hypnotic activity. For example, chloral hydrate, $CCl_3CH (OH)_2$ prepared by the action of chlorine on alcohol at ordinary temperature, was the first synthetic hypnotic used by Liebrien in 1869.

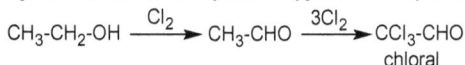

$$CH_3\text{-CH}_2\text{-OH} \xrightarrow{Cl_2} CH_3\text{-CHO} \xrightarrow{3Cl_2} CCl_3\text{-CHO}$$
chloral

In small doses (0.3-0.5 gm) it causes a sedative effect, but in large doses (2-3 gm) it induces a deep sleep. A few other halogenated aldehydes have been examined for sedative action, out of which only the butyl chloral hydrate gained importance. It is prepared by the chlorination of acetaldehyde.

$$CH_3CHO + Cl_2 \longrightarrow CH_2Cl.CHO + HCl$$

$$CH_3CHO + CH_2Cl.CHO \longrightarrow CH_3CH=CCl.CHO$$

$$CH_3CH=CCl.CHO \xrightarrow{Cl_2} CH_3CHCl.CCl_2.CHO$$

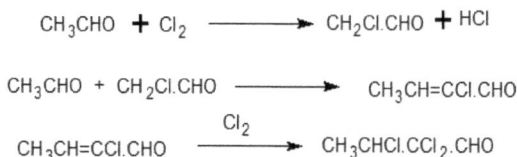

The hypnotic action of butyl chloral hydrate is less powerful than the chloral and its generally used for treating nuuralgic pain and migraine.

Most of the ketones have hypnotic properties e.g. diethyl ketone, $C_2H_5CO.C_2H_5$, is a strong hypnotic but could not be used longer because of its insolubility and unpleasant taste. Simple aromatic ketones, like benzophenone have a slight hypnotic action; but on the other hand, mixed aromatic and aliphatic ketones have more marked properties, e.g. acetophenone is a fairly strong hypnotic and used under the name of hypnotic. The introduction of ethyl group in the ketone further increases its hypnotic properties e.g. phenyl ethyl ketone, is more powerful hypnotic than acetophenone.

Sulphones are prepared by the condensation of a ketone with mercaptans, and substituent oxidation of the condensation product so formed.

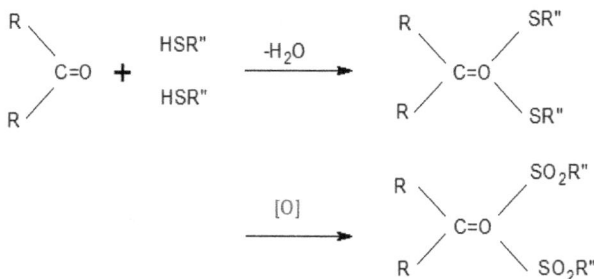

Sulphones

The hypnotic properties of sulphones were investigated by Baumann and Kasi in 1888, who observed that sulphonal has a strong hypnotic action on animals. Sulphonal is prepared from acetone and ethyl mercaptan according to the general method of their synthesis.

$$\begin{matrix} CH_3 \\ \\ CH_3 \end{matrix} C=O \quad + \quad \begin{matrix} HSC_2H_5 \\ \\ HSC_2H_5 \end{matrix} \quad \xrightarrow[HCl]{ZnCl_2} \quad \begin{matrix} CH_3 \\ \\ CH_3 \end{matrix} C \begin{matrix} SC_2H_5 \\ \\ SC_2H_5 \end{matrix}$$

$$\xrightarrow[HCl]{KMnO_4} \quad \begin{matrix} CH_3 \\ \\ CH_3 \end{matrix} C \begin{matrix} SO_2C_2H_5 \\ \\ SO_2C_2H_5 \end{matrix}$$

sulphonal

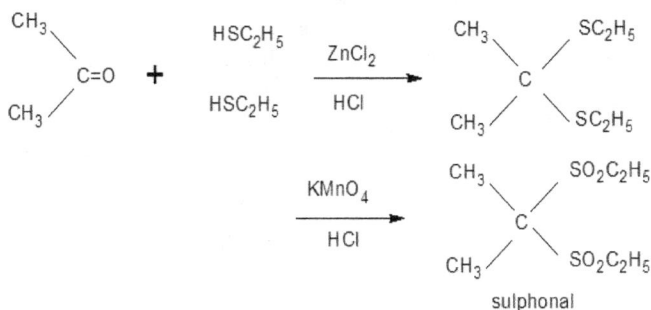

It has been observed that increase in the number of ethyl groups increase the hypnotic action, e.g., trional is more active than sulphonal, and tetronal is more active even than the trional.

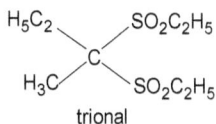

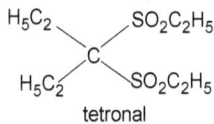

$$\begin{matrix} H_5C_2 \\ \\ H_3C \end{matrix} C \begin{matrix} SO_2C_2H_5 \\ \\ SO_2C_2H_5 \end{matrix}$$
trional

$$\begin{matrix} H_5C_2 \\ \\ H_5C_2 \end{matrix} C \begin{matrix} SO_2C_2H_5 \\ \\ SO_2C_2H_5 \end{matrix}$$
tetronal

On the other hand, decrease in the number of ethyl groups, decreases in the hypnotic's action of the sulphones.

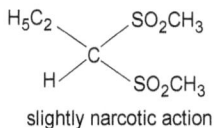

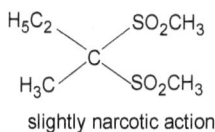

$$\begin{matrix} H_5C_2 \\ \\ H \end{matrix} C \begin{matrix} SO_2CH_3 \\ \\ SO_2CH_3 \end{matrix}$$
slightly narcotic action

$$\begin{matrix} H_5C_2 \\ \\ H_3C \end{matrix} C \begin{matrix} SO_2CH_3 \\ \\ SO_2CH_3 \end{matrix}$$
slightly narcotic action

Moreover, only those sulphones are found to be active in which the two SO_2R group are attached to the same carbon atom which in turn is having at least on alkyl group, e.g. ethylene diethyl sulphone I, and methylene diethylsulphone II are inactive, whereas ethylidenediethylsulphone, III has an action similar to that of sulphonal.

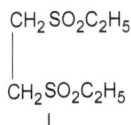

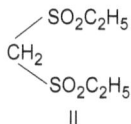

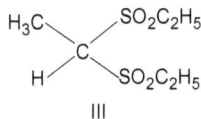

$$\begin{matrix} CH_2\,SO_2C_2H_5 \\ | \\ CH_2\,SO_2C_2H_5 \end{matrix}$$
I

$$CH_2 \begin{matrix} SO_2C_2H_5 \\ \\ SO_2C_2H_5 \end{matrix}$$
II

$$\begin{matrix} H_3C \\ \\ H \end{matrix} C \begin{matrix} SO_2C_2H_5 \\ \\ SO_2C_2H_5 \end{matrix}$$
III

Lastly, further increase in length or size of one or more of the sulphone group diminishes the hypnotic properties.

(c) Urethanes, Amides and Ureas: Urethane ($C_2H_5OCONH_2$), a well known anaesthetic for man, was introduced by Schmiedeberg in 1885 as a narcotic for dogs; but in human therapy it is rarely used as a narcotic. Although, urethanes itself is not an important narcotic for man. Its analogues have been successfully used as sedatives.

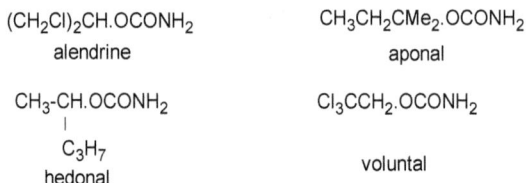

$(CH_2Cl)_2CH.OCONH_2$
alendrine

$CH_3CH_2CMe_2.OCONH_2$
aponal

$CH_3\text{-}CH.OCONH_2$
|
C_3H_7
hedonal

$Cl_3CCH_2.OCONH_2$
voluntal

Urethanes are obtained by the following two methods.

(a) By treating the corresponding alcohol with carbonyl chloride followed by reaction with ammonia in pyridine or quinoline.

$$ROH + COCl_2 \longrightarrow ROCOCl \xrightarrow{NH_3} ROCONH_2$$

(b) By treating the corresponding alcohol with carbamic acid. This method is cheaper and gives the purer products.

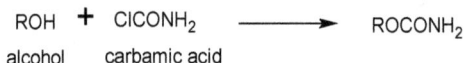

$$ROH + ClCONH_2 \longrightarrow ROCONH_2$$
alcohol carbamic acid

The amide and urea derivatives are not so powerful that they can produce hypnotic action and they are manly used as sedatives. The sedative amides are mainly of three types.

(i) Dialkylamines of aliphatic acid. The only important member is Falyl, α-bromo-N, N-diethylisovaleramide, $Me_2CHCHBr\ CONEt_2$.

(ii) Amide having free-$CONH_2$ group. The important members of group are described below.

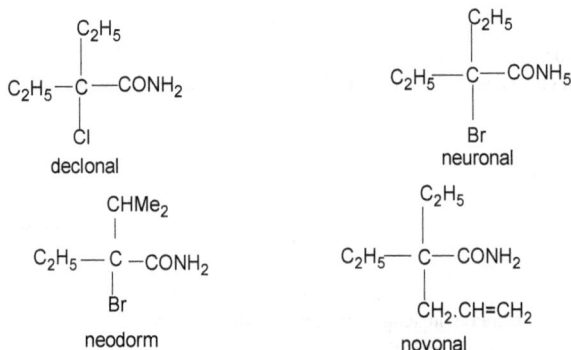

C_2H_5
|
$C_2H_5\text{—}C\text{—}CONH_2$
|
Cl
declonal

C_2H_5
|
$C_2H_5\text{—}C\text{—}CONH_5$
|
Br
neuronal

$CHMe_2$
|
$C_2H_5\text{—}C\text{—}CONH_2$
|
Br
neodorm

C_2H_5
|
$C_2H_5\text{—}C\text{—}CONH_2$
|
$CH_2.CH{=}CH_2$
novonal

These, amide except novonal, are prepared by heating the corresponding malonic acid derivative with thionyl chloride followed by liquid ammonia, e.g.

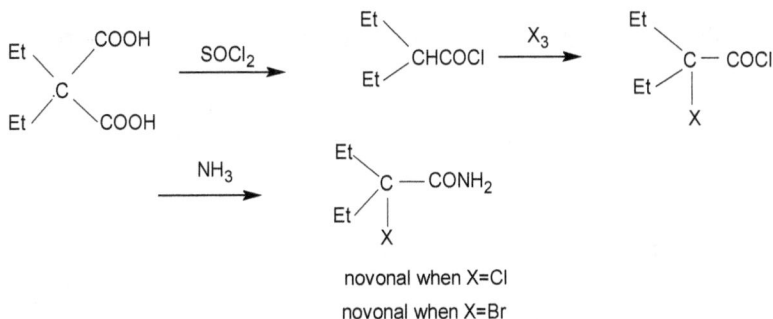

novonal when X=Cl
novonal when X=Br

Novonal, diethylacitamide is prepared from diethylcyano acetic ester, according to the following series or reactions.

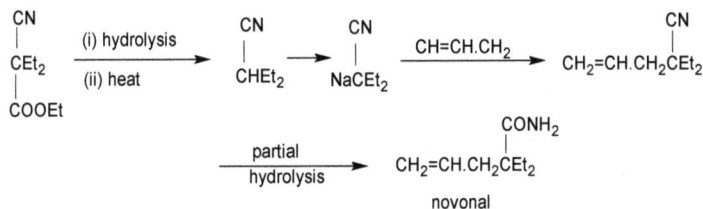

novonal

(iii) Cyclic amides Persedon, 3, 3-diethyl-2, 4-dioxotetrahydropuridine, is the most important member among cyclic amides used as a powerful hypnotic action. It is prepared from ethyl formate and diethylacetoacetic ester in the following manner.

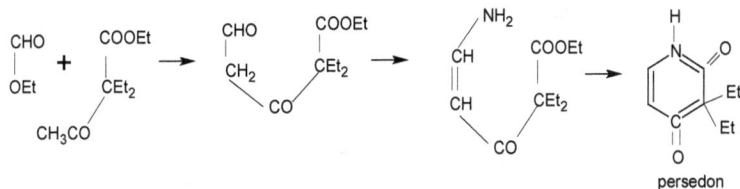

persedon

The dihydro and diallyl analogues of persedon are also used as hypnotics. The various alkyl substituted urea have been used as bromoisovaleryl bromide with mercurous isocyanate followed by treatment with ammonia.

$$\text{Me}_2\text{CH.CHBr.COBr} \xrightarrow{\text{HgCNO}} \text{Me}_2\text{CH.CHBrCO.NCO} \xrightarrow{\text{NH}_2} \text{Me}_2\text{CH.CHBr.CONHCONH}_2$$

bromural

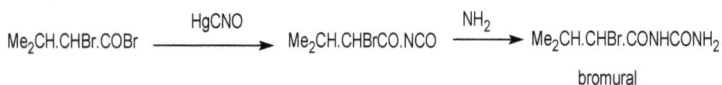

But commercially bromural is prepared by condensing the corresponding acid bromide with urea.

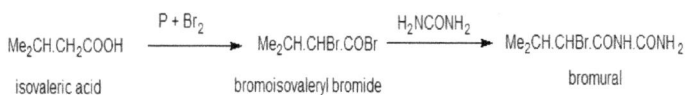

$$\text{Me}_2\text{CH.CH}_2\text{COOH} \xrightarrow{\text{P + Br}_2} \text{Me}_2\text{CH.CHBr.COBr} \xrightarrow{\text{H}_2\text{NCONH}_2} \text{Me}_2\text{CH.CHBr.CONH.CONH}_2$$

isovaleric acid   bromoisovaleryl bromide   bromural

(ii) Carbromal. Commecially it is Prepared by either of the following two method.

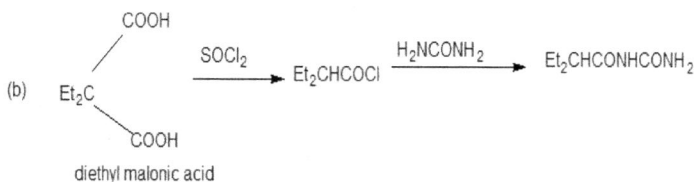

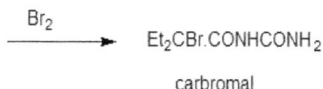

(a) $\text{Et}_2\text{CHCOOH} \xrightarrow{\text{P + Br}} \text{Et}_2\text{CBr.COBr} \xrightarrow{\text{H}_2\text{NCONH}_2} \text{Et}_2\text{CBr.CO.NHCONH}_2$

carbromal

(b)

$$\underset{\text{diethyl malonic acid}}{\text{Et}_2\text{C}\begin{cases} \text{COOH} \\ \text{COOH} \end{cases}} \xrightarrow{\text{SOCl}_2} \text{Et}_2\text{CHCOCl} \xrightarrow{\text{H}_2\text{NCONH}_2} \text{Et}_2\text{CHCONHCONH}_2$$

$$\xrightarrow{\text{Br}_2} \text{Et}_2\text{CBr.CONHCONH}_2$$

carbromal

(iii) Sedormid: It is prepared by the hydrolysis of the corresponding barbituric acid derivative with boiling dilute ammonia.

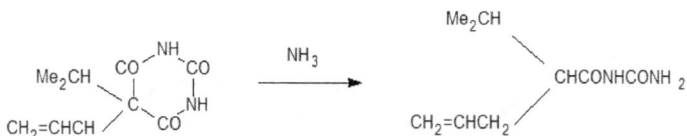

Sedormid

Since the elongation of the CONH-chain from simple amide to urea enhances the sedative action the elongation of the urea chain also increases the activity of the compound with low toxicity (Hill and Deguan). The acyl biurates are prepared by condensing a piperidourea, II, obained by the action of urea chloride on piperidine with diethyl acetyl isocyanate, I.

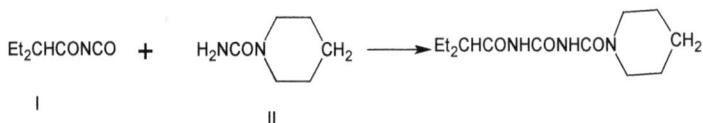

$$Et_2CHCONCO \quad + \quad H_2NCON{\Large\bigcirc}CH_2 \longrightarrow Et_2CHCONHCONHCON{\Large\bigcirc}CH_2$$

I

II

(d) Barbiturates: Veronal (barbitone) the diethyl derivative of barbituric acid was the first barbiturate introduced into therapy by Fischer and Merling, in 1903. It is prepared by the condensation of diethyl malonic ester with urea in the presence of sodium ethoxide.

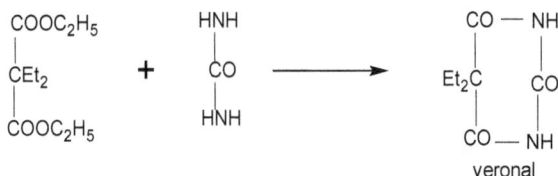

$$\begin{array}{ccc} COOC_2H_5 & HNH & CO — NH \\ | & | & | \quad\quad | \\ CEt_2 \quad\quad + & CO \longrightarrow & Et_2C \quad\quad CO \\ | & | & | \quad\quad | \\ COOC_2H_5 & HNH & CO — NH \\ & & \text{veronal} \end{array}$$

It is a powerful hypnotic and causes natural and dreamless sleep within 15-30 minutes. The barbitone has the drawback of having low therapeutic index (approximately 2) which means fatal overdoses is easily achieved. To overcome this defect attempts were made in changing the structural details of barbitone to find a compound with equal or greater hypnotic activity with the high therapeutic index. Of these the most commonly used are given below.

(i) Phenobarbitone (phenobarbital): It is prepared from benzyl cyanide by the following method.

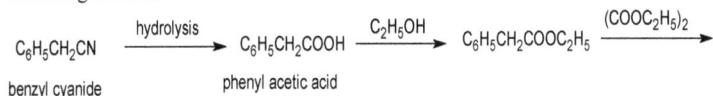

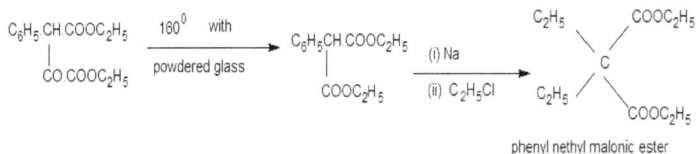

$$C_6H_5CH_2CN \xrightarrow{\text{hydrolysis}} C_6H_5CH_2COOH \xrightarrow{C_2H_5OH} C_6H_5CH_2COOC_2H_5 \xrightarrow{(COOC_2H_5)_2}$$

benzyl cyanide   phenyl acetic acid

$$C_6H_5.CH.COOC_2H_5 \xrightarrow[\text{powdered glass}]{160^0 \quad \text{with}} C_6H_5CH.COOC_2H_5 \xrightarrow[\text{(ii) } C_2H_5Cl]{\text{(i) Na}} \begin{array}{c} C_2H_5 \quad\quad COOC_2H_5 \\ \diagdown \quad\diagup \\ C \\ \diagup \quad\diagdown \\ C_2H_5 \quad\quad COOC_2H_5 \end{array}$$
$$| \quad\quad\quad\quad\quad\quad\quad\quad\quad\quad\quad\quad | $$
$$CO.COOC_2H_5 \quad\quad\quad\quad\quad\quad\quad COOC_2H_5$$

phenyl nethyl malonic ester

$$\xrightarrow[\text{C}_2\text{H}_5\text{ONa}]{\text{NH}_2\text{CONH}_2}$$

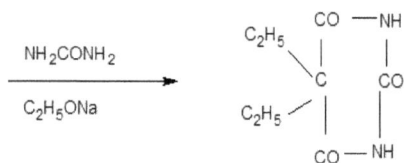

It is used both as a sedative and hypnotic.

(ii) Methylphenobarbitone: It is the N-methyl derivative of phenobarbitone and prepared from phenyl ethyl malonic ester and dicyndiamide in the following manner.

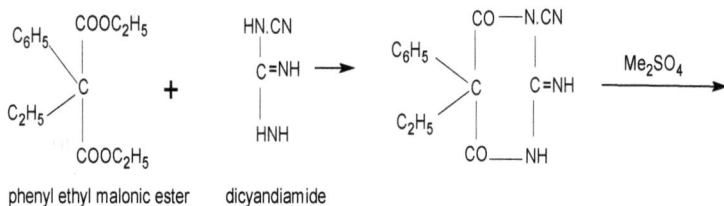

phenyl ethyl malonic ester   dicyandiamide

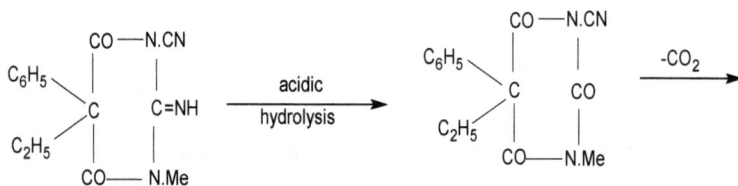

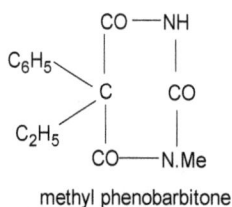

methyl phenobarbitone

(iii)   Cyclobarbitone: Cyclobarbitone is prepared from cyclohexanone and cyanoethyl acetate.

cyclobarbitone

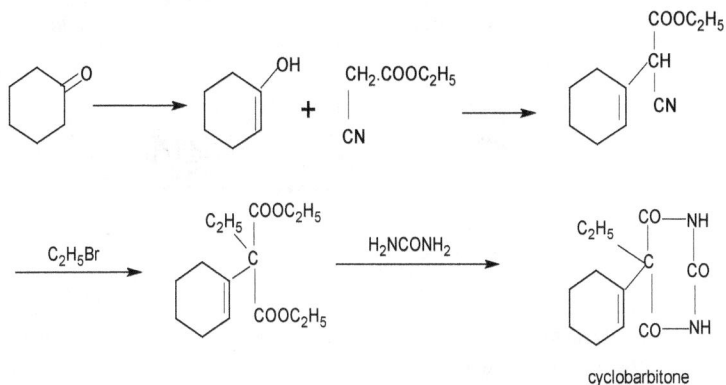

(iv) Hexobarbitone (evipan): It is obtained from cyclohexenyl methyl malonic ester and dicyanamide.

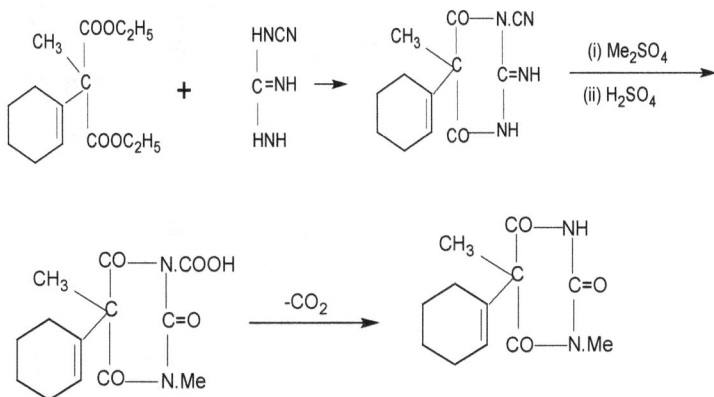

(v) Pentobarbitone: It is prepared by the condensation of ethyl (1-methyl) butyl 1 malonic ester with urea.

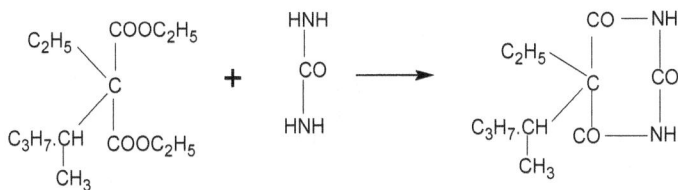

(vi) Dial(allobarbitone): It is prepared by the condensation of barbituric acid with allyl bromide in the presence of sodium ethoxide.

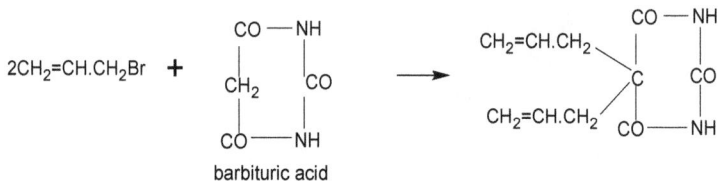

$$2CH_2=CH.CH_2Br \quad + \qquad \underset{\text{barbituric acid}}{\begin{array}{c} CO-NH \\ | \qquad | \\ CH_2 \quad CO \\ | \qquad | \\ CO-NH \end{array}} \qquad \longrightarrow \qquad \begin{array}{c} CH_2=CH.CH_2 \\ \diagdown \\ \qquad\qquad C \\ \diagup \\ CH_2=CH.CH_2 \end{array} \begin{array}{c} CO-NH \\ | \qquad | \\ CO \\ | \\ CO-NH \end{array}$$

(vii) Quinalbarbitone: It is obtained by the condensation of monoallyl barbituric acid with 1-methyl-butyl bromide.

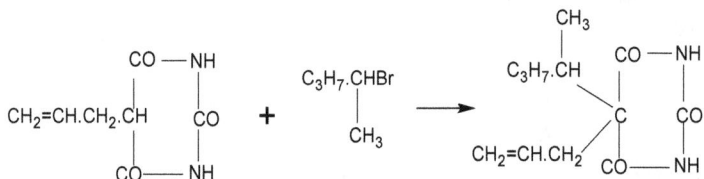

$$\underset{}{\begin{array}{c} CO-NH \\ | \qquad | \\ CH_2=CH.CH_2.CH \quad CO \\ | \qquad | \\ CO-NH \end{array}} \quad + \quad \begin{array}{c} C_3H_7.CHBr \\ | \\ CH_3 \end{array} \quad \longrightarrow \quad \begin{array}{c} CH_3 \\ | \\ C_3H_7.CH \\ \diagdown \\ \qquad C \\ \diagup \\ CH_2=CH.CH_2 \end{array} \begin{array}{c} CO-NH \\ | \qquad | \\ CO \\ | \\ CO-NH \end{array}$$

(viii) Thiobaebitone: It is sulphur analogue to the barbitone. and prepared in the same manner using thiourea as barbitone.

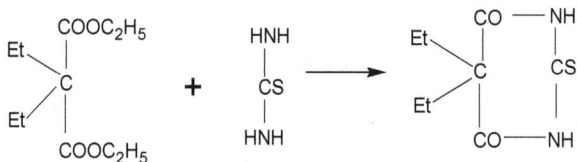

$$\begin{array}{c} Et \\ \diagdown \quad COOC_2H_5 \\ \qquad C \\ \diagup \quad | \\ Et \quad COOC_2H_5 \end{array} \quad + \quad \begin{array}{c} HNH \\ | \\ CS \\ | \\ HNH \end{array} \quad \longrightarrow \quad \begin{array}{c} Et \\ \diagdown \qquad CO-NH \\ \qquad C \qquad | \\ \diagup \qquad CS \\ Et \qquad | \\ \qquad CO-NH \end{array}$$

6.13 Tranquillisers: Although tranquillisers are primarily used for the treatment of symptoms in mental disease, their general influence is to free the mind from disturbance or passion and thus calm the mind i.e., they produce sedation without sleep. The entire tranquilliser can be grouped in any of the following four headings.
(a) Reserpine and its analogues
(b) The diphenyl diol family
(c) The diphenylmethane compounds
(d) Phenothiazine compounds

Reserpine and its analogues: Reserpine, an alkaloid from rauwolfia so., and its various naturally occurring analogues, e.g., deserpidine (11- desmethoxy reserpine) and rescinnamine (trimethoxy cinnamoylmethylreserpine) , have been used as teanquillisers.

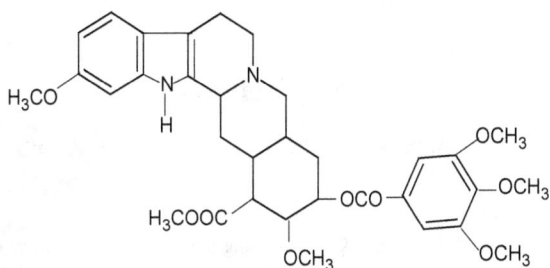

reserpine

Reserpine has been used in treating anxiety, tension states, insomnia, and hypertention. The side effects produced by the use of resepine include lethargy, nasal congestion, nausea, vomiting and sometimes serious depressions.

Miller and Weinberg in 1956 observed that even the simple tertiary amines having the trimethoxy benzoyl group shows the reserpine activity, e.g. the (3-diethylaminopropyl) ester of 3,4,5-trimethoxybenzoic acid is having one-third the activity of reserpine. It is prepared by condensing 3, 4, 5- trimethoxy benzoyl chloride with 3-diethylamino propanol.

7. Surfactants

7.1 Introduction

Surfactants and water have been reported as being among the major causes of hand eczema due to occupational exposure. In many work situations it is difficult to avoid wet work, constant hand washing and use of surfactants. To prevent hand eczema it is important to identify the risk factors. In the present study one of the most common classes of chemicals used in such situations i.e. in detergents, ethoxylated non-ionic surfactants, was investigated. Their ability to be oxidized when exposed to air and form oxidation products with allergenic properties was investigated. An allergenic activity due to a change in their composition can contribute to and aggravate occupational contact dermatitis.

7.2 Surfactants: chemistry and property

Chemical composition of surfactants, the name amphiphile is sometimes used synonymously with surfactant. The term relates to the fact that all surfactant molecules consist of at least two parts, one part which is soluble in a specific liquid (the lyophilic part) and one which is insoluble (the lyophobic part). When the liquid is water one usually talks about the hydrophilic and hydrophobic parts, respectively. The hydrophilic part is often referred to as the polar head group and the hydrophobic part as the tail. The primary classification of surfactants as cationic, anionic, non ionic and zwitterionic (amphoteric) is made on the basis of the charge of the hydrophilic group, which is either ionic or nonionic. The hydrophobic group is normally a hydrocarbon chain and the majority of these are linear to meet the demands for biodegradability.

The word surfactant is an abbreviation for surface active agent. A surfactant is characterized by its tendency to adsorbent surfaces and interfaces. Examples of interfaces involving a liquid phase include suspension (solid-liquid), emulsion (liquid-liquid) and foam (liquid-vapour). In many formulated products several types of interfaces are present at the same time. Another general and fundamental property of surface active agents is that monomers in solutions tend to form aggregates, called micelles. Micelles form already at very low surfactant concentrations in water. The concentration at which micelles start to form is called critical micelle concentration (CMC). Micelle formation, or micellization, can be viewed as an alternative mechanism to adsorption at the interfaces for removing hydrophobic groups from contact with the water, thereby reducing the free energy of the system. It is an important phenomenon since surfactant molecules behave very differently depending on whether they are present in micelles or as free monomers. The micelles behave as large molecules and influence the solubility of organic hydrocarbons and oils in aqueous solution and also influence the viscosity. The size of the micelle is measured by the aggregation number which is the number of surfactant molecules associated with a micelle. Only surfactant monomers contribute to surface and interfacial tension lowering. Wetting and foaming are governed by the concentration of free monomers in solution. The micelles can be seen as a reservoir for surfactant monomers. At higher concentrations of the surfactant other aggregates are formed. Different phase structures give very different physio chemical properties. To understand the physicochemical behaviour of a surface active compound over a concentration range at different temperatures, phase diagrams of the surfactants in water are constructed. Depending on the temperature there is a different solubility region of the surfactant-water system, shown in the phase diagram as a homogenous or a single phase system and heterogeneous systems of two or more phases.

7.2.1 Amphiphiles

The word amphiphile was coined by Paul Winsor 50 years ago. It comes from two Greek roots. First the prefix *amphi* which means "double", "from both sides", "around", as in amphitheater or amphibian. Then the root *philos* which expresses friendship or affinity, as in "philanthropist" (the friend of man), "hydrophilic" (compatible with water), or "philosopher" (the friend of wisdom or science).

An amphiphilic substance exhibits a double affinity, which can be defined from the physico-chemical point of view as a polar-apolar duality. A typical amphiphilic molecule consists of two parts: on the one hand a polar group which contents hetero atoms such as O, S, P, or N, included in functional groups such as alcohol, thiol, ether, ester, acid, sulfate, sulfonate, phosphate, amine, amide etc… On the other hand, an essentially apolar group which is in general an hydrocarbon chain of the alkyl or alkyl benzene type, sometimes with halogen atoms and even a few non ionized oxygen atoms.

The polar portion exhibits an strong affinity for polar solvents, particularly water, and it is often called hydrophilic part or hydrophile. The apolar part is called hydrophobe or lipophile, from Greek roots *phobos* (fear) and *lipos* (grease). The following formula shows an amphiphilic molecule which is commonly used in shapoos (sodium dodecyl sulfate).

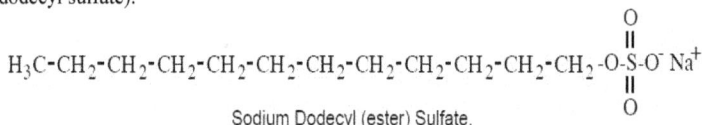

$$H_3C-CH_2-CH_2-CH_2-CH_2-CH_2-CH_2-CH_2-CH_2-CH_2-CH_2-CH_2-O-\overset{\displaystyle O}{\underset{\displaystyle O}{\overset{\|}{\underset{\|}{S}}}}-O^-\ Na^+$$

Sodium Dodecyl (ester) Sulfate.

7.2.2 Tension lowering agent Vs surfactant

Because of its dual affinity, an amphiphilic molecule does not feel "at ease" in any solvent, be it polar or non polar, since there is always one of the groups which "does not like" the solvent environment. This is why amphiphilic molecules exhibit a very strong tendency to migrate to interfaces or surfaces and to orientate so that the polar group lies in water and the apolar group is placed out of it, and eventually in oil.

In the following the word *surface* will be used to designate the limit between a condensed phase and a gas phase, whereas the term *interface* will be used for the boundary between two condensed phases. This distinction is handy though not necessary, and the two words are often used indifferently particularly in American terminology. In English the term surfactant (short for *surface-active-agent*) designates a substance which exhibits some superficial o interfacial activity. It is worth remarking that all amhiphiles do not display such activity; in effect, only the amphiphiles with more or less equilibrated hydrophilic and lipophilic tendencies are likely to migrate to the surface or interface. It does not happen if the amphiphilic molecule is too hydrophilic or too hydrophobic, in which case it stays in one of the phases.

In other languages such as French, German or Spanish the word "surfactant" does not exist, and the actual term used to describe these substances is based on their properties to lower the surface or interface tension, e.g. *tensioactif* (French), *tenside* (German), *tensioactivo* (Spanish). This would imply that surface activity is strictly equivalent to tension lowering, which is not absolutely general, although it is true in many cases.

Amphiphiles exhibit other properties than tension lowering and this is why they are often labeled according to their main use such as: *soap, detergent, wetting agent, dispersant, emulsifier, foaming agent, bactericide, corrosion inhibitor, antistatic agent,* etc. In some cases they are konwn from the name of the structure they are able to build, i.e. *membrane, micro emulsion, liquid crystal, liposome, vesicle* or *gel.*

7.3 Classification of surfactants
From the commercial point of view surfactants are often classified according to their use. However, this is not very useful because many surfactants have several uses, and confusions may

arise from that. The most accepted and scientifically sound classification of surfactants is based on their dissociation in water. The figures in page 4 show a few typical examples of each class. **(a) Anionic Surfactants** are dissociated in water in an amphiphilic anion, and a cation, which is in general an alkaline metal (Na^+, K^+) or a quaternary ammonium. They are the most commonly used surfactants. They include alkyl benzene sulfonates (detergents), (fatty acid) soaps, lauryl sulfate (foaming agent), di-alkyl sulfosuccinate (wetting agent), lignosulfonates (dispersants) etc. Anionic surfactants account for about 50 % of the world production. Anionic surfactants have the following general property;

- The hydrophilic part of the molecule has a negative charge – often found as sodium salts
- Common general purpose cleaners
- Often alkaline in solution

(b) Nonionic Surfactants come as a close second with about 45% of the overall industrial production. They do not ionize in aqueous solution, because their hydrophilic group is of a non dissociable type, such as alcohol, phenol, ether, ester, or amide. A large proportion of these nonionic surfactants are made hydrophilic by the presence of a polyethylene glycol chain, obtained by the poly condensation of ethylene oxide. They are called poly ethoxylated nonionics. In the past decade glucoside (sugar based) head groups, have been introduced in the market, because of their low toxicity. As far as the lipophilic group is concerned, it is often of the alkyl or alkyl benzene type, the former coming from fatty acids of natural origin. The poly condensation of propylene oxide produces a polyether which (in opposition to polyethylene oxide) is slightly hydrophobic. This polyether chain is used as the lipophilic group in the so-called poly EO poly PO block copolymers, which are most often included in a different class, e.g. polymeric surfactants, to be dealt with later, it possess following general properties;

- Have no charge on the hydrophilic part of the molecule – solubility is conferred by
  Hydrogen bonding
- milder cleaning action
- rarely used on their own

(c) Cationic Surfactants are dissociated in water into an amphiphilic cation and an anion, most often of the halogen type. A very large proportion of this class corresponds to nitrogen compounds such as fatty amine salts and quaternary ammoniums, with one or several long chain of the alkyl type, often coming from natural fatty acids. These surfactants are in general more expensive than anionics, because of a high pressure hydrogenation reaction to be carried out during their synthesis. As a consequence, they are only used in two cases in which there is no cheaper substitute, i.e. (1) as bactericide, (2) as positively charged substance which is able to absorb on negatively charged substrates to produce antistatic and hydrophobant effect, often of great commercial importance such as in corrosion inhibition. When a single surfactant molecule exhibit both anionic and cationic dissociations it is called *amphoteric* or *zwitterionic*. This is the case of synthetic products like betaines or sulfobetaines and natural substances such as amino acids and phospholipids.

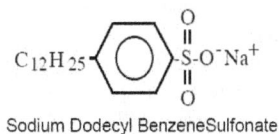

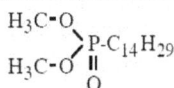

$C_{12}H_{25}$—⬡—$\overset{\overset{O}{\|}}{\underset{\underset{O}{\|}}{S}}$-O⁻Na⁺

Sodium Dodecyl BenzeneSulfonate

Abietic Acid

Dimethyl Ether of
Tetradecyl Phosphonic

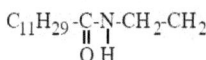

$C_{11}H_{29}$-$\overset{\overset{O}{\|}}{C}$-$\overset{\underset{H}{|}}{N}$-$CH_2$-$CH_2$

Lauryl Mono-Ethanol

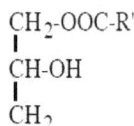

C_8H_{17}—⬡—O$\left[CH_2\text{-}CH_2\text{-}O\right]_n$H

Polyethoxylated Octyl Phenol

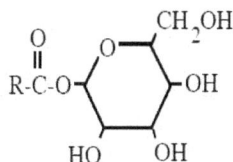

CH_2-OOC-R'
|
CH-OH
|
CH_2

Glycerol Diester (diglyceride)

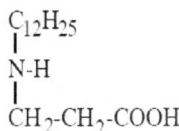

Sorbitan Monoester

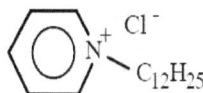

$C_{12}H_{25}$
|
N-H
|
CH_2-CH_2-COOH

Dodecyl Betaine

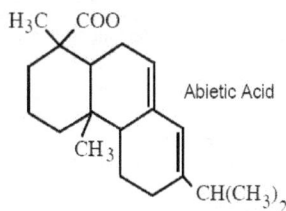

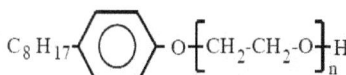

N-Dodecyl Piridinium Chloride

A few commonly used surfactants

Some amphoteric surfactants are insensitive to pH, whereas others are cationic at low pH and anionic at high pH, with an amphoteric behavior at intermediate pH.

Amphoteric surfactants are generally quite expensive, and consequently, their use is limited to very special applications such as cosmetics where their high biological compatibility and low toxicity is of primary importance.

The past two decades have seen the introduction of a new class of surface active substance, so-called polymeric surfactants or surface active polymers, which result from the association of one or several macromolecular structures exhibiting hydrophilic and lipophilic characters, either as separated blocks or as grafts. They are now very commonly used in formulating products as different as cosmetics, paints, foodstuffs, and petroleum production additives. It possess following general property;

- the hydrophilic part of the molecule has a positive charge
- Amphoteric surfactants have both negative and positive charged regions
- not generally used as cleaning agents
- can act as biocides

7. 4 Raw materials

Many kinds of surfactant structures are today available on the market and their price range from 1 $/lb to 20 times more. The raw materials are extremely varied and come from diverse origins, with a transformation ranging from a simple hydrolysis to multistep high pressure synthesis processes. With the single exception of rosin and tall oils the surfactant raw material market does not depends significantly on the surfactant manufacturing business. A consequence of this is that raw material costs can vary considerably because of factors external to the surfactant business. This volatile situation has produced changes and altered competitive margins in the surfactant industry.

For the sake of simplicity the raw materials for surfactant manufacturing are classified according to their origin (natural or synthesized from a petroleum cut). The following paragraphs mostly deal with the lipophilic group, since it is where the variety comes from. In effect, with the exception of ethylene and propylene oxides, the raw materials used in the hydrophilic groups (nitrogen, oxygen, sulfur and phosphorus compounds) are chemicals whose production is unrelated with the surfactant business.

This classification also takes into account the chronology of events.

7.4.1 Natural oil and fats : Triglycerides

Most oils and fats from animal or vegetal origin are *triglycerides,* i.e., triesters of glycerol and fatty acids, as for instance the structure indicated in the following formula.

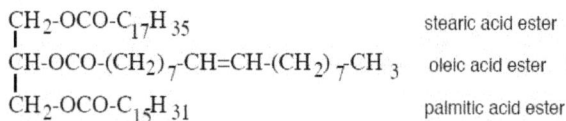

$$CH_2\text{-}OCO\text{-}C_{17}H_{35} \qquad \text{stearic acid ester}$$
$$CH\text{-}OCO\text{-}(CH_2)_7\text{-}CH=CH\text{-}(CH_2)_7CH_3 \qquad \text{oleic acid ester}$$
$$CH_2\text{-}OCO\text{-}C_{15}H_{31} \qquad \text{palmitic acid ester}$$

2-Oleyl palmytyl stearine

In some cases esterification is uncomplete, leading to mono and diglycerides. Some natural products include poly alcohols which are more complex than glycerol as for instance in C-5 and C-6 mono-sugar compounds. In all cases, the hydrolysis reaction allows the separation of the polyalcohol from the fatty acids.

Natural triglycerides contain the five most common fatty acids in various proportions: palmitic acid (symbolized as C16:0, i.e. 16 carbon atoms, no double bound) and the 4 main acids containing 18 carbon atoms: stearic (C18:0), oleic (C18:1), linoleic (C18:2) and linolenic (C18:3), with 0, 1, 2 and 3 double bounds, respectively.

The IUPAC (International Union for Pure and Applied Chemistry) nomenclature of acids starts with the name of the hydrocarbon and follows with suffix "-oic".

Alkane in C12: DODECANE   C12:0 dodecanoic acid
Alkane in C16: HEXADECANE   C16:0 hexadecanoic acid

When there is one (or more) double bond, the location is indicated in the formula:

Alkene in C18: OCTADECENE   C18:1 9-Octadecenoic acid
Diene in C18: OCTADECADIENE   C18:2 9,12-Octadecadienoic acid

In fact this nomenclature is rather cumbersome and in most cases the common names, which come from the triglyceride natural origin, are used instead. Butyric acid (C04:0) is found in butter, caproic (C06:0), caprilic (C08:1) etc., capric (C10:0) acids is found in milk, particularly from goats (*Capra* in Latin). Acid C16:0 has two common names coming from different origins: palmitic because it is one of the principal component of palm oil, and cetylic because it is also found in the liver oil of cetaceans such aswhales.C18:1 acids, mostly the 9-octadecenoic or oleic acid, are encountered in large proportions in most animal and vegetable oils and fats. A high proportion of C18:2 (linoleic) and C18:3(linolenic) acids are found in low viscosity vegetable oils such as corn, peanut, linseed, soya, and sunflower oils, in which a lower viscosity indicate a higher amount of double bounds in the acids. The next table indicates the proportions of different acids in most common natural oils and fats.

It is worth noting that natural oils and fats contain an even number of carbon atoms, and that they are linear with the acid group at one end. Natural oils exhibit an uncommon conformation, i.e., most of the C=C unsaturation are of the *cis* type, and in poly unsaturated chains the double bonds are not conjugated, whereas the *trans* conformation and the double bond conjugation are more stable from the thermodynamic point of view.

Fatty acids in the C_{12}-C_{18} range, particularly those from natural origin, are quite important in the manufacture of soaps and personal care specialties, because they carry alipophilic group which is completely biocompatible and well adapted to the preparation of surfactants for cosmetics, pharmaceuticals or foodstuffs.

OIL / FAT → ACID ↓		COCONUT	ALMOND	PEANUT	SOYA	OLIVE	CORN	PALM	PORK FAT	BEEF FAT	BUTTER
Caprilic	C08:0	07	04	-	-	-	-	-	-	-	01
Capric	C10:0	08	04	-	-	-	-	-	-	-	03
Lauric	C12:0	48	50	-	-	-	-	-	-	-	04
Myristic	C14:0	17	16	-	-	-	-	01	01	02	12
Palmitic	C16:0	09	08	11	11	14	12	46	26	35	29
Stearic	C18:0	02	02	03	04	03	02	04	11	16	11
Oleic	C18:1	06	12	46	25	68	27	38	49	44	25
Linoleic	C18:2	03	03	31	59	13	57	10	12	02	02
Linolenic	C18:3	-	-	02	08	-	01	-	01	-	-

Table:1 Fatty acid composition (%) of some Triglycerides

7. 4. 2 Other natural substances

(a) Wood oils

Some trees like pine and other conifer species contain esters of other carboxylic acids and glycerol (or other alcohols). They are called *rosin* oils and *tall* oils. It is worth noting that *tall* is not related with tallow, but with pine (in Swedish). During the wood digestion to make pulp, most esters are hydrolyzed and the acids are released. In a typical conifer wood digestion, fatty acids accounts for about 50%, while other acids are more complex substances such as abietic acid and its derivatives.

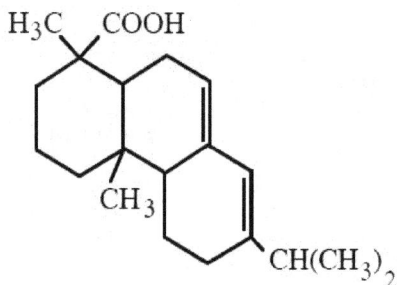

Abietic acid

(b) Lignin and derivatives

Lignin has been said to be the most common polymer on Earth. It accounts for approximately 30 % of dry wood weight. Lignin is a 3D polymer based on 3-hydroxy-4-methoxy-phenyl-propane (guanacyl, coniferyl and similar) units which can reach a high molecular weight (500,000-1,000,000). During wood digestion lignin is fragmented into small pieces and hydrophilic groups (-OH, -COOH, -SO$_3$-) are produced to make it water soluble, particularly at the high pH (11-12) of the pulping liquor. Lignin derivatives are polymeric surfactants of the grafted type, as will be discussed later. They are dispersants for solid particles, as in drilling fluids, among other uses. The figure indicates a likely structure for lignin.

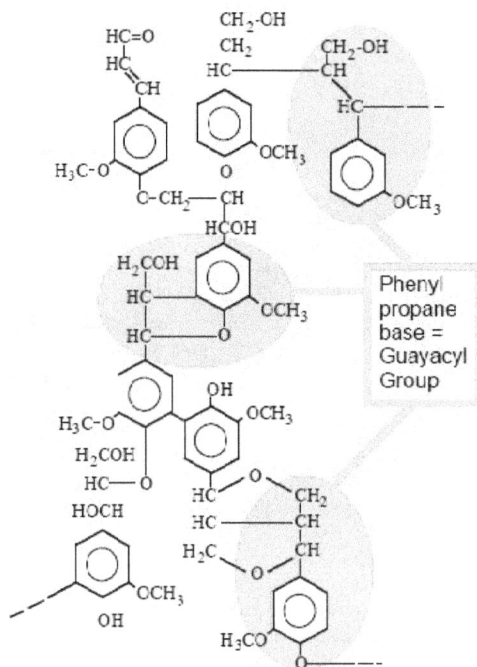

HC=O
HC
CH

CH₂-OH
CH₂
HC————CH

CH₂-OH
CH
HC———--

H₃C-O
C-CH₂——CH
HĊOH

OCH₃
O

OCH₃

H₂COH
HC
HC————O

OCH₃

Phenyl propane base = Guayacyl Group

H₃C-O
H₂COH
HC——O
HOCH

OH
OCH₃

HC———O
HC————CH
H₂C
O

CH₂
CH
CH

OCH₃
OH

H₃CO
O———--

7. 4. 3 Raw materials from petroleum

Other sources of lipophilic materials such as petroleum refining were considered in order to lower the cost, particularly for detergents. A proper lipophilic group exhibits a hydrocarbon chain containing from 12 to 18 carbon atoms. Such substances are found in light cuts (gasoline and kerosene) coming from atmospheric distillation and catalytic cracking. It is also possible to make such a chain by polymerization of short chain olefin, particularly in C_3 and C_4.

(a) Alkylates for alkylbenzene production

After World War II catalytic cracking and reforming processes were developed to produce high octane gasoline. They essentially consist in breaking an alkane chain to produce an α-olefin and to reform molecules in a different way, because of Markovnikov's rule. There formation happens with the attachment at the second carbon atom of the α-olefin, thus resulting in branching, which is the structural characteristic that confers a high octane number. These plants were producing short chain olefins which had no use in the early 1950's, particularly propylene, which was thus quite an inexpensive raw material to produce a surfactant lipophilic chain by polymerization. Because of the 3 carbon atoms difference between the *n*-mer and the *n+1*-mer, it is easy to separate by distillation the tetramer, with some amount of trimer and pentamer, to adjust the required chain length.

$$3 \ H_3C\text{-}CH=CH_2 \ \text{------>} \ H_3C\text{-}CH\text{-}CH_2\text{-}CH\text{-}C=CH_2$$

with CH₃ branches labeled "Branching":

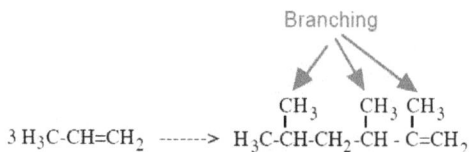

Trimerization of propylene

The alpha-olefin resulting from polymerization is used as an alkylate in a Friedel-Crafts reaction that ends in an alkyl-benzene. By sulfonation and neutralization, an alkyl benzene sulfonate of the detergent type is produced at a low cost, much lower than soap from natural oil and fat origin. However, alkylate is branched (see Figure), and this is quite an inconvenient because it is much more difficult to biodegradate than the linear counterpart. As a consequence, this kind of called hard alkylate have been banned by legislation in most countries, to be replaced by their linear equivalents.

(b) Linear paraffins, olefins and alkylates

Linear alkylates are produced either by separation from a petroleum cut containing a mixture of linear and isomerizes substances or by synthesis through ethylene oligomerization. Extraction of linear paraffins from refinery cuts can be carried out by two methods. The first one uses molecular sieves of the zeolite type. For instance zeolite Y exhibits a cage with a 5 Å diameter, in which a 4.7 Å diameter n-paraffin can enter, whereas iso-paraffins or cycloparaffin cannot. In practice the mixture is contacted with the solid zeolite powder, so that the linear compounds are able to adsorb. After drainage of the liquid the paraffins are recovered by evaporation, an operation which cost energy, from which an extra cost. Several commercial processes are found: MOLEX (UOP), ENSORB (EXXON), ISOSIEVE (Union Carbide) etc.

The second extraction method is based on F. Bengen discovery that urea is able to produce crystallized addition compounds with n-paraffins, but not with non-linear ones. These crystallized compounds (see Figure below) are relatively stable at ambient temperature and can be separated by filtration. On the other hand they are discomposed by heating around 80°C, temperature at which the n-paraffin can be separated from an urea aqueous solution.

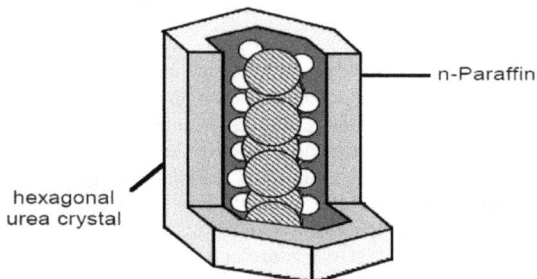

Fig: 1 Crystalline structure of urea/n-paraffin addition compound

On the other hand, a linear chain can be produced by polymerizing ethylene, since Markovnikov's rule does not apply to this two carbon olefin. In effect, the second carbon is the first on the other side. This is done through the so-called Ziegler oligomerization process which consists in forming a chain by poly condensation of ethylene on an organometallic template of the triethyl-aluminum type (see Figure below), and then to cut the oligomerized chain to recover the linear hydrocarbon.

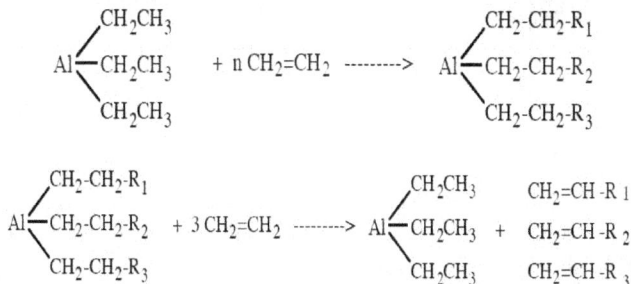

$$Al\left\langle\begin{array}{l} CH_2CH_3 \\ CH_2CH_3 \\ CH_2CH_3 \end{array}\right. + n\ CH_2=CH_2 \ \dashrightarrow\ Al\left\langle\begin{array}{l} CH_2\text{-}CH_2\text{-}R_1 \\ CH_2\text{-}CH_2\text{-}R_2 \\ CH_2\text{-}CH_2\text{-}R_3 \end{array}\right.$$

$$Al\left\langle\begin{array}{l} CH_2\text{-}CH_2\text{-}R_1 \\ CH_2\text{-}CH_2\text{-}R_2 \\ CH_2\text{-}CH_2\text{-}R_3 \end{array}\right. + 3\ CH_2=CH_2 \ \dashrightarrow\ Al\left\langle\begin{array}{l} CH_2CH_3 \\ CH_2CH_3 \\ CH_2CH_3 \end{array}\right. + \begin{array}{l} CH_2=CH\text{-}R_1 \\ CH_2=CH\text{-}R_2 \\ CH_2=CH\text{-}R_3 \end{array}$$

Ziegler oligomerization to produce n-alkenes

(c) Aromatics

Benzene, toluene and xylene are not found in crude oil. They come from dehydrogenation and dehydrocyclization reactions taking place in catalytic reforming and steam cracking plants. The most valuable substance is benzene and there are several methods to dealkylate toluene and xylene which are often carried out in the so-called BTX separation unit.

Benzene enters in the synthesis of the alkyl-benzene sulfonate, the most common surfactant in powdered detergents. It is also used in the synthesis of isopropyl benzene or cumene, which is an intermediate to produce both acetone and phenol by peroxidation. Alkyl phenols are synthesized by a Friedel-Craft reaction just as alkyl-benzene. In the 70's and 80'sethoxylated alkyl-phenols were the most popular surfactants for liquid dishwashing applications as well as many other. However, in the past few years, toxicity issues have cut down the production of such surfactants, which are likely be displaced by more environmentally friendly alcohol substitutes, although these later are not as good surfactants. Another surfactant application of alkyl-phenol is likely to stay around for a long time however. It is the production of ethoxylated phenol-formaldehyde resins, i.e. low MW bakelite type resins which are the current fashionable additives for crude oil dehydration (see polymeric surfactants).

7. 4. 4 Intermediate chemicals

(a) Ethylene oxide

Ethylene oxide was discovered by Wurst more than 100 years ago. However it is only after WWI that it was prepared by direct oxidation of ethylene by air on a silver catalyst (300 °C, 10 atm.). It is a very unstable gas, very dangerous to manipulate, because its triangular structure (see following formula) is submitted to extreme tension. The figure indicates the angle and bond distance (in between single and double)

61° angle
59° angle
C-C bond 1.47 Å

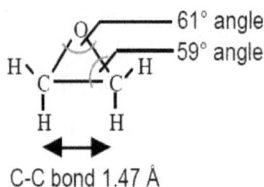

normal C-O-O angle : 111°
normal C-C-C angle : 91°
C-C normal bond: 1.55 Å
C=C normal bond: 1.35 Å

Ethylene oxide noted **EO** in formulas

$$EO = H_2C \overset{O}{-} CH_2$$

As a consequence, the molecule reacts very easily with any substance which is able to release a proton, according to:

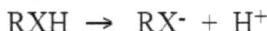

$$RXH \rightarrow RX^- + H^+$$

Where R is any hydrocarbon radical and X is a heteroatom capable of producing a negative ion (O, S). The reaction with the first mole of ethylene oxide can be written:

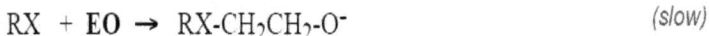

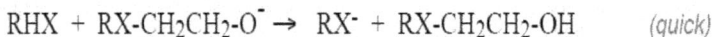

$$RX + EO \rightarrow RX\text{-}CH_2CH_2\text{-}O^- \qquad (slow)$$

$$RHX + RX\text{-}CH_2CH_2\text{-}O^- \rightarrow RX^- + RX\text{-}CH_2CH_2\text{-}OH \qquad (quick)$$

If other ethylene oxide molecules are available, they will react either with the remaining RX^-, or with the ethoxylated ion $RX\text{-}CH_2\text{-}CH_2\text{-}O^-$, which also display the RX^- structure. Everything depends on the relative reactivity of RX- and $RX\text{-}CH_2\text{-}CH_2\text{-}O^-$

(i) First case

RX^- is more acid than $RX\text{-}CH_2\text{-}CH_2\text{-}O^-$ as for instance with alkyl phenols $R\text{-}C_6H_4\text{-}OH$, mercaptans or thiols RSH, or carboxylic acids RCOOH. In this case the ethylene oxide molecule exhibits a stronger affinity for radicals RX^-, because they are more negative. As far as the kinetic point of view is concerned, this means that each RX^- radical react with one EO mole before poly condensation is able to start. The first reaction (to be completed) is (for instance with an alkyl phenol R-Ø-OH):

$$R\text{-}Ø\text{-}O^- + EO \rightarrow R\text{-}Ø\text{-}O\text{-}CH_2\text{-}CH_2\text{-}O^-$$

Afterward, when all R-Ø-O- species have reacted, polycondensation can take place according to:

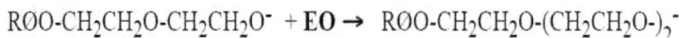

$$ROO\text{-}CH_2CH_2O^- + EO \rightarrow ROO\text{-}CH_2CH_2O\text{-}CH_2CH_2O^-$$

$$ROO\text{-}CH_2CH_2O\text{-}CH_2CH_2O^- + EO \rightarrow ROO\text{-}CH_2CH_2O\text{-}(CH_2CH_2O\text{-})_2^-$$

$$ROO\text{-}CH_2CH_2\text{-}O\text{-}(CH_2CH_2O\text{-})_2^- + EO \rightarrow ROO\text{-}CH_2CH_2O\text{-}(CH_2CH_2O\text{-})_3^-$$

etc., which can be summarized as:

$$\text{R\O O-CH}_2\text{CH}_2\text{O}^- + x \ \textbf{EO} \rightarrow \text{R\O O-CH}_2\text{CH}_2\text{O-(CH}_2\text{CH}_2\text{O-)}_x^-$$

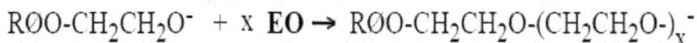

During the polycondensation, each EO molecule has the same probability to react with any already ethoxylated molecule, whatever its degree of ethoxylation. In other words all previous reactions have the same probability factor, independently of x. Consequently, the result is an oligomer distribution according to a Poisson law with mean m:

$$\% \text{ with x EO moles on R\O O-CH}_2\text{-CH}_2\text{-O}^- = \frac{e^{-m} m^x}{x!} \qquad x = 1,2,3,4,\ldots\ldots$$

The actual number of EO groups in the R\O OH molecule is $n = x+1$, and its mean ethoxylation degree is $\mu = m+1$, often called ethylene oxide number EON.

$$\% \text{ with with n EO moles on R\O O}^- = \frac{e^{-\mu+1}(\mu-1)^{n-1}}{(n-1)!} \qquad n = 2,3,\ldots\ldots$$

(ii) Second case

If RX$^-$ ions display the same acidity than R\O O(-CH$_2$-CH$_2$-O-)$_n^-$ ions as with water(H$_2$O), alcohols (R-OH) or amides (RCONH$_2$), both radicals compete from the first EO mole and the oligomer distribution is also a Poisson law but in n instead of (n-1).

(iii) Third case

If RXH is not acid enough to release a proton at alkaline pH, as it is the case with amines, then the reaction has to be carried out in two steps. During the first step the first EO mole is added at acid pH, so that the amine is transformed in ammonium. The reactions produce the mono-, di- and tri-ethanol amines.

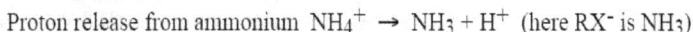

Proton release from ammonium $NH_4^+ \rightarrow NH_3 + H^+$ (here RX$^-$ is NH$_3$)

Then, the three condensation reactions:

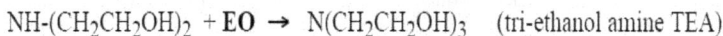

$$NH_3 + \textbf{EO} \rightarrow NH_2CH_2CH_2OH \quad \text{(mono-ethanol amine MEA)}$$

$$NH_2CH_2CH_2OH + \textbf{EO} \rightarrow NH(CH_2CH_2OH)_2 \quad \text{(di-ethanol amine DEA)}$$

$$NH\text{-}(CH_2CH_2OH)_2 + \textbf{EO} \rightarrow N(CH_2CH_2OH)_3 \quad \text{(tri-ethanol amine TEA)}$$

With an alkyl amine, first the alkyl ammonium ion is formed and it is deprotonated:

$$RNH_3^+ \rightarrow RNH_2 + H^+ \qquad \text{(here RX}^- \text{ is RNH}_2)$$

$$RNH_2 + \textbf{EO} \rightarrow RNH\text{-}CH_2CH_2OH \quad \text{(mono-ethanol alkyl amine)}$$

$$RNH\text{-}CH_2CH_2OH + \textbf{EO} \rightarrow RN(CH_2CH_2OH)_2 \quad \text{(di-ethanol alkyl amine)}$$

Once the ethanol amine is attained, the EO polycondensation is carried out at alkaline pH as previously. In many instance the first ethoxylation is stopped when the monoethanol alkyl amine is formed in order to avoid the polycondensation in more than one chain.

(b) Ethoxylated alcohols

Linear alcohols in C_{12}-C_{16} are used to prepare the alkyl-ester-sulfates used as detergents or foaming agents in shampoos, tooth pastes and hand dishwashing products. Ethoxylated alcohols tend to displace ethoxylated alkyl phenols, which are fading away because of their toxicity. Alcohols can be made by controlled hydrogenation of natural fatty acids. However, this is a costly way and in most cases they are rather produced by one of two available synthetic routes, as follows:

(i) The first one consists in oxidizing the Ziegler tri-alkyl aluminium complex (see section7.3.2) and to hydrolyse the resulting ether. This is called the ALFOL (alpha-olefin-alcohol) process.

(ii) The second so-called OXO process consists in the hydroformylation of an olefin. It is the most important process from the industrial point of view. It produces a mixture of primary and secondary alcohols.

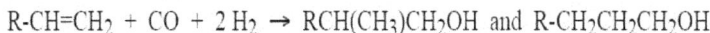

$$R\text{-}CH{=}CH_2 + CO + 2\,H_2 \rightarrow RCH(CH_3)CH_2OH \text{ and } R\text{-}CH_2CH_2CH_2OH$$

Note that if the olefin comes from the reduction of fatty alcohol (with an even number of carbon atoms) the OXO alcohol and the resulting ethoxylate would contain an odd number of carbon atoms. The most employed alcoyl group is the so-called tridecanol, which is often a mixture ranging from C_{11} to C_{15}.

7.5 Anionic surfactants

7.5.1 Soap

(a) The Chemical nature of soap

Soap is the result of the breakdown of fats or oils when contacted with alkali. The reaction known as hydrolysis and is the result of a reaction between the fat and water in the presence of alkali. Fats and oils are formed from fatty acids and glycerol to produce an ester. An ester is the result of a reaction between an alcohol and an organic acid. In the case of fats and oils, the acid is a carboxylic acid which has the general formula RCOOH, where R represents a hydrocarbon chain. The two simplest carboxylic acids are

- Formic acid: HCOOH, used in preserving biological specimens.
- Ethanoic acid: CH_3COOH, also known as acetic acid and the major constituent of vinegar.

Alcohols have the general formula $R_n(OH)_m$. Alcohols with 1, 2 and 3 OH groups are known.

Some common alcohols are

- Methanol: CH_3OH, found in methylated spirit and highly poisonous
- Ethanol: C_2H_5OH, the alcohol in alcoholic drinks
- Ethylene glycol: $CH_2OH\text{-}CH_2OH$, used in car anti-freeze
- Glycerol: $CH_2OH\text{-}CHOHCH_2OH$, found in animal and vegetable fats and oils

The reaction between carboxylic acids and alcohols to form esters is briefly as follows

$$R_1COOH + R_2OH \rightarrow R_1COOR_2 + H_2O$$

Thus the reaction between a fatty acid and an alcohol results in ester plus water. The reverse of this process is called hydrolysis and is the reaction between ester and water to produce alcohol and carboxylic acid. If an alkali is present, such as sodium or potassium hydroxide, and an alkali is often necessary to promote the reaction the resulting product is usually the sodium or potassium salt of the carboxylic acid.

In the case of animal or vegetable fats and oils, the alcohol is glycerol and the carboxylic acids are fatty acids. Fatty acids are acids like acetic acid, but the "R" part of the molecule is a long hydrocarbon chain containing mainly 12 – 18 carbon atoms but, in some cases, up to 22 carbon atoms. The reaction is

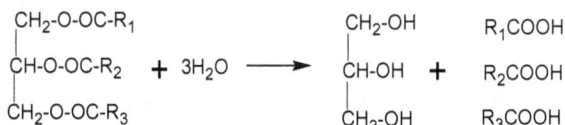

$$
\begin{array}{lcl}
CH_2\text{-O-OC-R}_1 & & CH_2\text{-OH} \qquad R_1COOH \\
CH\text{-O-OC-R}_2 \ + \ 3H_2O \ \longrightarrow & & CH\text{-OH} \ + \ R_2COOH \\
CH_2\text{-O-OC-R}_3 & & CH_2\text{-OH} \qquad R_3COOH
\end{array}
$$

The fatty acids react with alkali to produce a salt.

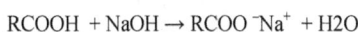

$$RCOOH + NaOH \rightarrow RCOO^-Na^+ + H2O$$

It is the sodium salt of the fatty acid that is the soap. When salt (NaCl) is added, the sodium salt of the fatty acid is precipitated out. This is because, not all the RCOONa is ionised. There is equilibrium

$$RCOONa \rightarrow RCOO^- Na^+$$

By adding salt, which is fully ionised in solution, an excess of Na+ ions is created. To try and remove this excess, some of the sodium ions combine with carboxyl ions (RCOO⁻) ions to form more RCOONa, resulting in the solution becoming saturated in RCOONa and the soap precipitating out as solid soap.

(b) Soap and other carboxylates

Strictly speaking the term *soap* refers to a sodium or postassium salt of a fatty acid. By extension the acid may be any carboxylic acid, and the alkaline metal ion may be replaced by any metallic or organic cation.

(i) Soap manufacture

Soaps are prepared by **saponification** of triglycerides from vegetal or animal source. For instance with a triglyceride containing 3 stearic acid (C18:0) units, the reaction with sodium hydroxide produces 3 moles of sodium stearate and 1 mole of glycerol.

$$3\ NaOH + (C_{17}H_{35}COO)_3\ C_3H_5 \rightarrow 3\ C_{17}H_{35}COONa + CH_2OH\text{-}CHOH\text{-}CH_2OH$$

This type of reaction has been used for centuries to manufacture soap from palm oil, olive oil (from which the brand name "Palmolive") etc, and mostly from tallow. The current process takes place in two steps. First the triglyceride is hydrolyzed at high pressure (240 °C, 40 atm.) with a ZnO catalyst, which is alkaline but not water soluble, and thus does not react with the acids. At the end of the hydrolysis, acids (oil phase) and glycerol (aqueous phase) are separated. Acids are then distilled under vacuum to separate too short and too long species, to keep the proper cut (C_{10}-C_{20}) and fractionate it into its components, particularly the C_{12}-C_{14} acids which are scares and more valuable than their C_{16}-C_{18} counterparts. This process allows formulating soaps with the proper mixture of acids, and with the desired hydroxide.

(a) Selection acids according to soap use

Luxury soap bars, at least in the past, were made only with vegetable oils, as implied by brand names like "Palmolive". However, it is seen from a previous table that tallow (beef fat) has a composition very close to a C_{16}-C_{18} mixture of vegetable oils, with a large proportion of unsaturated C_{18}. Consequently a similar but cheaper soap is obtained by saponification of tallow

("Marseille" soap) or of a mixture of tallow with vegetable oils. C_{16}-C_{18} soaps do not produce skin irritation, but they are not very water soluble and they produce whitish deposits (of calcium saps) with hard water. C_{12}-C_{14} soaps are often added in a small proportion (25%) to increase foam ability and tolerance to divalent cations (calcium and magnesium).

Transparent soaps are made by saponification of castor oil which contains a high proportion (80 %) of ricinoleic (12-hydroxy-oleic) acid. Sweet soaps are produced by leaving a certain amount of the produced glycerol.

Soap bars typically contains 30% water, and the actual structure is that of a liquid crystals, which is attained by kneading the soap according to a complex process that confers to the final product the right water solubility, without being too quick to dissolve.

(c) Cations

The hydroxide which is used to neutralize the acid is of great importance, because of the hydrolysis reaction which takes place in water. With very alkaline hydroxides, e.g. NaOH or KOH, the pH of the soap aqueous solution is very high. This will enhance the cleansing power but will result in irritation of biological tissues. Selecting the soap cation is a way to control the balance of cleansing action and solubility. The use of organic hydroxides such as ammonia, amine, amide, or ethanol amine, results in a less alkaline and less aggressive soap, although less water soluble. For instance triethanolamine oleate is a common soap used in cosmetic as well as in dry cleaning formulas. Calcium and magnesium soaps are oil soluble and are used as detergents or corrosion inhibitors in non polar media. Pb, Mn, Co and Zn soaps are used in paints because they accelerate drying. Cu soap exhibits fungicidal properties. Zn stearate is found in makes up. Lithium and aluminum soaps form fibrous mesophases with oils and are used as jellifying agent in lubricant greases.

7.5.2 Sulphonation and sulfatation

(i) Sulphonation mechanisms

Sulfonation of an aromatic ring takes place according to an electrophilic substitution, to produce an intermediate sigma complex that rearranges as an alkyl benzene sulfonic acid :

$$Ar\text{-}H + X \rightarrow X\text{-}Ar\text{-}H \rightarrow Ar\text{-}X\text{-} H^+$$

Where, Ar-H represents the aromatic ring an X electrophilic group: SO_3, H_2SO_4, etc. Symbol Ar-X- H+ is used because the sulfonic acid is a strong acid, i.e., completely dissociated, even at low pH. With alkyl benzene R-OH the reaction will be:

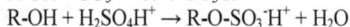

$$R\text{-}OH + SO_3 \rightarrow R\text{-}O\text{-}SO_3^-H^+$$
$$R\text{-}OH + H_2SO_4H^+ \rightarrow R\text{-}O\text{-}SO_3^-H^+ + H_2O$$

There exist other mechanisms, such as the addition on the double bond of an olefin or an unsaturated acid, or the nucleophilic substitution (SN^2) in α- position of a carboxylic acid.

(ii) Sulpatation mechanisms

Sulfatation is the esterification of an alcohol by one of the two acidities of sulfuric acid or anhydride. It results in an alkyl ester monosulfiric acid.

$$R\text{-}OH + SO_3 \rightarrow RO\text{-}SO_3^- H^+$$

$$R\text{-}OH + H_2SO_4H^+ \rightarrow RO\text{-}SO_3^-H^+ + H_2O$$

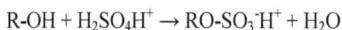

As for sulfonates, the salt (sulfate) is obtained by neutralization with an hydroxide. The product is called alkyl-sulfate. However, this is misleading term, and it is better to name it alkyl ester-sulfate in order to remember the existence of the ester bound, particularly because it is the one which is likely to break by hydrolysis, especially at acid pH. This is quite a difference with the sulfonates in which the C-S bound is quite resistant.

It is worth remarking that since the esterification-hydrolysis reaction is equilibrated, a small amount of alcohol will be always present, even at alkaline pH. This is why the most employed alkyl-sulfate, e.g., lauryl sulfate, always contains at least traces of dodecanol, which affects its properties. As a matter of fact an ultrapure lauryl sulfate is a poor foamer, and it is well known that the traces of lauryl alcohol produce a considerable foam boosting effect.

7.5.3 Sulphates

Alkyl sulfates were introduced just after WWII and except soaps, they are the oldest surfactants. They are excellent foaming and wetting agents, as well as detergents, and they are included in many different products for domestic and industrial use.

(a) Alkylsulphate (or better Alkyl-Ester-Sulfate)

They are very common, particularly the dodecyl (or lauryl) sulfate, as a sodium, ammonium or ethanolamine salt, which is the foaming agent found in shampoos, tooth paste, and some detergents. They are prepared by neutralization of the alkyl-ester-sulfuric acid by the appropriate base.

$$R\text{-}O\text{-}SO_3^- H^+ + NaOH \rightarrow R\text{-}O\text{-}SO_3^-Na^+$$

The sodium lauryl sulfate is an extremely hydrophilic surfactant. Lesser hydrophilicity can be attained with a longer chain (up to C16) or by using a weaker hydroxide (ammonia, ethanolamine).

(b) Alkyl sulphates (or better Alkyl-Ethoxy-Ester-Sulfate)

They are similar to the previous ones, but this time the sulfatation is carried out on an slightly ethoxylated (2-4 EO groups) alcohol.

For example: sodium laureth sulfate $C_{12}H_{25}\text{-}(O\text{-}CH_2\text{-}CH_2)_3$ $\text{-}O\text{-}SO_3^-Na^+$

The presences of the EO groups confer some nonionic character to the surfactant, and a better tolerance to divalent cations. They are used as *lime soap dispersing agents* (LSDA) in luxury soap, bath creams and shampoos. The ethoxylation step results in a mixture of oligomers, and the final product contains species having from 0 to 5 EO groups. This allows for a more compact packing of the polar heads at the air-water surface, in spite of the charge a characteristic which is associated with the excellent foaming ability of these surfactants.

(c) Sulphated alkanolamides

A similar result is attained by sulfating alkanol amides, particularly those in $C_{12}\text{-}C_{14}$ (cocoamide). In the following example dodecyl-amide sulfuric acid is neutralized by monoethanol amine, resulting in a foam booster used in shampoos and bubble bath products.

$$C_{11}H_{23}\text{-}CONH\text{-}CH_2\text{-}CH_2OH + SO_3 \rightarrow C_{11}H_{23}\text{-}CONH\text{-}CH_2\text{-}CH_2O\text{-}SO_3^-H^+$$

$$+ NH_2CH_2\text{-}CH_2OH \rightarrow C_{11}H_{23}\text{-}CONH\text{-}CH_2\text{-}CH_2O\text{-}SO_3^- {}^+NH_3\text{-}CH_2\text{-}CH_2OH$$

These surfactants have a large hydrophilic group and do not irritate the skin. They are used as LSDA and foam stabilizers in soap bars and shampoos. In general only 80-

90% of the alkyl amide is sulfated, so that the remaining unsulfated alkyl-amide can play a foam booster role.

(d) Glyceride sulphates and other sulphate

Alkyl sulfates are often prepared by starting with the hydrolysis of a glyceride to produce the fatty acid, which is then reduced into the alcohol. If a glyceride is hydrolyzed in presence of sulfuric acid, both the alcohol and the sulfate can be produced at the same time. The following example illustrates the case of a diglyceride which is both hydrolyzed and sulfated:

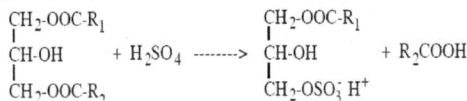

$$
\begin{array}{lll}
CH_2\text{-OOC-}R_1 & & CH_2\text{-OOC-}R_1 \\
| & & | \\
CH\text{-OH} \quad + H_2SO_4 \text{ -------> } & CH\text{-OH} & + R_2COOH \\
| & & | \\
CH_2\text{-OOC-}R_2 & & CH_2\text{-OSO}_3^- \ H^+
\end{array}
$$

This double reaction is carried at a low cost, but precaution is required to control the conditions and avoid side reactions. Sulfated monoglycerides which are neutralized by an ethanolamine are excellent foaming agents, even with a C_{18} chain. This is remarkable since alkyl-sulfates are foaming agents only with short C_{12}-C_{14} chain, i.e., a lipophilic group which comes from coconut oil, and thus a raw material much more expensive than tallow (C_{16}-C_{18}).A mole of sulfuric acid can be added on a double bond of one of the acid of a glyceride. The sulfated acid can be separated (by hydrolysis) or stay in the glyceride, to result in an emulsifying agent.

The sulfate of ricinoleic acid (12-hydroxy-9-octadecenoic acid) which comes from castor oil is used as a fixer of Turkey red dye (alizarine) on wool. Turkey red oil, a mixture of sulfated castor oil compounds, was one of the first attempt (in 1875) to produce a soap with some insensitivity to calcium ions.

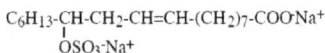

$$
\begin{array}{l}
C_6H_{13}\text{-CH-CH}_2\text{-CH=CH-(CH}_2)_7\text{-COO}^-Na^+ \\
\quad\quad | \\
\quad\quad OSO_3^-Na^+
\end{array}
$$

Disodium ricinoleate sulfate

7. 5. 4 Sulphonates

(a) A bit of history about petroleum sulphonates

Lubricating oils are made from lateral cuts of the vacuum distillation unit, i.e. high MW hydrocarbons in the 30-40 carbon atom range, containing (normal, iso, and cyclo) paraffins and aromatics, often polyaromatics. The first step in manufacturing lubricating oil is to remove the aromatics which are not acceptable for two reasons: they are likely to react at high temperature and their viscosity index is not appropriate. Today a liquid-liquid extraction with furfural or phenol is used to separate the aromatics, but during the first part of the XX century the extraction of aromatics was based on a sulfonation reaction that attached a sulfonic acid group on the aromatic ring. These acids were then removed from the oil by a liquid-liquid extraction with an alkaline solution. The aromatic species were thus obtained as alkyl-aryl sulfonates, so called *mahogany* sulfonates because of their reddish color.

R-Ar-H $+ SO_3 \rightarrow$ R-Ar-SO$_3$-H   (oil soluble)

R-Ar-SO$_3$H (in oil) $+ NaOH$ (aqueous solution) $\rightarrow$ R-Ar-SO$_3^-$Na$^+$ (aqueous solution)

In the previous reaction R-Ar-H stands for an alkyl-aromatic hydrocarbon which typically contains at least one aromatic ring and an alkyl chain, as in the following figure.

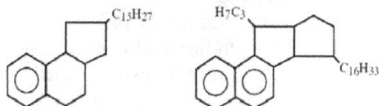

Alkyl aromatic structures found in lube oil

Nowadays the sulfonation reaction is carried out on the appropriate cut of the extracted aromatic stream, to make the so-called petroleum sulfonates. The MW of their sodium salt typically ranges from 400 to 550 Daltons. Care is taken to add only one sulfonate group, in general by reducing the sulfonic agent concentration below stoichiometry requirement. As a consequence the final product often contains a large proportion of unsulfonated oil. These sulfonates represent about 10% of the total production of sulfonated products. They are used in many industrial products as emulsifiers, dispersants, and tension lowering agents, detergents and floatation aids. Calcium salts, which are oil soluble, are used in lubricating oils and dry cleaning products. They are the main candidates for the enhanced oil recovery processes by surfactant flooding, because they allow the attainment of ultralow interfacial tensions ($0.1 \mu N/m$) and they are the cheapest surfactants available on the market.

(b) Dodecyl benzene sulphonate and synthetic detergents

During WWII, the catalytic cracking processes were developed to produce high octane aviation gasoline. As shown in the following reaction, the cracking of a paraffin results in the formation of shorter paraffin and an α-olefin, in the present case a propylene molecule.

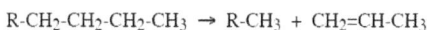

$$R\text{-}CH_2\text{-}CH_2\text{-}CH_2\text{-}CH_3 \rightarrow R\text{-}CH_3 + CH_2\text{=}CH\text{-}CH_3$$

In 1945 propylene was a by-product with little use, since the plastic era had not started yet. By controlled polymerization a low cost propylene tetramer was obtained:

$$4\,CH_3\text{-}CH\text{=}CH_2 \rightarrow CH_3\text{-}\overset{\underset{\textstyle CH_3}{|}}{C}H\text{-}CH_2\text{-}\overset{\underset{\textstyle CH_3}{|}}{C}H\text{-}CH_2\text{-}\overset{\underset{\textstyle CH_3}{|}}{C}H\text{-}CH_2\text{-}CH\text{=}CH_2$$

Because of stereochemical reasons (Markovnikov's rule) propylene polymers are branched alpha-olefins. It is worth remarking that the tetramer can be produced with a high degree of purity, since impurities are the other polymers, i.e. trimer and pentamer species, whose MW is quite different. It was thus possible to manufacture a cheap alkyl benzene sulfonate by a series of easy to carry reactions, e.g., Friedel-Crafts alkylation, sulfonation and neutralization. The commercial alkyl benzene sulfonate product so-called ABS, contained an alkylate with an average number of carbon atoms around C12 coming from various origins, particularly propylene tetramer, whose synthesis resulted in a branched "tail".

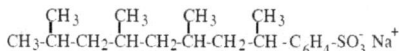

$$CH_3\text{-}\overset{\underset{\textstyle CH_3}{|}}{C}H\text{-}CH_2\text{-}\overset{\underset{\textstyle CH_3}{|}}{C}H\text{-}CH_2\text{-}\overset{\underset{\textstyle CH_3}{|}}{C}H\text{-}CH_2\text{-}\overset{\underset{\textstyle CH_3}{|}}{C}H\text{-}C_6H_4\text{-}SO_3^-\,Na^+$$

In the late 1940 and early 1950 synthetic detergents displaced soaps in domestic washing particularly in washing machine use, because they displayed several advantages, such as a better tolerance to hard water, a better detergency, and a cheaper price. Production and use rose quickly. However, they had a major drawback that industrialized countries soon noticed in the areas of high population density, may be as one of the first major ecological warnings. Wastewaters carried ABS to lakes and rivers which were being covered by a layer of persistent foam. It was shown that the culprit was not the detergent by itself, but the fact that the alkylate was branched, which made it much more difficult for micro-organisms to degradation it. By 1965 most industrialized countries had passed laws banning the use of branched alkylate, and detergent manufacturers turned to linear alkyl benzene sulfonates (LAS) which were still relatively inexpensive, in spite of the extra production cost.

These LAS are still around and account for a very large proportion of the powdered detergents. They have an alkyl chain in the C_{10}-C_{16} range with a benzene ring which is attached in any position of the linear chain, not necessarily at the end. Since there are many possibilities of attachment, the commercial product is in general a mixture of oligomers, the most common (from the statistical point of view) being the ones in which the benzene is attached at 3-6 carbon atoms from the extremity, as for instance in the following species.

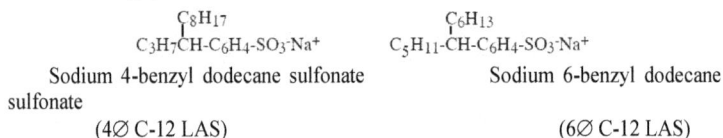

$$\overset{\displaystyle C_8H_{17}}{\underset{\displaystyle C_3H_7\text{-}CH\text{-}C_6H_4\text{-}SO_3^-Na^+}{|}} \qquad\qquad \overset{\displaystyle C_6H_{13}}{\underset{\displaystyle C_5H_{11}\text{-}CH\text{-}C_6H_4\text{-}SO_3^-Na^+}{|}}$$

Sodium 4-benzyl dodecane sulfonate   Sodium 6-benzyl dodecane sulfonate

(4Ø C-12 LAS)   (6Ø C-12 LAS)

LAS are as good detergents as ABS, much better than alkane sulfonates and other intented substitutes without a benzene ring; however, there are not as good as ABS as foaming agents or emulsifiers. LAS sodium salts are water soluble up to 1Ø C-16, but the maximum detergency is attained with C_{12}-C_{13}. LAS in C_9-C_{12} are wetting agents, whereas those in C_{15}-C_{18} are used as tension lowering agents and emulsifiers. Today domestic detergent formulas, either in powdered or liquid forms, contain a high proportion of LAS, as seen in the following table.

Table: 2 Typical detergent formulation

	Powder for machine	Dishwashing liquid	Fine fabric hand wash (liq)
Surfactants	14% C12 LAS	24% C12 LAS	15% C12 LAS
	3% Alcohol + 6EO	5% C12 Sulfate	10% C12 ether Sulf.
Foaming agent		5% Coco amide	5% C12 DEamide
Antifoaming	3% C18 soap		
Hydrotope		5% Xylene sulfonate	
Builder	48% STP		15% C12 Sulfobetaine
Alkaline	10% Na Silicate		
Salts	13% Na Sulfate		
Antiredeposition	0,3% CMC		0,5% CMC
Other		1% latex	
Water	ex	60% with 5% ethanol	55% with 4% urea

Domestic uses account for about 50% of the LAS production. Industrial uses include emulsion polymerization (polystyrene, polymethacrylate, PVC and other resins), agricultural self emulsifying concentrates for seed and crop phytosanitary protection, production of elastomer of solid foams, emulsified paints, industrial cleaning and cleansing, petroleum production, dry cleaning etc.

(c) Short tail alkyl benzene sulphonates hydrotropes

Hydrotropes (from Greek **tropos** "turn") are substances which help other to become compatible with water. For instance, it is well known that short alcohols and urea are able to cosolubilized organic compounds such as perfumes. Hydrotropes are non-surfactant amphiphiles, which enter the micelles as cosolubilizing agents and introduce disorder in any mesophases structure. For cheap commodity products such as liquid detergents, hydrotropes are alkylbenzenesulfonates with very short alkyl chain e.g., toluene, xylene, ethyl or propyl benzene sulfonates. Hydrotropes are used in powdered detergents to reduce hygroscopy, in pastes to reduce viscosity, and in dishwashing and fine fabric hand washing liquids to avoid precipitation at low temperature.

(d) α- Olefin sulphonates

Since most linear alkylates are often α-olefins, which can be sulfonated, it is worth asking the question: why α-olefin sulfonates have not displaced alkyl benzene sulfonates, since the later exhibit an expensive and potentially toxic benzene ring?

The principal problem is that the sulfonation of an α-olefin results in various compounds, such as the α-olefin sulfonate (60-70 %), the hydroxyalkane sulfonate (20 %), and even some amount of β-olefin sulfonate and sulfate of hydroxyalkane sulfonate.

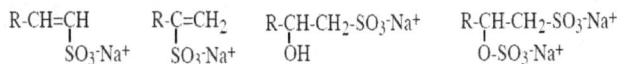

$$R-CH=CH \quad\quad R-C=CH_2 \quad\quad R-CH-CH_2-SO_3^-Na^+ \quad\quad R-CH-CH_2-SO_3^-Na^+$$
$$\quad\; | \quad\quad\quad\quad\quad | \quad\quad\quad\quad\quad | \quad\quad\quad\quad\quad\quad\quad\quad\quad | $$
$$\quad SO_3^-Na^+ \quad\quad SO_3^-Na^+ \quad\quad OH \quad\quad\quad\quad\quad\quad\quad\quad O-SO_3^-Na^+$$

α-olefin sulfonate β-olefin sulfonate hydroxyl alkane sulfonate hydroxyl alkane sulfate

α-olefin sulfonates display a better hard water tolerance than LAS, but they are not as good detergents; they are used as additives, particularly in low phosphate formulas: C_{12}-C_{14} in liquids, C_{14}-C_{18} in powders.

(e) Lignosulphonates

Lignosulfonates come from the reaction of wood lignin with bisulfite or sulfate ions during the wood digestion reaction to make the pulp. It has been seen in section 2 that lignin tridimensional polymer containing numerous aromatic rings as well as hydroxyl methyl-ether functions.

A sulfonating agent is able to add asulfonate group on aromatic rings or to sulfate a hydroxyl group. In both cases the resulting sulfonate or sulfate increases the hydrophilicity of the polymer and can turn it water soluble. This solubilization in the so-called black liquor at alkaline pH is the way to separate lignin compounds from insoluble cellulose fibers. A typical commercial lignin compound contents lignin chunks with MW ranging from4000 Daltons (about 8 aromatic ring units) to 20.000 or more. Lignosulfonates are used as clay dispersants in drilling fluids.

Lignin calcium salts are non water-soluble and are used as dispersant in non-aqueous media. Alkaline (sodium, ammonium, potassium) salts are polyelectrolites which are used as heavy metal ions sequestrants or protein agglutinant for granulated food, waste water treatment.

(f) Sulpho carboxylic compounds

These compounds display at least 2 hydrophilic groups: the sulfonate group and one or two carboxylic group(s) as carboxylate or ester. The mono carboxylic compounds are not very important in practice, and the best known product of this category is the sodium lauryl sulfoacetate which is found in tooth pastes, shampoos, cosmetics and slightly alkaline soaps.

$$NaOOC-(CH_2)_{11}-SO_3^-Na^+$$

C_{18} compounds are used in butter and margarine as anti-splattering agents, because they are able to fix the water in food emulsions so that the evaporation is not explosive when they are heated in a pan.

On the other hand sulfodicarboxylic compounds, such as sulfosuccinates and sulfosuccinamates are well known and used in many applications. Succinic acid is a di acid which should be named: 2-butene 1, 4-dicarboxylic. The diesters (succinate) is produced by direct reaction of maleic anhydride with an alcohol, followed by a sulfonation.

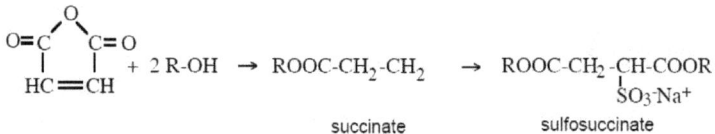

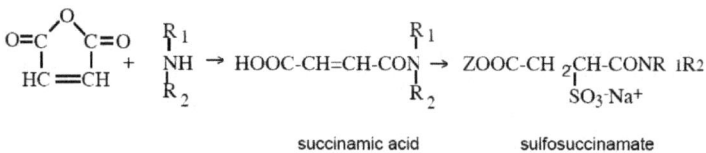

$$O=C \overset{O}{\diagup \diagdown} C=O$$
$$\underset{HC=CH}{| \quad |} + 2\,R\text{-}OH \rightarrow ROOC\text{-}CH_2\text{-}CH_2 \rightarrow ROOC\text{-}CH_2\text{-}\underset{SO_3^-Na^+}{\overset{|}{CH}}\text{-}COOR$$

succinate   sulfosuccinate

$$O=C \overset{O}{\diagup \diagdown} C=O$$
$$\underset{HC=CH}{| \quad |} + \underset{R_2}{\overset{R_1}{\underset{|}{\overset{|}{NH}}}} \rightarrow HOOC\text{-}CH=CH\text{-}\underset{R_2}{\overset{R_1}{\underset{|}{\overset{|}{CON}}}} \rightarrow ZOOC\text{-}CH_2\text{-}\underset{SO_3^-Na^+}{\overset{|}{CH}}\text{-}CONR_1R_2$$

succinamic acid   sulfosuccinamate

Where Z represents a sodium atom, an alkyl group, or another amide (it will be then a sulpho succinamide).The best known compound is sold as Aerosol OT by American Cianamide; it is the dioctyl (actually di-bis-ethyl-hexyl) sulfosuccinate which is prepared from secondary octanol. A similar compound is attained with hexanol.

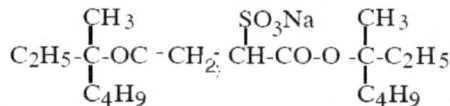

$$C_2H_5\text{-}\underset{C_4H_9}{\overset{CH_3}{\underset{|}{\overset{|}{C}}}}\text{-}OC\text{-}CH_2\text{-}\overset{SO_3Na}{\overset{|}{CH}}\text{-}CO\text{-}O\text{-}\underset{C_4H_9}{\overset{CH_3}{\underset{|}{\overset{|}{C}}}}\text{-}C_2H_5$$

2-bis (ethyl-hexyl) sodium sulfosuccinate

These surfactants are the best wetting agents to be known; they are also foaming agents, dispersants and emulsifiers. They are used in emulsion polymerization, pigments dispersion in paints and latex, shampoos, cosmetics etc. However their used is limited by their price which is high in the anionic category.

7.5.5 Other anionic surfactants

(a) Organo phosphored surfactants

The chemistry of phosphorus is particularly complex. It is just necessary to point out here that the double valency (3 and 5) leads to three acids:

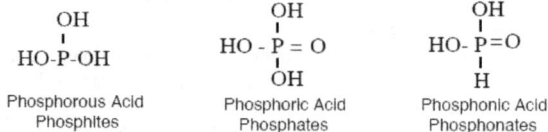

$$\begin{array}{ccc} OH & OH & OH \\ | & | & | \\ HO\text{-}P\text{-}OH & HO\text{-}P=O & HO\text{-}P=O \\ & | & | \\ & OH & H \end{array}$$

Phosphorous Acid   Phosphoric Acid   Phosphonic Acid
Phosphites   Phosphates   Phosphonates

The mono and di esters of fatty alcohols (R-OH), whose extra acid H's are neutralized by an alkaline hydroxide or a short amine, do not hydrolyze and exhibit a good tolerance to electrolytes. They are used in agrochemical emulsions, particularly when the aqueous phase contains fertilizers.

$$
\begin{array}{cc}
\text{O-Na} & \text{O-R} \\
| & | \\
\text{R-O-P=O} & \text{Na-O-P=O} \\
| & | \\
\text{O-Na} & \text{O-R}
\end{array}
$$

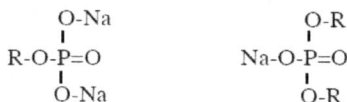

If an ethoxylated alcohol is used, the monoestar can exhibit a good water solubility even at acid pH at which the other acid H's are not neutralized.

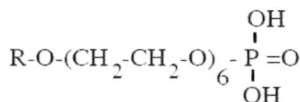

$$
\text{R-O-(CH}_2\text{-CH}_2\text{-O)}_6 \text{ - } \overset{\text{OH}}{\underset{\text{OH}}{\text{P}}} \text{=O}
$$

As a final comment of this section, it is worth remarking that most of the structural elements of biological membranes are amphoteric amphiphiles so-called phospholipids, as for instance the following lecithin.

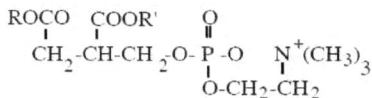

$$
\begin{array}{l}
\text{ROCO} \quad \text{COOR'} \qquad \text{O} \\
\quad | \qquad \quad | \qquad \qquad \| \\
\text{CH}_2\text{-CH-CH}_2\text{-O- P -O} \quad \text{N}^+\text{(CH}_3)_3 \\
\qquad \qquad \qquad \quad | \\
\qquad \qquad \qquad \text{O-CH}_2\text{-CH}_2
\end{array}
$$

Diglyceride aminophosphate

(b) Sarcosides or alkyl amino acids

The root product of this surfactant family is sarcosine, or methyl glycine, a cheap synthetic amino acid.

$$\text{CH}_3\text{NH-CH}_2\text{-COOH}$$

The acid acylation is carried out with a fatty acid chloride, to result in a surfactant that displays a fatty amide group as the lipophilic tail.

$$\text{R-CO-Cl + CH}_3\text{-NH-CH}_2\text{-COONa} \rightarrow \text{RCO-N(CH}_3)\text{-CH}_2\text{-COONa}$$

This reaction can take place with many different amino acids, particularly those which come from the hydrolysis of proteins. This results in the so-called sarcosides, whose structure is very similar to our biological tissues.

The most used synthetic product is lauryl sarcosinate, which is both a strong bactericide and a blocking agent of hexokinase (putrefaction enzyme). Since it is not cationic, it is compatible with anionic surfactants, and it is used in tooth paste and "dry" shampoos for carpets and upholstery.

7.6 Non ionic surfactants

(a) Types of non ionic surfactants

During the last 35 years, nonionic surfactants have increased their market share, to reach about 40 % of the total surfactant production worldwide. Nonionic surfactants do not produce ions in aqueous solution. As a consequence, they are compatible with other types and are excellent candidates to enter complex mixtures, as found in many commercial products. They are much less sensitive to electrolytes, particularly divalent cations, than ionic surfactants, and can be used with high salinity or hard water. Nonionic surfactants are good detergents, wetting agents and emulsifiers.

Some of them have good foaming properties. Some categories exhibit a very low toxicity level and are used in pharmaceuticals, cosmetics and food products.

Nonoionic surfactants are found today in a large variety of domestic and industrial products, such as powdered or liquid formulations. However the market is dominated by polyethoxylated products, i.e., those whose hydrophilic group is a polyethylenglycol chain produced by the polycondensation of ethylene oxide on a hydroxyl or amine group.

Table: 3 Main nonionic surfactants

Surfactant Type	% Total
Ethoxylated Linear Alcohols	40
Ethoxylated Alkyl Phenols	15
Fatty Acid Esters	20
Amine and Amide Derivatives	10
Alkylpolyglucosides	---
Ethleneoxide/Propyleneoxide Copolymers	---
Polyalcolols and ethoxylated polyalcohols	---
Thiols (mercaptans) and derivates	---

It was seen in section2 that ethylene oxide can be polycondensated on various types of molecules with general formula RXH that can ionize as RX⁻.

Depending on the relative acidity of the RXH molecule and its ethoxylated counterpart $RX-CH_2-CH_2-OH$, the polycondensation leads to a different result. However, in all cases the commercial product contains a mixture of different oligomers with a distribution of ethylene oxide number (EON), a characteristic which can be an advantage or a drawback. At least 4-5 ethylene oxide groups are needed to insure a good solubility in water with a lipophilic group such as a C_{13} alkyl. However, for some applications, the ethoxylation degree can reach EON=20 and even 40.It is worth noting that, though the polyethylene-oxide chain is globally hydrophilic, each EO group contains 2-methylene ($-CH_2-$) units which are hydrophobic. This duality becomes evident when it is known that the polypropylene-oxide chain, i.e. the three carbon atoms counterpart, is globally hydrophobic. It can be said that the hydrophilicity conferred by the oxygen atom is thus compensated by approximately 2.5-methylene groups.

This remark is quite important, because it clearly indicates that the poly EO hydrophilic group is not an extremely hydrophilic group, a characteristic that explains why this kind of surfactant is soluble in organic solvents. Moreover, any change in formulation or temperature that affects the interaction between the poly EO chain and the water/oil physicochemical environment is likely to affect the behavior of this kind of surfactant.

(b) Ethoxylated alcohols and alkyl phenols

(i) Ethoxylated linear alcohols

Alcohols come from various origins, but today the main point is to select those with linear alkyl groups. Primary alcohols have their -OH group at the end of the chain. They are generally prepared by moderate hydrogenation of fatty acids, so called catalytic hydrogenolysis (150 °C, 50 atm H_2, Copperchromite (catalyst):

$$R\text{-}CO\text{-}OH + H_2 \rightarrow R\text{-}CH_2\text{-}OH + H_2O$$

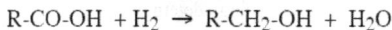

They can be prepared too by Ziegler hydroformylation of olefins (OXO process) or controlled oxidation of paraffins. Secondary alcohols, which have their hydroxyle group attached on the second carbon atom of the alkyl chain, are produced by hydration of α-olefins in sulfuric medium:

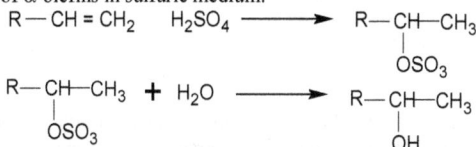

$$R\text{—}CH = CH_2 \quad H_2SO_4 \longrightarrow R\text{—}\underset{\underset{OSO_3}{|}}{CH}\text{—}CH_3$$

$$R\text{—}\underset{\underset{OSO_3}{|}}{CH}\text{—}CH_3 + H_2O \longrightarrow R\text{—}\underset{\underset{OH}{|}}{CH}\text{—}CH_3$$

The polycondensation of ethylene oxide on anhydrous alcohol is carried out in presence of an alkaline catalyst (NaOH, KOH, Na metal), in absence of air and with caution. Since the probability of condensation is the same for the unreacted alcohol as for already ethoxylated molecules, there is a large distribution of oligomers.

The most used alcohol is the so-called tridecanol, actually a C_{12}-C_{16} mixture. The ethoxylation degree ranges from EON = 6-10 for detergents, EON > 10 for lime soap dispersants, wetting agents and emulsifiers. Foaming ability passes through a maximum for a proper ethoxylation degree. For dodecanol, it is EON = 30.

(ii) Ethoxylated alkyl phenols

Phenol (or hydroxybenzene) is mostly prepared as a sub product of acetone manufacturing via the peroxidation of cumene (isopropyl benzene).Ethoxylated alkyl-phenols are produced by two ways, depending on the available raw material. The first method consists in alkylating the phenol according to a classical Friedel-Crafts reaction:

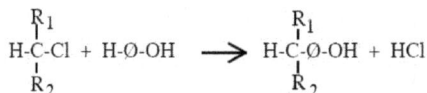

$$H\text{-}\underset{\underset{R_2}{|}}{\overset{\overset{R_1}{|}}{C}}\text{-}Cl + H\text{-}\varnothing\text{-}OH \longrightarrow H\text{-}\underset{\underset{R_2}{|}}{\overset{\overset{R_1}{|}}{C}}\text{-}\varnothing\text{-}OH + HCl$$

The second method consists in adding an alpha-olefin such as propylene trimer or tetramer, or isobutylene dimer, on an aromatic ring. This technique results in nonyl, dodecyl and octyl phenols, with branched, thus non biodegradable alkylates. One of the most common alkyl phenol has been for many years the 3^0 octyl-phenol produced by the Friedel-Crafts alkylation of phenol by isobutylene dimer. As seen in the following formula this substance exhibits two tertiary carbon atoms which are a challenge to biodegradation.

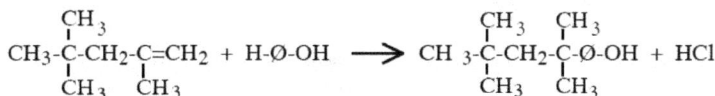

$$CH_3\text{-}\underset{\underset{CH_3}{|}}{\overset{\overset{CH_3}{|}}{C}}\text{-}CH_2\text{-}C=CH_2 + H\text{-}\varnothing\text{-}OH \longrightarrow CH_3\text{-}\underset{\underset{CH_3}{|}}{\overset{\overset{CH_3}{|}}{C}}\text{-}CH_2\text{-}\underset{\underset{CH_3}{|}}{\overset{\overset{CH_3}{|}}{C}}\text{-}\varnothing\text{-}OH + HCl$$

Common commercial products are the octyl, nonyl and dodecyl-phenol with a degree of ethoxylation ranging from 4 to 40. Octyl and nonyl-phenols with EON = 8-12 are used in detergents. With EON < 5 the attained products are antifoaming agents or detergent in non aqueous media. With EON ranging from 12 to 20, they are wetting agents and O/W emulsifiers. Beyond EON = 20 they exhibit detergent properties at high temperature and high salinity. The main use of alkyl phenols was and still is as ingredients for domestic and industrial detergents, particularly for high electrolyte level: acid solution for metal cleaning, detergents for dairy plants, agrochemical emulsions, styrene polymerization etc.

Since branched alkylates are not readily biodegradable, the trend has been in the past decades to go into more linear products. However the additional cost has restrained that trend, which has been substituted in the past decade by another way to cut price and toxicity alike, i.e. the elimination of the benzene ring altogether, e.g. the substitution by ethoxylated linear alcohols. The dilemma is that alcohol ethoxylates are not as good detergents as their counterpart phenol compounds, just as it is the case with alkyl benzene sulfonates versus alkane or olefin sulfonates.

(iii) Ethoxylated thiols

Thiols (the alcohol structure in which the O atom is replaced by a S atom) can be ethoxylated just as alcohols or phenols. The corresponding products are excellent detergents and wetting agents, which are used in industry only, since the possibility of releasing stinking mercaptans bar them from domestic use.

Ter-dodecyl mercaptans with EON = 8-10 exhibits a good solubility in both water and organic solvents. Moreover it is an excellent industrial detergent. It is use in raw wool treatment and agrochemical emulsions, in which its wetting ability enhances the cleansing action.

(c) Fatty acid esters

The esterification of a fatty acid by a -OH group from polyethylene oxide chain tip or poly alcohols generates an important family of nonionic surfactants, not only for their market share (20 % of all nonionics), but also because of their compatibility with biological tissues, which make them suitable for pharmaceuticals, cosmetics and food stuffs.

(i) Acid ethoxylated fatty acids (polyethoxy-esters)

The condensation of ethylene oxide on a carboxylic acid takes place just as it is for alkyl phenols. Polyethoxy esters are produced which are identical to those attained by the esterification of the acid by the polyethylene glycol.

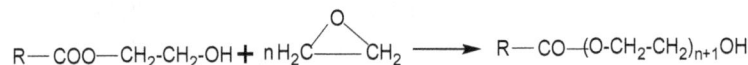

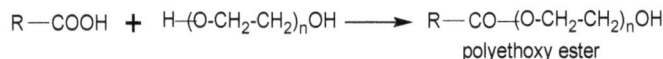

$$R-COOH + H_2C \overset{O}{\diagup\diagdown} CH_2 \longrightarrow R-COO-CH_2\text{-}CH_2\text{-}OH$$

$$R-COO-CH_2\text{-}CH_2\text{-}OH + n H_2C \overset{O}{\diagup\diagdown} CH_2 \longrightarrow R-CO-(O\text{-}CH_2\text{-}CH_2)_{n+1}OH$$

$$R-COOH + H-(O\text{-}CH_2\text{-}CH_2)_nOH \longrightarrow R-CO-(O\text{-}CH_2\text{-}CH_2)_nOH$$
polyethoxy ester

Polyethoxy esters of fatty acids and other natural carboxylic acids, e. g. abietic acid, are among the cheapest nonionic surfactants. However, they are not very good detergents and they don't foam. They are used to cut the cost of dish and fabric washing formulas, and many others, though with a limitation, i.e., they cannot be used at alkaline pH because they would hydrolyze.

(ii) Glycerol esters

Triglycerides which are found in most vegetal and animal oils and fats are triesters of glycerol (propane-triol) and are not hydrophilic enough to be water soluble. Contrariwise glycerol mono and diesters, so-called mono and diglycerides can exhibit surfactant properties. They can be synthesized by the reaction of glycerol with fatty acids, but the typical industrial way is fat and oil alcoholisation, which consists in reacting a triglyceride with an excess of glycerol in alkaline conditions.

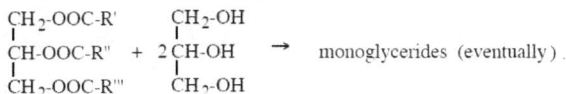

$$CH_2\text{-OOC-R'} \qquad CH_2\text{-OH}$$
$$CH\text{-OOC-R''} \quad + \quad 2\ CH\text{-OH} \qquad \rightarrow \qquad \text{monoglycerides (eventually)}$$
$$CH_2\text{-OOC-R'''} \qquad CH_2\text{-OH}$$

The size of the hydrophilic part can be increased by using a poly glycerol, which is obtained by dehydratation of glycerol.

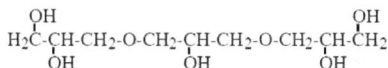

$$\overset{OH}{H_2C\text{-CH-CH}_2\text{-O-CH}_2\text{-CH-CH}_2\text{-O-CH}_2\text{-CH-CH}_2}$$
$$\overset{}{OH} \qquad\qquad OH \qquad\qquad OH$$

Glycerol trimer

Glycerol esters and derivatives are used in the conditioning of food stuffs, bread, dairy emulsions and foams such as beverages, ice creams, margarine, butter etc. They are used in pharmaceuticals as emulsifiers, dispersants and solublizing agents.

(iii) Esters of hexitols and cyclic anhydro hexitols

Hexitols are hexahydroxy hexanes obtained by the reduction of hexoses or monosaccharides. The most common is sorbitol, attained by the reduction of D-glucose.

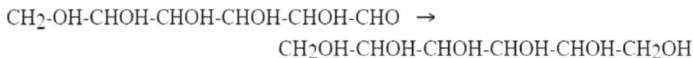

$$CH_2\text{-OH-CHOH-CHOH-CHOH-CHOH-CHO} \rightarrow$$

$$CH_2OH\text{-CHOH-CHOH-CHOH-CHOH-CH}_2OH$$

Monosaccharides can form a cycle or ether loop called hemi-acetaldehyde. The same happens to hexitols when they are heated at acid pH. Two hydroxyl groups merge to produce an ether link resulting in a 5 or 6 atom cycle called hydrosorbitol or sorbitan. In some cases a two cyclo bi-anhydro sorbitol product called isosorbide is produced.

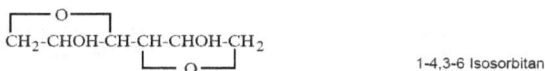

$$\overline{O}$$
$$CH_2\text{-CHOH-CHOH-CH-CHOH-CH}_2\text{-OH} \qquad \text{1-4 Sorbitan}$$

$$\overline{O}$$
$$CH_2\text{-OH-CH-CHOH-CHOH-CH-CH}_2\text{-OH} \qquad \text{2-5 Sorbitanne}$$

$$\overline{O}$$
$$CH_2\text{-CHOH-CHOH-CHOH-CH-CH}_2\text{-OH} \qquad \text{1-5 Sorbitan}$$

$$\overline{O}$$
$$CH_2\text{-CHOH-CH-CH-CHOH-CH}_2$$
$$\underline{O} \qquad\qquad \text{1-4,3-6 Isosorbitan}$$

The sorbitan ring exhibits 4 hydroxyl groups, whereas isosorbide bicycle has only 2.These -OH groups can be reacted with:
- Either fatty acids, to add to the molecule one or various lipophilic groups.
- A polyethylene oxide condensate, to increases hydrophilicity.

Because of these two possibilities, and the fine tuning that can be adjusted for each of them, it is feasible to prepared a "à-la-carte" surfactant molecule. Commercial sorbitan esters (SPAN brand or equivalent) and their ethoxylated counterparts (TWEEN brand or equivalent)can have a lipophilic group ranging from monolaurate (one C_{12}) to trioleate (3 C_{18}). In the ethoxylated products the ethylene oxide groups (often 20) are disseminated on the different available -OH groups, before the esterification is carried out. The following figure indicates a likely formula for an isomer of sorbitan 20 EO monolaurate, sold as TWEEN 20, which obviously displays a hydrophilic part which is much bulkier than its lipophilic tail.

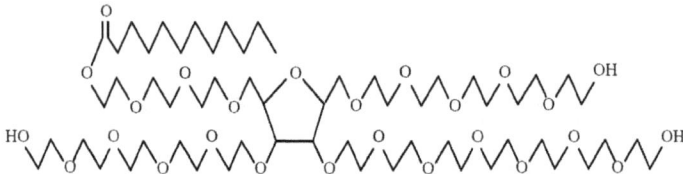

These molecules seem very complex. Nevertheless, they are quite easy to manufacture from commonly available natural raw materials, e.g. fat and sugar, which make them biologically compatible for food and pharmaceutical use.

It is very easy to adjust hydrophilicity of these surfactants, either my manipulating the EO/fatty acid ratio in a molecule, or by mixing different species. It is worth noting anyway that the ethoxylation and esterification do result in different species. Hence, the commercial product is always a mixture of many different substances. Mixtures are known to produce excellent emulsifying agents widely employed in food conditioning (creams, margarine, butter, ice cream, mayonnaise) as well as in pharmaceuticals and cosmetics. They have been used more recently in other applications, such as the preparation of micro emulsions, with for instance the sorbitan 20 EO trioleate (TWEEN 85) specie.

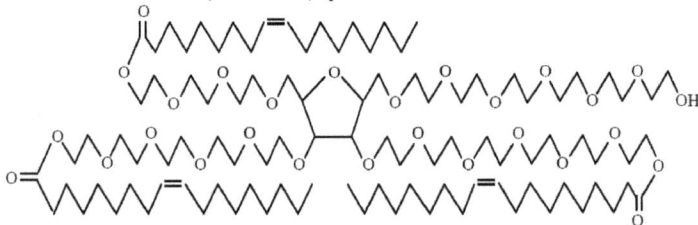

7.6.4. Nitrogenated non ionic surfactants

It has been said previously that amines and amides can be ethoxylated. The first EO group must be added at acid pH, whereas the other ones (from the second group on) are added at alkaline pH (see section 7.6.1(c)).

Products such as ethoxylated amines consist in a fatty amine with one or two polyethylene glycol chains. Those with only 2-4 EO groups behave as cationic surfactants at acid pH. They are used as corrosion inhibitors and emulsifiers with a better water solubility than most cationics. Imidazoles, i.e. cyclic alkyl-diamines, are ethoxylated to produce fabric softeners for machine washing, which also provide an anticorrosion protection for the hardware.

Ethoxylated alkyl amides are good foamer, which are used as additives. Because of their partially cationic character, they also provide antistatic and anticorrosion effects.

Ethoxylated and acylated urea also result in fabric softening substances. The same is attains with an imides.

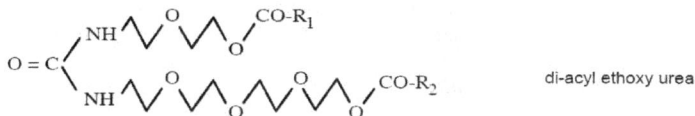

di-acyl ethoxy urea

ethoxylated imide

All these products are biocompatibilizing agents. They exhibit a favorable effect on the skin tissues and are used as additives in hand dish-washing products and liquid soaps. Tertiary amine oxides are used as foam boosters. It is worth noting that the polarization of the N-O bond in which the nitrogen atom provides both required electrons, results in a negative oxygen atom which is able to capture a proton in aqueous solution. As a consequence the actual form of an amine oxide foam booster is the cationic hydroxylamine.

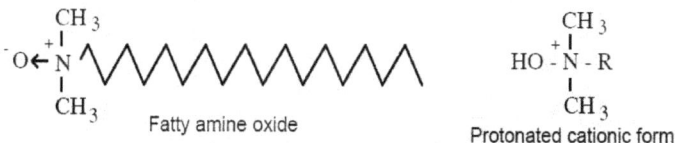

Fatty amine oxide

Protonated cationic form

Commercial amine oxides contain one long alkyl chain and two short (usually methyl) alkyls as for instance the oxide of stearyl dimethyl ammonium. Some products include two amine oxide groups, with the amine H often replaced by ethanol groups as in the following foam booster used in bubble bath, hand dish washing detergents and baby shampoos.

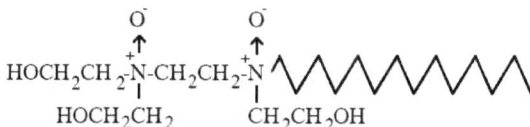

7.7 Cationic surfactants

Cationic surfactants account for only 5-6% of the total surfactant production. However, they are extremely useful for some specific uses, because of their peculiar properties. They are not good detergents nor foaming agents, and they cannot be mixed in formulations which contain anionic surfactants, with the exception of non quaternary nitrogenated compounds, or when a cationic complex synergetic action is sought. Nevertheless, they exhibit two very important features.

First, their *positive charge* allows them to adsorb on negatively charged substrates, as most solid surfaces are at neutral pH. This capacity confers to them an antistatic behavior and a softening action for fabric and hair rinsing. The positive charge enables them to operate as floatation collectors, hydrophobating agents, corrosion inhibitors as well as solid particle dispersant. They are used as emulsifiers in asphaltic emulsions and coatings in general, in inks, wood pulp dispersions, magnetic slurry etc.

On the other hand, many cationic surfactants are *bactericides*. They are used to clean and as peptize surgery hardware, to formulate heavy duty disinfectants for domestic and hospital use, and to sterilize food bottle or containers, particularly in the dairy and beverage industries.

7.7.1 Linear alkyl amines and alkyl ammoniums

(a) Nomenclature

Most used cationic surfactants are fatty amines, their salts, and quaternary derivatives. Actually fatty amines are not cationic but anionic surfactants. However, they are generally classified with cationics because they are mostly used at acid pH, in which their salts are cationic.

The amine is labeled as primary, secondary or tertiary respectively when the nitrogen is linked with 1, 2 our 3 alkyl groups. If the nitrogen possesses 4 bonds with C atoms, the compound is called a quaternary ammonium.

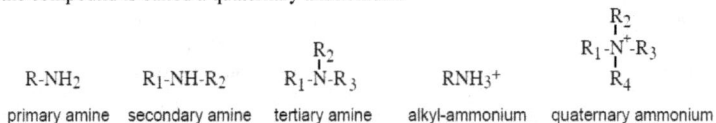

$$R\text{-}NH_2 \qquad R_1\text{-}NH\text{-}R_2 \qquad R_1\text{-}\overset{\overset{\displaystyle R_2}{|}}{N}\text{-}R_3 \qquad RNH_3^+ \qquad R_1\text{-}\overset{\overset{\displaystyle R_2}{|}}{\underset{\underset{\displaystyle R_4}{|}}{N}}{}^+\text{-}R_3$$

primary amine   secondary amine   tertiary amine   alkyl-ammonium   quaternary ammonium

In an ammonium structure, the nitrogen $C_{14}H_{29}NHCH_3$ Tetradecyl methyl amine atom gives two electrons to ensure the fourth bond, and thus remains with a positive charge. Alkyl-ammonium ions are produced in acid medium by the reaction of a proton with the amine. The resulting salt (in general chloride or bromide) is soluble in water thanks to the cation solvatation.

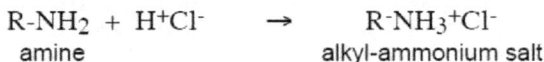

$$R\text{-}NH_2 \ + \ H^+Cl^- \qquad \rightarrow \qquad R\text{-}NH_3^+Cl^-$$
amine   alkyl-ammonium salt

Fatty amines come from fatty acids; hence their chain is linear with a even number of carbon atoms. The IUPAC nomenclature uses common names.

$C_{12}H_{25}NH_2$ Dodecyl amine or Lauryl amine, or even coco (C_{12}-C_{14}) amine

When the substance contains more than one long alkyl group, the longest is named first as in the following substances.

$C_{14}H_{29}NHCH_3$ Tetradecyl methyl amine

$C_{16}H_{33}\text{-}N^+(CH_3)_3 \ Br^-$ Cetyl trimethyl ammonium bromide (CETAB)

                                  or Hexadecyl trimethyl ammonium bromide (HTAB)

Another common product is formed with a long alkyl chain and different short substituent such as two methyl and one benzyl group.

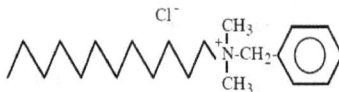

Benzalkonium or alkyl dimethyl benzyl-ammonium chloride

(b) Fatty amine synthesis

There are several methods. The mostly used in practice starts from a fatty acid and transforms it into amide, nitrile, primary amine, secondary amine, tertiary amine and finally quaternary ammonium.

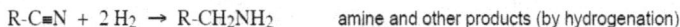

$R\text{-}COOH + NH_3 \rightarrow RCONH_2 + H_2O$ amide (by dehydration)

$R\text{-}CONH_2 \rightarrow RC\equiv N + H_2O$ nitrile (by heating/dehydration)

$R\text{-}C\equiv N + 2 H_2 \rightarrow R\text{-}CH_2NH_2$ amine and other products (by hydrogenation)

If the primary amine is transformed in secondary amine by catalytic removal of ammonia, the two alkyl groups are the same.

$$2 R\text{-}NH_2 \rightarrow R\text{-}NH\text{-}R + NH_3$$

In general, only one long alkyl group is required. Hence, the secondary amine is obtained by methylation, either by reaction of methyl amine on nitrile or by reductive methylation with formaldehyde.

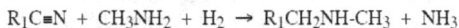

$$R_1C\equiv N + CH_3NH_2 + H_2 \rightarrow R_1CH_2NH\text{-}CH_3 + NH_3$$

$$R\text{-}NH_2 + 2 HCHO + 2 H_2 \rightarrow R\text{-}N(CH_3)_2 + 2 H_2O$$

(c) Preparation of quaternary alkyl ammoniums (QUATS)

Primary and secondary amines are quaternized by exhaustive methylation with methyl chloride, with removal of produced HCl in order to displace the reaction.

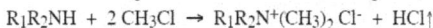

$$R_1R_2NH + 2 CH_3Cl \rightarrow R_1R_2N^+(CH_3)_2\ Cl^- + HCl\uparrow$$

Another way is to reacts an alkyl bromide with a tertiary amine. This is the usual way to prepare the cetyltrimethylammonium bromide (CETAB).

$$C_{16}H_{33}Br + N(CH_3)_3 \rightarrow C_{16}H_{33}\text{-}N^+(CH_3)_3\ Br^-$$

If a sulfate anon is required, the quaternization of the tertiary amine is carried out with dimethyl or diethylsufate.

$$R_1R_2CH_3 + (CH_3)_2SO_4 \rightarrow R_1R_2N^+(CH_3)_2 + CH_3\ SO_4^-$$

All these methods result in alkyl ammoniums displaying different alkyl groups.

(d) Uses of cationic surfactants

It can be concluded from the previous sections that the attainment of an amine o alkyl ammonium surfactant requires a chain of chemical reactions which are more or less selective and not necessarily complete. Consequently, only a small part of the original raw material ends up as the desired product. This is why cationic surfactants are in general more expensive than anionics such as sulfonates or sulfates. Hence, cationic surfactants are used only in applications in which they cannot be substituted by other surfactants, i.e. those which require a positive charge or a bactericide action.

They are found as antistatic agents in fabric softeners and hair rinse formulas. They are used in textile manufacturing to delay dye adsorption. In this application they compete with dye and thus slow down their adsorption and help attaining an uniform coloration. Their action as corrosion inhibitor in acid environment is similar, but in this case they compete with H+ ions. Collectors for mineral floatation are often ammonium salts or quarts. Asphalts emulsions for roadway pavement and protective coatings and paints are often stabilized by fatty amine salts (at acid pH) or quarts (at neutral pH).

Benzalalconium and alkyltrimethyl ammonium chloride or bromides are used as antiseptic agents, disinfectants and sterilizing agents. They are also incorporated as additive in non-ionic detergents formulation for corrosion inhibition purposes, and (in very small quantity) in anionic powdered formulas to synergize detergency.

7.7.2. Other cationic surfactants

There are many other cationic surfactants, but only a few are used in practice, in the following applications.

Linear diamines contain 2 or 3 methylene groups between the two amine groups, which are not equivalent, since one is much less dissociated that the other. Depending on the pH, only one or the two groups are quaternized, and thus carry a positive charge, a feature which can because in practical application, such as dispersant for asphalt emulsion or hydrophobant for earth roads.

Another important cationic class contains aromatic or saturated heterocycles including one or more nitrogen atoms. This is the case of a well used n-dodecyl pyridinium chloride, which is prepared by reacting dodecyl chloride on pyridine, an aromatic amine similar to benzene, in which a CH is replaced by nitrogen, e.g., C_5H_5N. N-dodecyl pyridinium chloride is used as bactericide and fungicide. If a second hydrophilic group is added (amide, ethylene oxide) the product is the both a detergent and a bactericide.

A cyclic compound with two nitrogen atoms, so-called imidazole is prepared from ethylene diamines. It has been used in the past two decade as a softener that is compatible with anionic formulas since it is not strictly cationic.

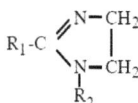

$$R_1\text{-}C \underset{\underset{\underset{R_2}{|}}{N-CH_2}}{\overset{N-CH_2}{\diagup}}\diagdown|$$

Where R_1 = long alkyl group and R_2 = short alkyl group, sometimes with -$CONH_2$ etc. All these substances can be quaternized to exhibit a positive charge. They are called imidazolium salts.

The last cationic class worth citing here are morpholine compounds. Morpholine is a saturated cycle containing both a oxygen and a nitrogen atom. The dialkylation of the nitrogen atom results in a salt called N, N-dialkylmorpholinium. When quaternization is carried out with dodecyl methyl sulfate, the resulting substance exhibits two surfactant ions, e.g. lauryl sulfate and N, N cetylmethyl morpholinium. These strange surfactants are called equiionic because they contain both anionic and cationic surfactant specie. They don't seem anyway to have any peculiar application.

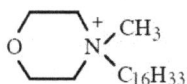

$$O\diagup\diagdown N^{+}\diagup^{CH_3}_{\diagdown C_{16}H_{33}}$$

$$C_{12}H_{25}\text{-}O\text{-}SO_3^{-}$$

N,N-Cetylmethyl morpholinium cation   Dodecyl sulfate anion

Other cyclic products have been prepared, particularly with three nitrogen atoms like triazoles, but their cost is prohibitive.

7.7.3. Nitrogenated surfactants with a second hydrophile

Cationic surfactants cannot be mixed in general with anionics, since they react with one another to produce insoluble catanionic compounds. This is quite a practical problem since most inexpensive formulas contain anionics of the sulfonate or sulfate ester type, and it would be desirable to add to them some cationic substance for softening purpose. The incompatibility problem is circumvented by using a nitrogenated surfactant of the amine type, i.e. with no actual positive ion, whose water solubility is enhanced by incorporating a second polar group.

(a) Amide- ester and ether- amines

It was previously mentioned that an amide can be produced by reacting a fatty acid with ammonia. Another way to prepare an amide is to react a fatty acid with a short chain alkyl amine or diamines, e.g. ethylene diamine or diethylene triamine. The resulting amide is used as textile softeners.

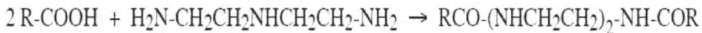

$$2 \text{ R-COOH} + \text{H}_2\text{N-CH}_2\text{CH}_2\text{NHCH}_2\text{CH}_2\text{-NH}_2 \rightarrow \text{RCO-(NHCH}_2\text{CH}_2)_2\text{-NH-COR}$$

The hydrophilicity of this substance (dialkylamido triamine) can be enhanced by quaternization or by adding a nonionic moiety such as polyethylenglycol. They are used as antistatic agents and bactericides in the textile industry. Ester amines are prepared by reacting fatty acid with an ethanol amine.

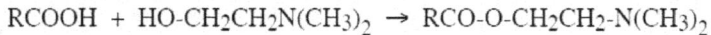

$$\text{RCOOH} + \text{HO-CH}_2\text{CH}_2\text{N(CH}_3)_2 \rightarrow \text{RCO-O-CH}_2\text{CH}_2\text{-N(CH}_3)_2$$

Whereas ether amines are prepared by condensing an alcohol on acrylonitrile double bond, followed by the hydrogenation of the nitrile.

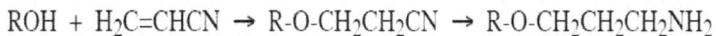

$$\text{ROH} + \text{H}_2\text{C=CHCN} \rightarrow \text{R-O-CH}_2\text{CH}_2\text{CN} \rightarrow \text{R-O-CH}_2\text{CH}_2\text{CH}_2\text{NH}_2$$

(b) Oxy and ethoxy amines

Amine oxides are prepared by reacting peroxide or a peracid with a tertiary amine. The amine oxide possesses a semi polar $\rightarrow$O bond in which the nitrogen atom provides the two electrons so that there is a strong electronic density on the oxygen atom. Amine oxides capture a proton from water to become quaternized cationic hydroxyamines at acid pH, and remain nonionic in neutral and alkaline conditions. They are among the best foam boosters available at neutral and alkaline pH, with additional corrosion inhibition properties at neutral pH.

$$\text{O} \leftarrow \text{NR(CH}_3)_2 + \text{H}^+ \rightarrow \text{HO-N}^+\text{R(CH}_3)_2$$

In section 4 it was seen that ethylene oxide can be poly condensed on amines to produce first an alacnol amine, then ethoxy amines that can be quaternized with an alkyl chloride.

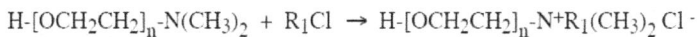

$$\text{H-[OCH}_2\text{CH}_2]_n\text{-N(CH}_3)_2 + \text{R}_1\text{Cl} \rightarrow \text{H-[OCH}_2\text{CH}_2]_n\text{-N}^+\text{R}_1\text{(CH}_3)_2 \text{ Cl}^-$$

Ethylene oxide can also be condensed on an amine in presence of water to produce an ethoxy-ammonium.

$$\text{RN(CH}_3)_2 + n \quad \text{H}_2\text{C}\overset{\text{O}}{-}\text{CH}_2 \rightarrow \text{H-(OCH}_2\text{CH}_2)_n\text{-N}^+\text{R(CH}_3)_2 \text{ Cl}^-$$

These surfactants are used in textile industry as untangling and softening agents, as well as corrosion inhibitors.

(c)Alkanol-amides

Fatty acid alkanol-amides are commonly used as foaming and wetting agents in hand dish-washing detergents, shampoos and bar soaps, particularly the diethanollauryl (coco) amide.

(d) Amino acids

Amino-acids are amphoteric surfactants which contain both an acid and an amine group. Depending on the substance and on the pH, they can exhibit anionic or cationic tendencies. They will be discussed in the next chapter.

(e) Quartenry ammonium salt and phospholipids

This means that primary, secondary and tertiary amines are not very effective surfactants. Apart from the fact that many amines have unpleasant odours. There is, however, another class of compounds related to amines. If we take a look at the ammonia molecule, the nitrogen atom possesses a lone pair of electrons which it can donate to a hydrogen ion, forming the ammonium ion. If one or more of the hydrogen atoms is replaced by an organic group, then this ability to donate the electrons to a hydrogen ion is weakened. However, if instead of a hydrogen ion, we introduce a fourth organic group we produce a compound which is a relatively strong base and which is not readily returned to the amine by the addition of alkali. Such compounds are known as quaternary ammonium compounds and are the basis of a number of cationic surfactants.

Quaternary ammonium compounds have the general formula

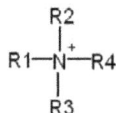

$$R_1 \overset{\overset{\displaystyle R_2}{|}}{\underset{\underset{\displaystyle R_3}{|}}{N^+}} R_4$$

Quaternary ammonium salt

The $R_1 - R_4$ represent hydrocarbon groups. If three of these are relatively short and one is much longer, then you have a cationic surfactant.

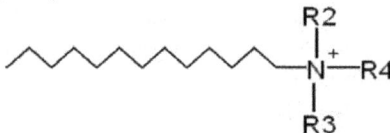

$$\wedge\wedge\wedge\wedge\wedge\wedge\wedge \overset{\overset{\displaystyle R_2}{|}}{\underset{\underset{\displaystyle R_3}{|}}{N^+}} R_4$$

Many such compounds have the advantage that they have antibacterial properties. This arises because the cell wall comprises compounds called "phospholipids"

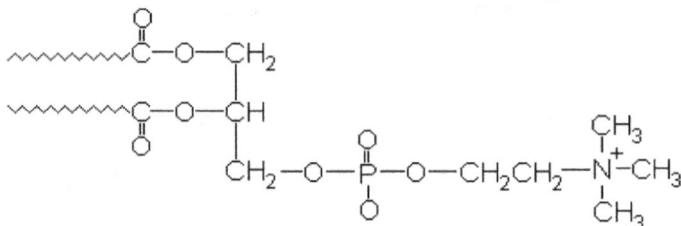

$$\text{Phospholipids structure}$$

Phospholipids

These compounds have a surfactant structure and form a double layer with the hydrocarbon tails adjacent.

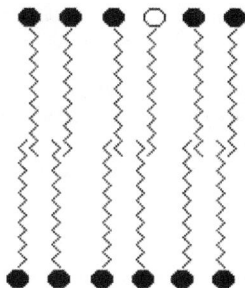

Phospholipid bilayer

As illustrated, quaternary ammonium compounds can replace phospholipids in the cell wall and ultimately cause disruption of the cell wall. Another application of cationic surfactants is in fabric softeners. Many fabrics have negatively charged groups on their surface which attract the positively charged cationic surfactants. The hydrocarbon "tail" on the molecule gives the surface of the fabric a soft feel to it.

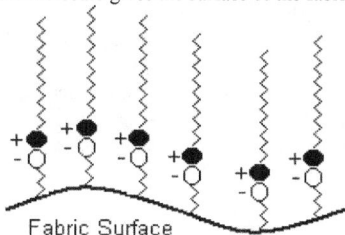

Fabric Surface

7.8 Other surfactants

This final chapter gathers all other surfactants. To start with, there are the last ones to be classified according to their ionization in water, e.g. amphoteric surfactants, which exhibit both a positive and a negative charge. On the other hand those which are different from conventional surfactants because of some peculiarity, even if they can be classified according to their ionization: silicon surfactants, fluorinated surfactants, and polymer surfactants.

7.8.1 Amphoteric surfactants

Amphoteric or zwitterionic surfactants have two functional group, one anionic and one cationic. In most cases it is the pH which determines which of the groups would dominate, by favoring one or the other ionization: anionic at alkaline pH and cationic at acid pH. Near the so called isoelectric point, these surfactants display both charges and are truly amphoteric, often with a minimum of interfacial activity and a concomitant maximum of water solubility. Amphoteric surfactants, particularly the amino acid ones are quite biocompatible, and are used in pharmaceuticals and cosmetics.

(a) Amino propionic acids

The general formula for amino propionic acids is RN^+-CH_2-CH_2-COO^-. Their isoelectric point is around pH = 4. They are soluble in acid or alkaline solutions. They adsorb on skin, hair, and textile fibers. They are used as antistatic and lubricants for hair and fabrics. The dodecyl amino propionic acid is used in cosmetics as wetting agent and bactericide. At high pH it is good detergent and foaming agent. However, due to their carboxylic acid groups, these surfactants are sensitive to divalent cations.

(b) Imido propionic acids

Their general formula is $HOOC-CH_2-CH_2-RN^+H-CH_2-CH^2-COO^-$. Their isoelectric point is around pH = 2-3. They are thus more water soluble than the previous ones. They are used as textile softeners. Dicarboxylic compounds of alkyl imidazole, in which the alkyl group is located on the carbon placed between the nitrogen atoms, are used in cosmetics and de lux soap bars.

(c)Quartenized compounds

Quaternized compound have similar structures. The most important are betaines and sulfobetaines or taurines, which have a single methylene group between the acid and the quaternary ammomium.

$$R-N^+(CH_3)_2-CH-COO^- \qquad \text{alkyl betaine}$$

$$R-N^+(CH_3)_2-CH_2-SO_3^- \qquad \text{alkyl sulfobetaine}$$

These surfactants are amphoteric at neutral and alkaline pH, and cationic at acid pH (at which the carboxylic acid is not ionized). Since the nitrogen atom is quaternized, these surfactants always display a positive charge. They tolerate a high salinity, particularly divalent cations, e.g. calcium and magnesium. They are the most used class of amphoteric surfactants.

They are found in softeners for textiles, hair rinse formulas, and corrosion inhibition additives. They are good foam boosters because of their cationic characteristics. Sulfobetaines have an anionic group (sulfonate) which tolerates calcium ions, and are used as lime soap dispersing agents (LSDA) in de lux soap bars and detergents.

7.8.2 Silicon surfactants

The hydrophobic character of silicon oil, particularly dimethylpolysiloxane is well known. The introduction of organosilicon group in a surfactant molecule tends to increase its hydrophobicity. Since Si is a heavier atom than C, a similar hydrophobicity is attained with less Si atoms than C atoms. Essentially all surfactant types can be made with silicon based hydrophobic tail by replacing several C atoms by one Si atom or one dimethylsiloxane group.

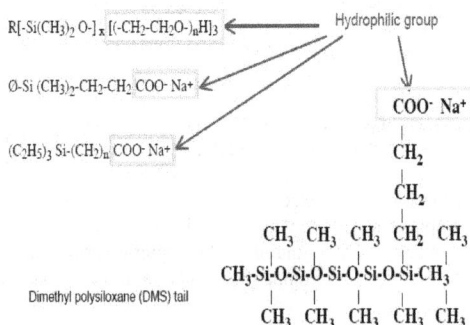

Most of these surfactants can be crystallized in acetone to attain a high purity degree. Some of them are used in pharmacy as anti-fluent surfactants since they are biologically inert.

7.8.3 Fluorinated surfactants

Hydrogen atoms of the surfactant hydrocarbon tail can be substituted by halogens, particularly F to produce fluorinated hydrophobe, which exhibit properties similar to polymerized tetrafluoro ethylene (PTFE), known under the commercial brand name Teflon, high chemical inertia, mechanical and thermal resistance, low surface energy, thus very high hydrophobicity. As silicon compounds, the fluorinated tails sometimes won't mix with hydrocarbons. Hence, the poly tetra fluorinated hydrophobe is also lipophobe with respect to hydrocarbons.

Per fluorinated surfactants are prepared by polymerizing tetra fluoroethylene in presence of methanol, so that the end product is an alcohol. The alcohol is then oxidized to produce a carboxylc acid or treated synthesize another hydrophilic group: sulfate, amine or phosphate.

$$H(CF_2CF_2)_n\text{-}CH_2\text{-}OH \qquad\qquad n = 2\text{-}4$$

Per fluorinated acid in C_5 ($n = 2$) gives a sodium salt with good surfactant properties. This is perfectly consistent with the fact that its molecular weight (increased by 8 F atoms) is close to sodium palmitate MW. Salts of per fluorinated carboxylic acids are surfactants when they possess from 5 to 9 carbon atoms. These salts are much more dissociated than their hydrocarbon counterparts and tolerate high salinity and divalent cations. However, they are much more expensive and their use is justified only by very special conditions, as for instance in fire extinction foams, in which their thermal resistance is imperative.

Per fluorinated carboxylates and sulfonates produce mono layers with less lateral interactions than their hydrocarbon counterparts. They are able to turn surface non-wettable to both water and organic solvents. They produce a superficial (air-aqueous solution) tension down to 15 mN /m, i.e. twice as low as the value reachable with the best tension reducing hydrocarbon surfactants.

7.8.4 Polymeric surfactants

A macromolecule can obviously exhibit an amphiphilic structure. Asphaltenes, which are natural compounds found in crude oils, have polar and non polar groups. However the location and segregation of these groups is often ill-defined, or at least less defined than in smaller molecules. There are two main configurations: *"block"* and *"graft"*, which are illustrated in the following scheme, where H and L represent hydrophilic and lipophilic monomer units, respectively.

Block type polymer 　　H-H-H-H-H-H-H-H-L-L-L-L-L-L-L-L-L-L

Graft type polymer 　　L-L-L-L-L-L-L-L-L-L-L-L-L
　　　　　　　　　　　　|　　　　|　　　　|
　　　　　　　　　　　　H　　　H　　　H

In the first case hydrophilic monomer units H are linked together to form an hydrophilic group, and lipophilic units L just do the same to form a lipophilic group. The result is a macromolecular surfactant with well defined and separated hydrophilic and lipophilic parts, which is just much bigger than a conventional surfactant molecule. The most used block polymeris the so called copolymer of ethylene-oxide and propylene oxide either with two or three blocks (as shown below). Although the hydrophilic and lipophilic parts are quite separated, the polymer polarity segregation is not that obvious since both groups are slightly polar, one (Poly EO) just barely more polar than the other (poly PO).

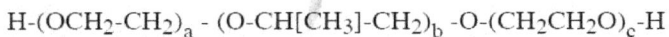

$$H-(OCH_2-CH_2)_a - (O-CH[CH_3]-CH_2)_b -O-(CH_2CH_2O)_c-H$$

These surfactants have many uses, in particular as colloid and nano emulsion dispersants, wetting agents, detergents and even additive to dehydrate crude oils. However, most polymeric surfactants belong to the second (graft) type, particularly synthetic products such as poly electrolytes which are not strictly surfactants or are not used for their surfactant properties. It is the case of hydrosoluble or hydrodispersible polyelectrolytes which are utilized for the entire position, dispersant and viscosity-enhancing properties such as carboxymethyl cellulose, polyacrylic acid and derivatives.

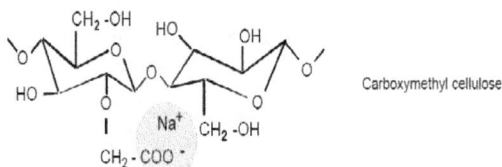

Carboxymethyl cellulose

There exist a wide variety of graft type polymeric surfactants. The main way to prepare them is to produce a lipophilic polymer with functional group where a hydrophilic group can be attached later, for instance the copolymer of maleic acid and styrene. The anhydride group readily reacts with water and any alcohol, amine, acid, etc... to produce a hydrophilic group.

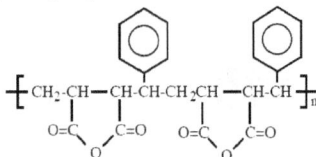

Polycarbonates are produced by polycondensation of an alkyl phenol with salicylic acid and formaldehyde.

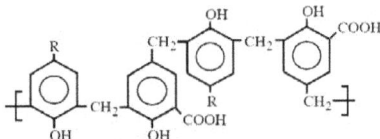

This kind of polymer can be sulfonated (on an aromatic ring) or sulfated (on a –OH group). The degree of sulfonation/sulfatation allows adjusting the hydrophilicity of the final compound. It is, for instance, the case of sulfonated polystyrene.

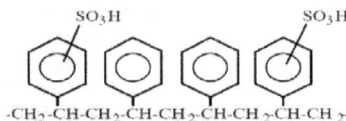

Alkyl phenol formlaldehyde resins can be turned hydrophilic by adding polyethylene oxide or sulfate or ether-sulfate groups. These products are currently in use as crude oil dehydration additives (to break water-in-oil emulsion at well head).

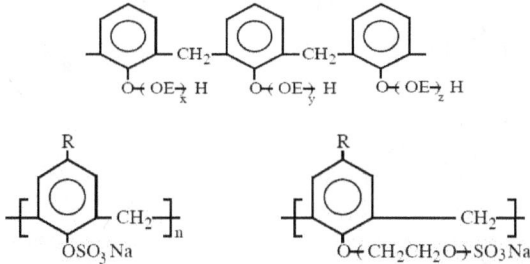

Similar substances can be prepared with cationic characteristics, such as polyvinyl pyridines or pyrolidones.

7.8.5 Association polymers

It is well known that most surfactants self-associate in solution to produce aggregates so called micelles, liquid crystals or micro emusions. Some surfactants, in particular double tail species, tend to associate as a bilayer instead. It is worth noting that this bilayer is the structural skeleton of many biological membranes in plants and animals, as those produced by phospholipids association:

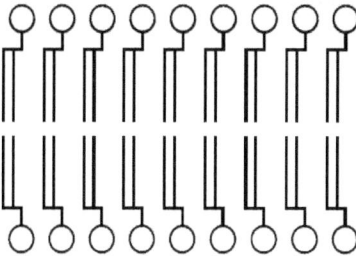

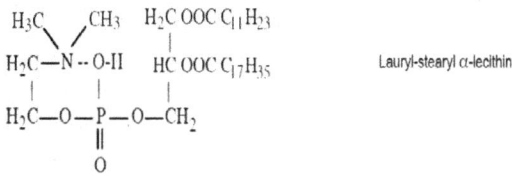

$$H_3C \diagdown CH_3 \qquad H_2C\ OOC\ C_{11}H_{23}$$
$$H_2C—N--O-II \qquad HC\ OOC\ C_{17}H_{35}$$
$$H_2C—O—P—O—CH_2$$
$$\overset{\parallel}{O}$$

Lauryl-stearyl α-lecithin

273

8. Dye and pigments

8.1 Introduction

A dye or dyestuff is usually a coloured organic compound or mixture that may be used for imparting colour to a substrate such as cloth, paper, plastic, lather or food in a reasonably permanent fashion. In other words, a dyed substrate should resistant to a normal laundry or cleaning procedures (wash fast) and stable to light.

In the olden days, when there was no industry on synthetic dyes the fibers were used to be dyed by the colouring matters obtained from plant and animal, e.g. a yellow dyestuff saffron from the dried flowers of the colour thistle, indigo (blue dye) from the leaves of indigo plants, alizarin or Turkey red or madder from the roots of madder plant, cochineal (red dye) from the dried bodies of the females of the insect Coccus cacti, Tyrian purple (red) from the snail Murex brandarin etc.

The first synthetic dyestuff was prepared in 1856 by an eighteen year old Englishman W.H. Perkin, assistant at the Royal College of Chemistry, while he was studying the action of oxidizing agent ($K_2Cr_2O_7$-H_2SO_4) on crude aniline in an attempt to synthesized quinine. Instead of quinine, he obtained a black resinous mass from which he extracted a lilac colouring matter by means of alcohol. He purified this and found that it had possibilities as a dye. Perkin's violet dye, commonly known as *mauve*, contained mainly N-phenylphenosafranine and its homologs.

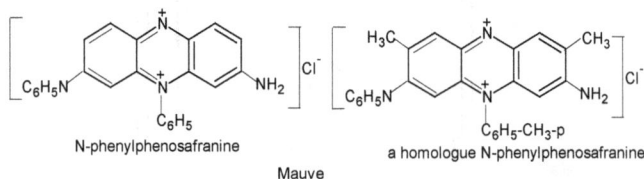

N-phenylphenosafranine   a homologue N-phenylphenosafranine

Mauve

In 1857, Perkin began the manufacture of mauve on commercial scale. The discovery of mauve stimulated the study and researches on the synthetic dyes, and the progress was very rapid. In 1859 was synthesized, in 1863, aniline yellow was synthesized and in 1869 Graebe and Libermann synthesized the first dyestuff alizarin. A great variety of other synthetic dyes, exceeding in purity, low cost, range and brilliancy of color, fastness to light etc. became soon available.

Since the most of synthetic dyestuff are synthesized from a few starting material, such as benzene, naphthalene, phenol, aniline, etc., which are obtained from coal-tar, the synthetic dyes are also known as coal-tar dyes.

Structural features of a dyes: For a substance to act as a dye there are two essential structural features:

(i) Presence of chromophores (Greek chroma-color, phoros-bearer): These are some groups the presence of which makes the compound colored. The important chromophores are: -N=O, -NO$_2$, -N=N-, -C=O, -C=S, -CH=CH- etc.

(ii) Presence of auxochromes (Greek auxo-increase, chroma-color): As already described that a dye should be fast to light, water and soap so it must be attached to the fibers by means of stable chemical bonds. These chemical bonds are formed by some groups which may be either acidic or basic in nature. Such groups are known as auxochromes, some of which are: -OH, -COOH, -SO$_3$H (acidic), -NH$_2$, -NHR, -NR$_2$ (basic).

These groups also serve to deepen or intensity the color even though they impart no color in the absence of chromophore, and hence they are known as auxochrome (color intensifying groups). Chromogens without an auxochrome (color intensifying groups) can never act as dye, e.g. azobenzene although red colored yet not a dye; on the other hand, p-aminoazobenzene is a dye (aniline yellow).

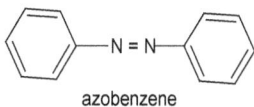

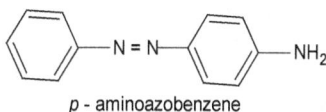

azobenzene

p - aminoazobenzene

Bathochromic and hypsochromic effects: A structural change which can shift the absorption towards higher wave lengths will deepen the color according to sequence yellow, orange, red, purple, violet, and blue, green, black. Any group or a factor that produces such change is known as bathochromic and effect, i.e. deepening of color is known as bathochromic effect. Controversially any group which shifts the absorption band of longer wave length to that of shorter wave length will lighten the color(i.e. in the order black to yellow described above), and such group and its effect produced are known as hypsochrome and hypsochromic effect, respectively.

The bathochromic and hypsochromic effects are produced by the presence of some groups, a summary of all of them may be given in the following points.

(i) The introduction of additional auxochromes such as –OH and –NH$_2$ results in a bathochromic effect, e.g. aniline yellow is changed to orange colored dye, chrysoidine having two –NH$_2$ groups.

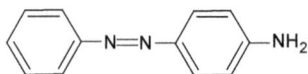 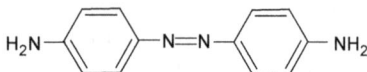

aniline yellow   chrysodine(orange)

It is important to remember all the dyes may not necessarily be colored substances. Therefore, optical brighteners or whiteners which may be called white dyes may be included in the term dye. Previously dyes were obtained from animal and vegetable sources. Today most of the available dyes prepared from aromatic compounds which are obtained from coal tar or petroleum.

(ii) Not only the auxochromes, but other o-p-directing groups when present on the aromatic ring also deepen the color, e.g. alkyl, aryl, and halogens all are bathochromes.

(iii) Alkylation of amino group produces bathochromic effect, on the other hand alkylation as well as acylation of phenolic group produces hypsochromic effect.

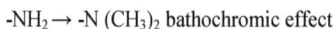

$$-NH_2 \rightarrow -N(CH_3)_2 \text{ bathochromic effect}$$

$$-OH \rightarrow -OCH_3 \text{ or } -OCOCH_3 \text{ hypsochromic effect}$$

The introduction of six methyl groups in p-nitrosaniline (red dye) nucleus to give crystal violet (violet color) is a good example of bathochromic effect.

(iv) Salt formation also produces bathochromic effect due to the more possibilities of resonance which is again due to the formation of dye ions.

$$-NH_2 \rightarrow NH_3^+Cl^-$$

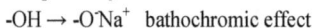

$$-OH \rightarrow -O^-Na^+ \text{ bathochromic effect}$$

8.2 Nomenclature of dyes

Of course, there is no definite nomenclature of dyes, and most of the dyes have several different names for the same dye because generally dye manufactures assign their own names to a single dye, e. g. the rose red-dye is generally named either by pararoseaniline, magenta or fuchsine. Moreover, the later used after some days indicate various factors.(i) generally they represents colours, e.g. Y for yellow, O for orange, R for red and B for blue; (ii) sometimes they represent the class of the dye, e. g. alizarin blue D indicates that the dye is a direct fuchsine S indicates that the dye is an acid *(sauer)*; (iii) sometimes they represent the property of a dye, e. g. F indicates that the dyes is fast to light.

To overcome these difficulties a **colour index** has been compiled by the Dyers and Colourists. In this colour index each dye is given its individual colour number (C. I. No.).

8.3 Requisites of a True Dye

All colored substances are not dyes. However, the requisites of a true dye are as follows:

(a) It must have a suitable color.

(b) It must have an attractive color.

(c) It must be able to attach itself to material from solution or to be capable for fixed on it.

For example azobenzene is colored but further, a dye may not be able to dye all types of substrates. For example picric acid to dye silk or wool a permanent yellow but not cotton. Thus, a dye either forms a chemical union with the substrate being dyed or it may get associate it in an intimate physical union.

(d) It must be soluble in water or must form a stable and good dispersion in water. Alternatively, it must be soluble in the medium other than water. However, it is to be remembered that pickup of the dye from the medium should be good.

(e) The substrate to be dyed must have natural affinity for an appropriate dye and must be able to absorb it from solution or aqueous dispersion, if necessary in the presence of auxiliary substances under suitable conditions of concentration, temperature and P^H.

(f) When a dye is fixed to a substrate, it must be fast to washing dry cleaning, perspiration, light, heat and other agencies. It must be resistant to the action of water, acids or alkalis, particularly the latter due to the alkaline nature of washing soda and washing soap. There is probably no dye which can be guaranteed not to alter shade under all condition.

(g) The shade and fastness of a given dye may vary depending on the substrate due to different interaction of the molecular orbital of the dye with the substrate, and the ease with which the dye may dissipate its absorbed energy to its environment without itself decomposing.

8.4 Classification of Dyes
The dyes used in various industries can be mainly classified by two ways:
(a) On the basis of the nature of chromophore group and
(b) On the basis of its mode of application
More accepted way of classification is according to their mode of application on to fabric.

(a) On the basis of the nature of chromophore group
On the basis of their chromophore groups dyes can be classified as azo, anthraquinone, nitro, nitroso, xanthenes, acridine, indigoid, thiazole etc. Amongst these different classes, azo dyes represent the largest and the most versatile class of synthetic dyes. These dyes are prepared by azo coupling with aromatic amines, phenols, naphthols or aliphatic enols. They absorb light in the visible spectrum due to their chemical structure, which is characterized by one or more azo groups (-N=N-). The major application of azo dyes is the dyeing and printing of cellulose fibers especially cotton.

Fig:1 Synthesis of azo dyes

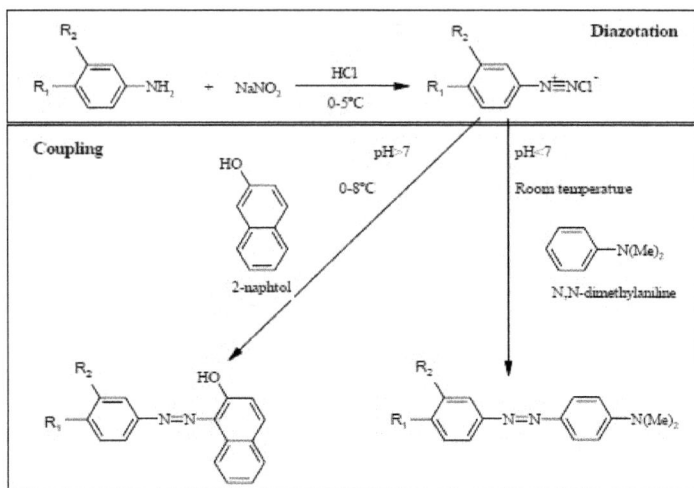

CLASS	CHROMOPHORE	EXAMPLE
Nitro dyes	$-N\begin{smallmatrix}O\\\\O\end{smallmatrix}$	C.I. Acid Yellow 24
Nitroso dyes	$-N=O$	Fast Green O
Azo dyes	$-N=N-$	Methyl Orange
Triphenyl methyl dyes		C.I. Basic Violet 3
Phthalein dyes		Phenolphtaleine
Indigoid dyes		C.I. Acid Blue 71
Anthraquinone dyes		C.I. Reactive Blue 19

(b) On the basis of its mode of application
(i) Reactive Dyes

These are dyes with reactive groups that form covalent bonds with OH-, NH_2-, or SH groups in fibers. Reactive dyes are extensively used in the textile industry because of their wide variety of color shades, ease of application and minimal energy consumption (Singh et al. 2007). The hydrolysis of the reactive groups is an undesired side reaction that lowers the degree of fixation. Thus, it is estimated 10-50% will not react with fabric and would remain hydrolyzed in the water phase. The problem of colored effluents is therefore identified mainly with use of reactive dyes. Reactive dyes are of two types

(i) In this type, dye first introduced in cyanuric chloride and then it is reacting with fibers possessing $-NH_2$ or $-OH$ in presence of alkali to form ether bond.

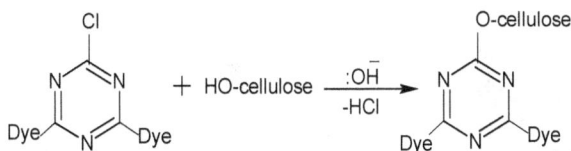

(ii) In this type, the reactive group is $SO_2-(CH_2)_2-SO_3H$ which gets hydrolysed to the vinyl sulphone ($SO_2-CH = CH_2$) in the presence of alkali and tends to form a covalent bond with – NH_2 or $-OH$ by addition reaction.

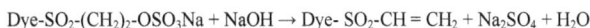

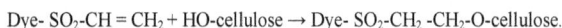

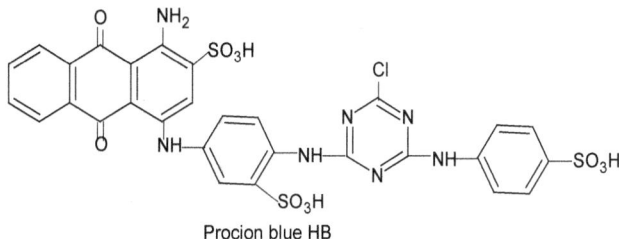

$$Dye-SO_2-(CH_2)_2-OSO_3Na + NaOH \rightarrow Dye- SO_2-CH = CH_2 + Na_2SO_4 + H_2O$$

$$Dye- SO_2-CH = CH_2 + HO-cellulose \rightarrow Dye- SO_2-CH_2 -CH_2-O-cellulose.$$

Procion blue HB

(ii) Acid Dyes

The largest class of dyes in the color index is referred as acid dyes (~ 2300 different acid dyes listed, ~ 40% of them are in current production). Acid dyes are anionic compounds which are mainly used for dyeing basic groups containing fabrics like wool, polyamide and silk and modified acryl. Application is generally made under acidic conditions which cause protonation of basic groups. The dyeing process is reversible and dyes are generally removed from fabrics during washing. Most of the acid dyes contain azo and anthraquinone or triarylmethane type of chromophore groups.

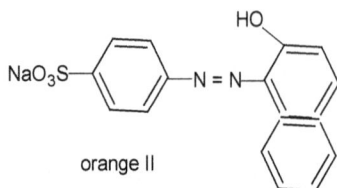

orange II

(iii) Direct Dyes

They are large, flat linear molecules. They have high affinity for cellulose fibers and they bind to the fiber through Van der Waals forces. The common salt or Glauber's salt is often used with direct dye to promote dyeing, because the presence of excess sodium ions favors establishment of equilibrium with the minimum content of dye. The dyeing process with direct dyes is reversible and exhibit poor wash fastness. A typical example of direct dye is Congo red.

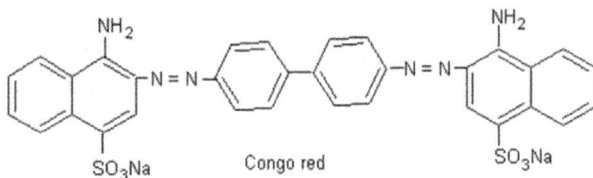

Congo red

(iv) Basic Dyes

They are cationic compound that are used for dyeing acid group containing fibers, usually synthetic fibers like modified polyacryl. Basic dyes represent ~5% of all the dyes listed in color index. They generally give intense and brilliant shades but have poor light fastness. They are used for dyeing silk and wool directly. Methyl Violet, Crystal Violet, Rhodamine, Magenta etc. are very common examples of basic dyes.

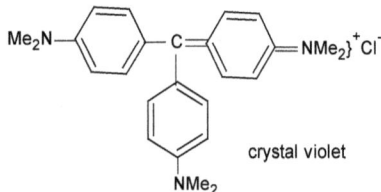

crystal violet

(v) Mordant Dyes

Mordants are the compounds which attach to the fiber and then combine with the dye to form an insoluble complex. About 600 different mordant dyes are listed in the color index but nowadays the use of mordant dyes is gradually decreased (only 23%) still, they are used for dyeing wool, leather, silk, paper and modified cellulose fibers. These dyes by themselves have poor affinity for the fiber. However, these dyes require a pretreatment of the fiber with mordants (usually metal salts such as chromium and iron salts). Most mordant dyes are azo, oxazine or triarylmethane compounds.

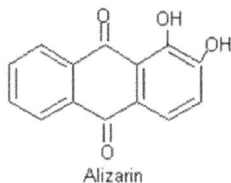
Alizarin

(vi) Disperse Dyes
They are less soluble dyes and are specifically used to dye synthetic fibers like cellulose acetate, polyester, polyamide, acryl, etc. Its diffusion requires swelling of the fiber, either by high temperature (>120C) or with the help of chemical softener. Disperse dyes form the third largest group of dyes in the color index and about 1400 disperse dyes are listed and out of which ~40% are currently manufactured. They are usually small azo or nitro compounds (yellow to red), anthraquinones (blue and green) or metal complex azo compounds (all colors).

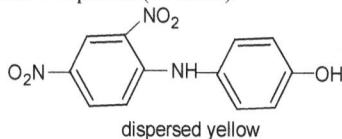
dispersed yellow

(vi) Vat Dyes
They are water insoluble dyes however their reduced forms are soluble. These dyes are applied in their reduced forms which are obtained by treating the compound with some reducing agent such as alkaline sodium dithionite. When the reduced dye is adsorbed on the fiber; the original insoluble dye is reformed upon oxidation with air or chemicals. Vat dyes offer excellent fastness, however, they are quiet expensive and must be applied with care. Such vat dyes contain indigo and anthraquinone type of chromophore group.

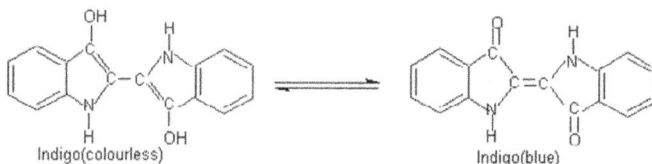

Indigo(colourless)   Indigo(blue)

(viii) Sulfur Dyes
They are heterocyclic S-ring containing polymeric aromatic compounds. Dyeing with sulfur dyes involve reduction and oxidation process, when reduced with sodium sulfide, they become soluble and exhibit affinity for cellulose and upon exposure to air they get oxidized to insoluble dye inside the fiber. They are mainly used for dyeing cellulose fibers. e.g., Nickel black

(ix) Solvent Dyes
Solvent dyes (lysochromes) are non ionic dyes that are used for dyeing substrates in which they can dissolve, e.g. plastics, varnish, ink, waxes and fats. Most solvent dyes are diazo compounds and have very less application in textile industry.

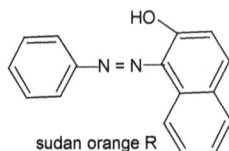

sudan orange R

8.5 Synthesis and uses of some important dyes

1. Congo red

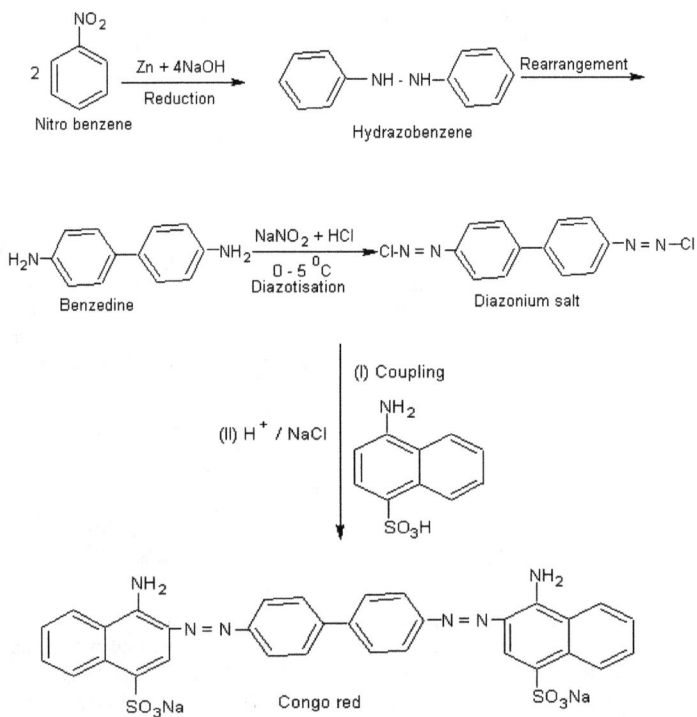

NO$_2$

2 Nitro benzene

$\xrightarrow[\text{Reduction}]{\text{Zn + 4NaOH}}$

—NH - NH—

Hydrazobenzene

$\xrightarrow{\text{Rearrangement}}$

H$_2$N——NH$_2$

Benzedine

$\xrightarrow[\substack{0 - 5\ ^0C \\ \text{Diazotisation}}]{\text{NaNO}_2 + \text{HCl}}$

Cl-N = N——N = N—Cl

Diazonium salt

(I) Coupling

(II) H$^+$ / NaCl

NH$_2$

SO$_3$H

NH$_2$ —N = N—— —N = N— NH$_2$

SO$_3$Na   Congo red   SO$_3$Na

Uses: (i) It is used as an indicator in acid-base titration.

(ii) It is used to making Congo red paper.

(iii) It is used to imparting red colour on cotton.

2. Indigo

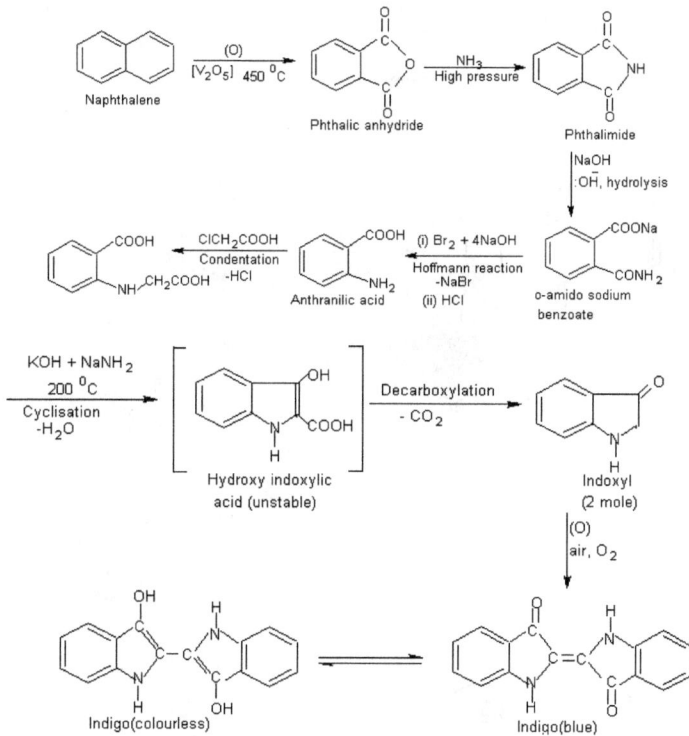

Naphthalene → Phthalic anhydride → Phthalimide → o-amido sodium benzoate → Anthranilic acid → ... Hydroxy indoxylic acid (unstable) → Indoxyl (2 mole) → Indigo(colourless) ⇌ Indigo(blue)

Uses: (i) Dyeing a blue colour on cotton

(ii) In printing

3. Alizarin

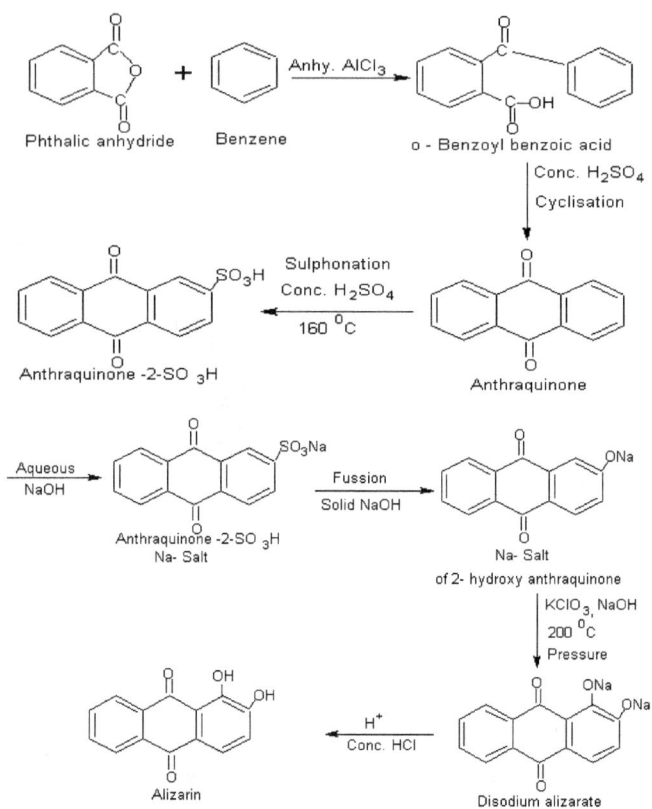

Phthalic anhydride + Benzene → (Anhy. AlCl₃) → o - Benzoyl benzoic acid

Conc. H₂SO₄ Cyclisation

Anthraquinone ← (Sulphonation Conc. H₂SO₄ 160 °C) Anthraquinone -2-SO₃H

Aqueous NaOH → Anthraquinone -2-SO₃H Na- Salt → (Fussion Solid NaOH) → Na- Salt of 2- hydroxy anthraquinone

KClO₃, NaOH 200 °C Pressure

Alizarin ← (H⁺ Conc. HCl) ← Disodium alizarate

Uses: (i) It is useful for dyeing wool and it gives different colour on wool with different mordent.

(ii) It is used in making printing ink.

(iii) It is used as staining agent and also making medicine.

4. Eosin

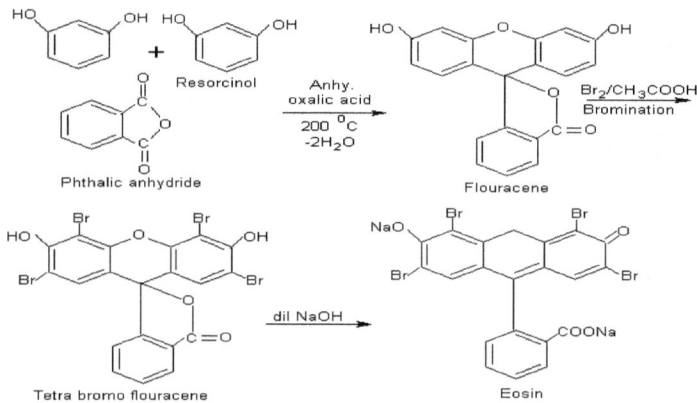

Phthalic anhydride + Resorcinol → Anhy. oxalic acid, 200 °C, -2H₂O → Flouracene → Br₂/CH₃COOH Bromination → Tetra bromo flouracene → dil NaOH → Eosin

Uses: (i) Eosin used to dye wool and silk a pure red, which a yellow fluorescence.

(ii) It is used to making red ink and refills for poster printing. It is also useful for dyeing

papers and production of lipsticks and nail polishes.

5. Methylene blue

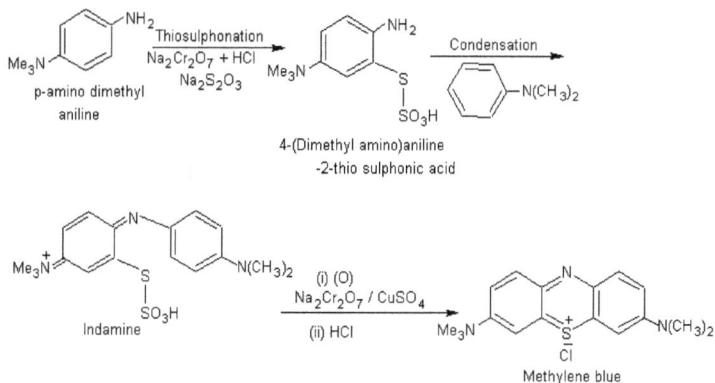

p-amino dimethyl aniline → Thiosulphonation Na₂Cr₂O₇ + HCl, Na₂S₂O₃ → 4-(Dimethyl amino)aniline-2-thio sulphonic acid → Condensation → Indamine → (i) (O) Na₂Cr₂O₇ / CuSO₄ (ii) HCl → Methylene blue

Uses: (i) It is used as an indicator in PH titration more over it is useful to determine toxicity of microorganism.

(ii) It is useful in calico printing and medicines.

(iii) It is used to dyeing cotton through tannin mordant.

285

6. Crystal violet

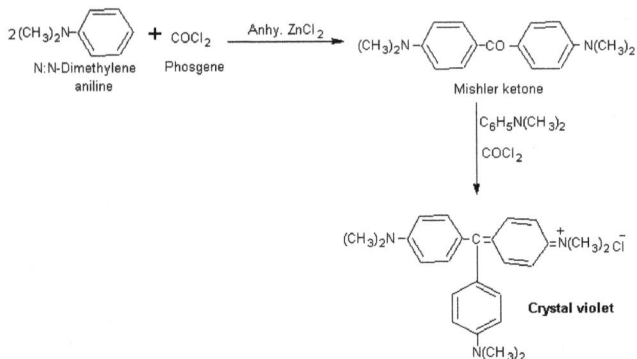

$$2(CH_3)_2N\text{-}\bigcirc + COCl_2 \xrightarrow{\text{Anhy. ZnCl}_2} (CH_3)_2N\text{-}\bigcirc\text{-}CO\text{-}\bigcirc\text{-}N(CH_3)_2$$

N:N-Dimethylene aniline   Phosgene

Mishler ketone

$$C_6H_5N(CH_3)_2$$
$$COCl_2$$

$$(CH_3)_2N\text{-}\bigcirc\text{-}C=\bigcirc=\overset{+}{N}(CH_3)_2\,\bar{Cl}$$

Crystal violet

$\overset{\cdot}{N}(CH_3)_2$

Uses: (i) It is used to dyeing silk and wool and dyeing cotton through tannin mordant gave purple flourescene.

(ii) It is used to making ink for stamp pad and pencils.

(iii) It is used as an indicator for determination of concentration of H^+ in solution.

7. Eriochrome black-T

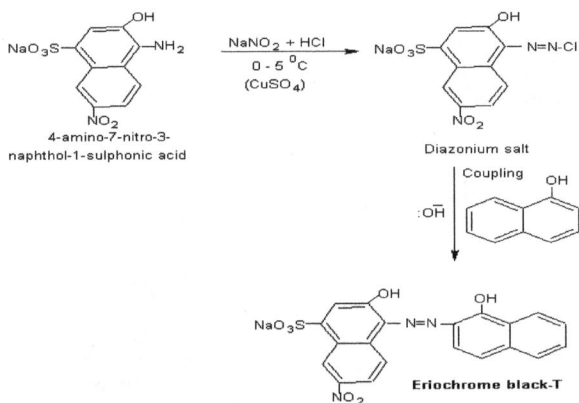

$$NaO_3S\text{-}\overset{OH}{\bigcirc}\text{-}NH_2 \xrightarrow[\substack{0-5\,^{\circ}C \\ (CuSO_4)}]{NaNO_2 + HCl} NaO_3S\text{-}\overset{OH}{\bigcirc}\text{-}N\text{=}N\text{-}Cl$$

$\overset{\cdot}{NO_2}$

4-amino-7-nitro-3-naphthol-1-sulphonic acid

$\overset{\cdot}{NO_2}$

Diazonium salt

Coupling

:OH

$$NaO_3S\text{-}\overset{OH}{\bigcirc}\text{-}N\text{=}N\text{-}\overset{OH}{\bigcirc}$$

Eriochrome black-T

$\overset{\cdot}{NO_2}$

Uses: It is very useful basic dye for dyeing cotton, wool, silk, lather and paper through tannin mordant. It has many applications in textile industries.

8. Rhodamine-B

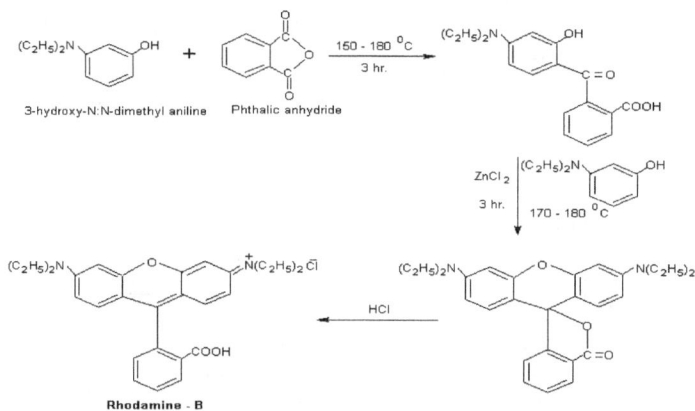

3-hydroxy-N:N-dimethyl aniline   Phthalic anhydride

Rhodamine - B

Uses: (i) It is generally used for dyeing paper.

(ii) It is a pigment and useful for making red and pink lake.

(iii) It is used for food adulteration (chilly).

(iv) It gives a faint colour on cotton, wool and silk through tannin mordant.

8.6 Colour and chemical constitution

Colour: The ordinary light is made up of rays varying wavelength which can be categorized in three headings.

Range of wave length of light (Angstrom unit)	Part of light
1000-4000	Ultraviolet
4000-7500	Visible
7500- 100,000,0	Infra red

But our eye is sensitive only to the spectral region from about $4000A$ to $7500A°$ and hence this region of ordinary light is the responsible region for producing a definite colour to a particular compound. Actually, colour is the psychological sensation produced when light of the above range of wavelengths reaches the eye. When white light falls on the substance, colour may be obtained in two ways. Firstly, if all the wavelengths of white light are absorbed except one single narrow band which is reflected band of light will be the colour of the substance, e.g. monochromatic light of wavelength $4500A°$ is blue. Secondly, if only a single band of white light of is absorbed, the substance will have the complementary colour of absorbed band, e.g. the blue colour is obtained when light of $5900A°$ region (of yellow colour) is absorbed because the composite (the sum of the other colours) of the remaining wavelengths (which are reflected) gives the blue appearance to eyes. Thus blue and yellow are known as *complementary* colours from light gives the other. Some of the common complementary are givens below.

2. Deepening or depth of colour must not be confused with the different *intensities* of the particular colour.

Wavelength absorbed, A^o	Coloured absorbed	Complementary (visible) colour
4000-4350	violet	Yellow -green
4350-4800	blue	yellow
4800-4900	Green-blue	orange
4900-5000	Blue-green	red
5000-5600	green	purple
5600-5800	Yellow-green	violet
5800-5950	yellow	blue
5950-6050	orange	Green-blue
6050-7500	red	Blue-green

In actual practice no dye gives a pure shade, i.e. does not reflect only one band of wave-length, e.g. malachite green reflects mainly green light, but also red, blue and violet to a small extent. Deepening of colour (bathochromic effect) is defined as the change in the colour in the order given above in the table, thus the following colours are in the order of increasing depth: yellow, orange, red, purple, violet, blue, green, black. Since visible colours are the complementary colours of the absorbed bands, the deepening of colour is produced by shifting the absorption band of the longer wave-length (red). Controversially, the opposite effect (hypsochromic effect) is produced by shifting the absorption band of shorter wave length. These changes are, actually, produced by the presence of a group which may said to be bathochromic or hypochromic group.

8.7 Relation between colour and constitution

Like the physical and chemical properties of organic compounds, there is a definite relationship between the colour and constitution, e.g. benzene is colourless, whereas isomer, fulvene is coloured; similarly, on reduction, certain compounds are converted into colourless compounds, which on oxidation give back the coloured compound (Graebe and Libermann 1868). Such relations between colour and constitution have been studied intensively because they have practical applications in the field of synthetic dyestuff. Various attempts have been made to formulate the laws governing the relationship between colour and constitution, but so far it had not been possible to predict the colour of a compound on the basis of its molecular structure. However, the following theories have been proposed to explain the observed general relationship exiting between colour and constitution.

8.7.1 Wilt's theory

In 1876 Wiltpur forward a theory according to which the colour of substance is mainly due to the presence of an unsaturated group known as *chromophores* (Greek *chroma*-colour, and *phores*-bearing).The important chromophores are:

$-N{=}O$	Nitroso	$\overset{H}{\underset{\vert}{-C}}{=}N$	azomethine

Structure	Name	Structure	Name
$\backslash C = S$	Thiocarbonyl	$-\overset{+}{N}\diagdown\overset{O}{_O}$	nitro
$-N = N-$	Azo	$\backslash C = O$	carbonyl
$-N = N \rightarrow O$	Azoxy	$\backslash C = C \diagup$	ethylene
$-N = N-NH$	Azoamine		o-quinoid
(p-quinoid ring structure)	p-quinoid	(o-quinoid ring structure)	

The compounds possessing chromophores are known as *chromogens*. The chromophoric groups are of the following two types:

(i) A single chromophore is sufficient to impart the colour to the compound. Examples are: NO,

NO_2, -N=N-, -N=N-NH, -N=N - O, p-quinoid, etc.

(ii) More than one chromophores is required to impart the colour, e.g. carbonyl, ethylenic etc. This can be field by various examples. Acetone (having one carbonyl as chromophore) is colourless, whereas acetylacetone (two carbonyl) is yellow.

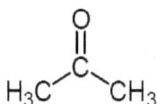

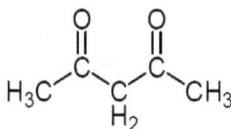

| Acetone | Acetyl acetone |

Another interesting example of this is series diphenylpolyenes, Ph (CH=CH)n Ph. If n= 1or 2 the compound is colourless, while n=3 a pale yellow colour is developed. Moreover, n is increased, the colour deepens, e.g. when n=5 the colour is orange, when n=7 the colour is copper-bronze and when n is 11 the colour is violet-black.

Witt also pointed out another type of groups which, while themsalves are unable to produce the colour, but can deepen it if the molecule possesses a chromophore. To such group be called *auxochrome* (Greek *auxein*- to increase, and *chroma*-colour). The important auxohromes are: (acidic) –OH,-SO₃H,-COOH; (basic) -NH₂ , -NHR, -NR₂ etc. the importance of auxochromoes as colour increasing groups may be illustrated by the example that the benzene, nitrobenzene (having only chromophore) and aniline(having only auxochrme) are colourless, whereas p-nitro aniline (having -NO₂ as cromophore and NH₂ as auxochrome) is a orange. Such groups which increase the depth of the colour are known as bathochromic groups, on the other hand, there are some groups which decrease the depth of colour (hypsochromic groups), e.g. acetylated phenolic or -NH₂ groups. Since the auxochromes are capable of forming salts either with a basic or acidic groups, their presence also converts a coloured compound (devoid of salt forming groups) into a dye which must fix permanently to the fiber, i.e. it must be fast to water, light, soap, and laundering, when fixed to the fiber. The permanent fixing of dye to the fiber is generally due to the formation of chemical bond between the fiber and the auxochrome. This can be explained by the following examples.

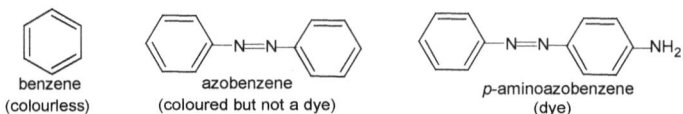

benzene
(colourless)

azobenzene
(coloured but not a dye)

p-aminoazobenzene
(dye)

8.7.2 Armstrong theory (Quinonoid theory)

Armstrong, in 1885, suggested that all colouring matters may be represented by quinonoid structures (*para or ortho*) and thus believed that if a particular compound can be formulated in a quinonoid form it is coloured, otherwise it is colourless. Some of the important compounds the colouring properties of which can be explained on the basis of this theory are given below.

(i) On the basis of this theory we can say that benzene is colourless, whereas benzoquinones are coloured.

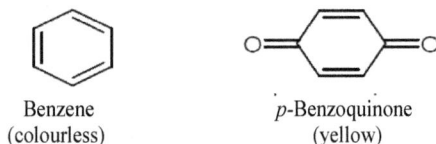

Benzene
(colourless)

p-Benzoquinone
(yellow)

(ii) Moreover, this theory explains very well the colouring properties of triphenylmethane dyes and phenolphthalein, e.g. phenolphthalein is coloured when it possess *p*-benzoquinone structure, while colourless when *p*-quinonoid structure is absent.

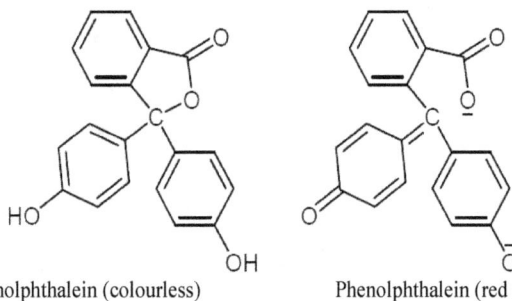

Phenolphthalein (colourless)   Phenolphthalein (red colour)

But, the quinonoid theory is not sufficient to account the colouring characteristics of all of the compounds. For example, iminoquinone and di-imonoquinone, both possess a quinonoid structure even then they are colourless.

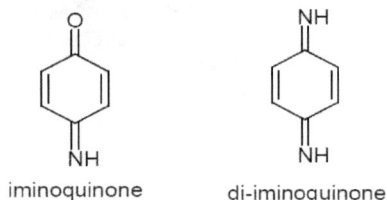

iminoquinone   di-iminoquinone

On the other hand, a number of coloured compounds are known their coloured cannot be explained by a quinonoid structure, e.g.

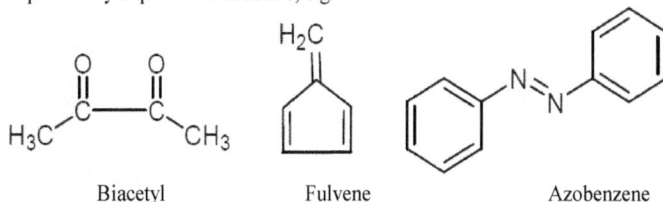

| Biacetyl | Fulvene | Azobenzene |

However, quinonoid theory has great importance because it stimulated further on the problem of colour and constitution. e. g. it was observed that the nitro phenols are coloured which can be explained easily on the basis of these theory because, the nitro phenols exist as a mixture of the following two tautomers. The existence of the phenomenon of tautomerism in nitro phenols was further proved by Hantzsch (1906) who prepared the two types of ethers (colourless and coloured). Moreover, quinonoid theory has also got practical application in the hands of dye chemist.

8.7.3. Modern theories

The above two theories discussing the relationship between colour and constitution are only empirical. The next to most important theories which explained the possibility of the relation between colour and constitution required somewhat theoretical background about the effect of absorbed light on the molecule. So, first of all, we will be dealing the effect of absorbed light on the molecule of the compound. We know that a compound appears coloured if it absorbs light energy in the visible spectrum only to which our eye is sensitive (describe earlier). The amount of light energy absorbed in visible spectrum is the only responsible factor for the shade of the colour. The most important function of this absorbed light is the molecule from its ground state to the exited state, and let us assume that E_1 and E_2 are the energies of a molecule before and after absorbing light or in the ground and exited states respectively, the difference in these two energies($E_2 - E_1$) is the absorbed energy(ΔE).The amount of ΔE_1 depends upon the constitution of the compound, e.g. if the electrons of a molecule are firmly bound, as in saturated compounds, no light of visible region will be absorbed(i.e. there will be no ΔE), but only of high energy(i.e. of ultra-violet region) will be absorbed, and hence the compound will appear colourless; on the other hand , if some of the electron are bound loosely, as in unsaturated compounds, light of wavelengths in the visible region will also be absorbed and the substance will appear coloured. Now the question arises how the ΔE is responsible for producing colour; we will discuss the answer of this question below;

Since the electron occupy in definite orbital, ΔE and hence the frequency of light absorbed must have definite value. Now each value of the frequency of the absorbed light is associated with a particular line in the spectrum of a compound, having large number of molecules and hence large number of exited states, will consist of a very large number of lines which finally appears to be just like a band (as the value of ΔE are very close to each other). Now it is existence of these bands in definite parts of the spectrum that produces the colour.

It must be noted that the dipole moment also plays an important role in the transition of the molecule. A molecule can absorb light only when its dipole moment changes, e.g. the more symmetrical the molecule, the smaller is the possibility of transition dipole, and hence the molecule will absorb light very slowly, similarly it has been observed that greater the transition dipole the greater will be the absorption of intensity. So, if the introduction of group reduces the symmetry of the molecule, the transition dipole and consequently the intensity absorption of light will be increased.

(a) Valence bond theory (resonance theory)

The resonance theory regards chromophores as groups of atoms the electrons of which may be transferred to energy rich states (or charged structure is developed) by the absorption of radiation, thus producing colour, and auxochromes as groups which enhance resonance by interacting the unshared pair of electrons of nitrogen or oxygen(the key atoms of auxochromes) atoms of the auxochromes with the unsaturation electrons of the ring which thus not only increase the intensity of absorption band to longer wavelength and hence deepen the colour. Thus we can say that increases in resonance of a compound must deepen the colour, and actually it is found to be so. The relation of colour and symmetry of molecule or we can say that transition dipole of the molecule may also be interpreted in the terms of resonance theory, because as the number of charged canonical structures increase the colour of the compound deepens. The more the possibility and longer the path for a charge to oscillate in a compound the more longer wave lengths of light will be absorbed and hence more deeper will be the colour of compound. Now let us visualize some of the important and common examples to exemplify our statement.

(i) The simplest and common example is that the benzene is colourless, nitrobenzene is pale yellow and nitro aniline is orange-red. It is explained by the fact that the charged structure of benzene(in exited state)contributes little to the resonance hybrid of the molecule and thus it absorbs light only in the ultra violet region, and further the absorption in weak due to the symmetry of the molecule; in nitrobenzene the contribution of the charged structure is larger than benzene and thus nitrobenzene absorbs light of longer wavelength producing a pale yellow colour, furthermore the intensity of absorption is also increased owing to the loss of symmetry of the molecule; lastly in nitro aniline the contribution of the charged structure is still larger, and hence the light of more longer wave-length is absorbed, and thereby deepening the colour of orange-red.

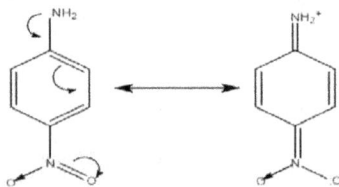

The reality high concentration of the charged structure to the resonance hybrid of p-nitro aniline is further indicated by dipole moments. The values of dipole moments for nitrobenzene, aniline and nitro aniline are 4.21D, 1.48D, ands 6.1D, respectively. The larger value (6.1D) for nitro aniline over the some fact that each amino and nitro groups co-operatively arguments the other in shifting the actual state of molecule in the direction of charged structure.

(ii) Pure p-nitro phenol is colourless but yellow in alkaline solution, it is due to fact that in alkaline solution phenol exists as phenoxide ion, in which only charged structures contribute to the resonance hybrid and thus the charged compound absorbs light waves of the higher wavelength.

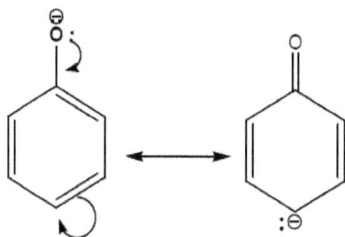

(iii) The colour of *para* and *ortho*-quinones may also explained by resonance among following charged structures.

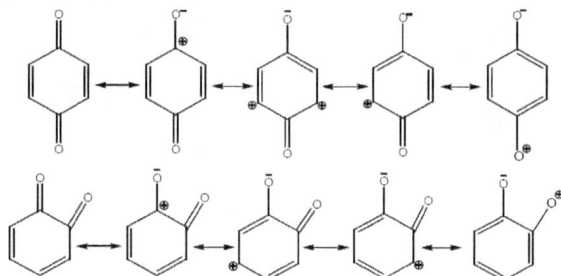

(iv) Aminoazobenzene is yellow, but in acidic solution it becomes violet. The deepening of colour in acidic solution is explained on the basis that in the yellow form it is a resonance hybrid of the two structures out of which only one is charged, whereas in the acid solution both the contributing structures are charged.

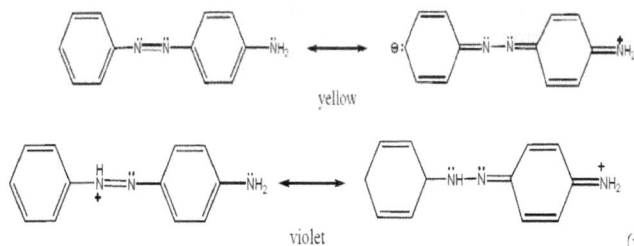

yellow

violet

(v) The much deeper colour of all the triphenylmethylene dyes is associated with delocalization of a unit electrical charge over a long path. For example,

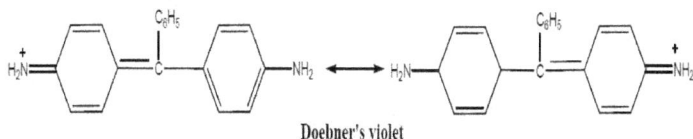

Doebner's violet

The corresponding compound in which one of the two amino groups is replaced by hydrogen atom is almost colourless, because in such cases there is no possibility for the oscillation of the positive charge.

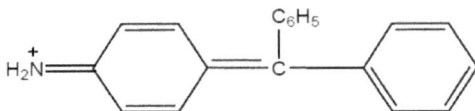

P-aminotriphenylmethane

But it must be noted that the amount of the charge also matters, the more the magnitude of the oscillating charge the lesser will be the frequency of the light absorbed and hence deeper will be the colour. For example, although there is more possibilities for the oscillation of the positive charge in crystal violet than in malachite green, yet the latter is more deeply coloured than the former.

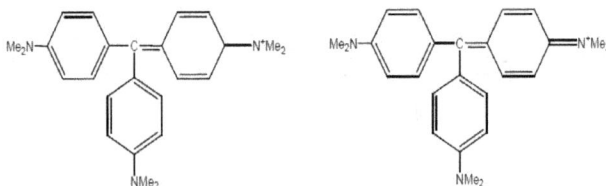

| Crystal violet | Malachite green |

It is due to the fact that the magnitude of the charge on each nitrogen atom in crystal violet is lesser than in malachite green, and hence it will absorb light of higher frequency than that of malachite green, and thus colour will be lighter as compared to malachite green.

(vi) *Effect of conjugation on resonance*

Since the conjugated system of double bond provides a long path for resonance, it plays an extremely important role in producing deep colour, the longer the conjugation in a molecule the deeper will be the colour. The isolated C=C group is extremely weak chromophore; e. g. ethylene and even the diphenyl ethylene are colourless compounds.

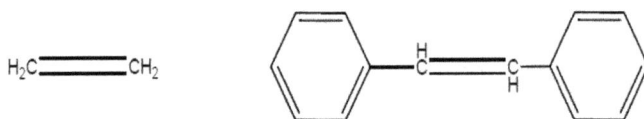

Ethylene
Absorption Max. 1800 A^0

1, 2-Diphenyl ethylene
Absorption Max. 3190 A^0

But as the number of C=C bonds in conjugated system is increased the absorption shifts progressively to longer and longer wavelengths, e.g. the following observations are made on a series of diphenylpolyenes.

Diphenyl polyene	Maximum absorption(A^0)	Colour
$C_6H_5(CH=CH)_2 C_6H_5$	3520	Colourless
$C_6H_5(CH=CH)_3 C_6H_5$	3770	Pale yellow
$C_6H_5(CH=CH)_4 C_6H_5$	4040	Greenish-yellow
$C_6H_5(CH=CH)_5 C_6H_5$	4240	Orange
$C_6H_5(CH=CH)_6 C_6H_5$	4450	Brownish-Orange

$C_6H_5(CH{=}CH)_7 C_6H_5$	4650	Copper-Bronze
$C_6H_5(CH{=}CH)_8 C_6H_5$	5700	Greenish-Pink

Lewis and Calvin observed that the effect of conjugation in deepening the colour is due to the increase in the number of electrons involved in the oscillation. The increase in conjugated in system also explains the deep colours of some compounds lacking aromatic nucleus, for example β-carotene possessing a conjugated system of all 11 ethylenic bonds is orange red in colour.

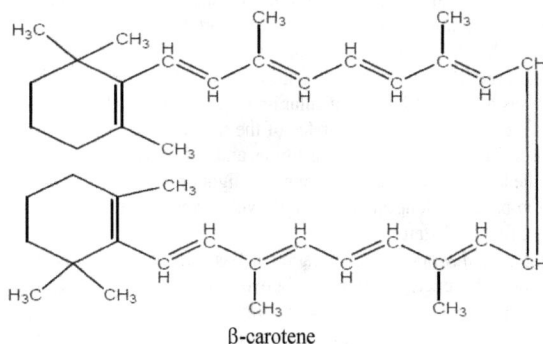

β-carotene

The colour of the cyanine and related dyes may be introduced in terms of extended conjugation. Lastly, when the conjugated system also possess atoms, such as N,S,O, etc., it absorbs light of longer wavelengths than the corresponding compound having the conjugated system of only the carbon atoms. It is due to two reasons; firstly the former conjugated system has a charge, and secondly because it is less symmetrical than the latter.

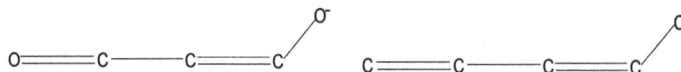

(b) M.O. Theory

In M.O. theory the excitation of a molecule means the transference of one electron from an orbital of lower energy to that of higher energy; but such transitions occur only between permitted orbital, e.g. s↔p, p↔d, etc. The M.O. theory explains beautifully the relation between colour and constitution from the quantitative point of view. According to this theory conjugated system is responsible for producing the colour. Let us take the simplest, although colourless, conjugated compound butadiene to study the relation between colour and constitution from the M.O. theory point of view. Since more than two electrons cannot occupy the same M.O. the four pz electrons of butadiene are present in two M. O.'s of different energy levels: two pz electrons in one M.O.

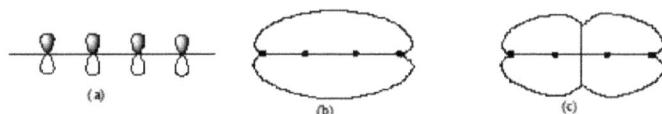

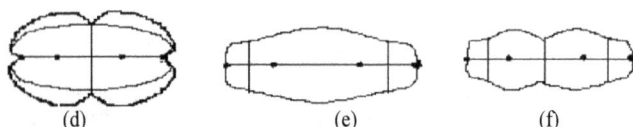

| (d) | (e) | (f) |

It must be noted that the orbital(c) is at higher energy level than that of (b) and similarly (f)is at higher energy level than that of (e).

The fig.(d) represents both the M.O.'s ;(b) of lower energy level and (c) of higher energy level; in one diagram i.e. it represents the ground state of the butadiene molecule. In the excitation of the butadiene there are four possibilities for an electron to transfer from lower energy level to higher energy level;(b) to (e), (b) to (f), (c) to (e) or (c) to (f); and if all these four transitions occur, four absorption band would be produced. It has been seen by the calculation of the energy differences (ΔE) between various possible exited and ground states of the molecule of the four transition (c) to (e) is lowest than all the other three transitions and in this transition state the molecule will absorb the longest possible wave-length of light or in other words we can say that the absorption bands of longest wavelength will correspond the transition of the one electron From (c) to (e) state.

By the experimental work and calculation it has been seen that as the conjugated system increase, the energy difference between highest level of ground state and lowest level of exited state(which corresponds to the lowest ΔE among all the possible transitions in the molecule) decreases and hence the absorption maximum of the compound in-creases, and when the absorption maximum of the compound reaches the visible part of the spectrum the compound appears to be coloured. For example, in polyenes $CH_3 (CH=CH)_n CH_3$, when n=6, the absorption band appears in the blue region of the spectrum, the compounds yellow (complementary colour of blue). This also explains the deepening of colour of diphenylpolyenes.

Now let us consider the aromatic compounds, the parent compound of which colourless benzene. Since benzene is a symmetrical molecule, all the carbon atoms have equal charges of unity and hence there is not transition dipole in the benzene molecule, and thus it is not absorb light. When a nitro group is introduced into the benzene molecule, its symmetry is lost and thus its (nitrobenzene) carbon atoms are unequally charged. The last property of the molecule produces a definite value of the dipole moment in the compound which is then responsible for the high transition dip-ole, and consequently absorption of longer wavelength of light is expected. Moreover, since by the introduction of nitro group in the benzene molecule its conjugation is extended, the energy difference between the highest level of the ground state and lowest level of the exited state is decreased, and hence the absorption band will appear at a longer wave-length than that of benzene, and thus the nitrobenzene is coloured (yellow) whereas benzene is colourless. Now let us consider the aniline molecule, the lone pair of which (on nitrogen) is in conjugation with the benzene ring, and hence the molecule has dipole moment and thus finally a transition dipole. On the whole, the aniline will absorb light of longer wave-length than benzene. However, in acid solution the absorption maximum of aniline is almost the same as that of benzene, which is due to the fact that the lone pair of electrons (which was responsible for dipole moment and hence transition dipole) has now been co-ordinate with a proton, and thus the conjugation destroyed. So conjugation is the important factor in producing colour, it can further be evidenced by the deeper colour(orange red) of nitro aniline in which the conjugation is more extended than in either nitrobenzene or aniline, which causes the greater separation of charge, and hence the absorption of light of longer wavelength. In general, any group which conjugates the benzene ring will shift the frequency of light to longer wavelength, and as the change in dipole moment of a compound increases colour becomes deeper and deeper. The last property is achieved at its best when the two polar groups, connected through a conjugated system, are present at the maximum possible distance.

8.8 Fluorescent Brightening Agents

It is found that white textile articles attain yellow colour long before they are worn out. This undesirable effect can be overcome in the following different ways:

(i) In the first method, a chemical bleaching agent such as hypochlorite or peroxide may be used. However, one should use such substances carefully because they spoil the coloured good and may damage the fiber.

(ii) In the second method, a small amount of blue colouring matter, e.g. ultramarine, may be used. This absorbs yellow light and thus the yellow light and thus the yellowed fabric appears white. However, its actual function is to extend the region of the spectrum over which light is absorbed and therefore, the treated article becomes pale-grey.

(iii) In third method, fluorescent brightening agents may be used. They are colourless compounds but they strongly absorb light of shorter wavelength the in the ultraviolet region (330-380 nm) and fluorescence or reemit lighter wavelength in the visible region of the spectrum (430-490 nm). Thus, they produce a brightening or whitening effect which is useful in making yellow products appear whiter and white materials brighter. Fluorescent brightening agents have in fact largely replaced ultramarine and other blueing agents. The large quantities are consumed annually as ingredients in soap and washing powders. The principal textile application is on cotton, trough fluorescent brightening agents suitable for wool, nylon and other synthetic fibers are of increasing importance.

8.8.1 Characteristic Properties of Fluorescent Brighteners

1) A brightening agent molecule must possess a conjugated system.

2) It must be planar.

3) It must contains electron donating groups such as, -OH and $-NH_2$.

4) It must be substantive non-toxic, compatibles with detergent action under all likely conditions.

5) On long exposure, it should not yield coloured decomposition products.

6) It must be sufficiently fast to light to remain effective for a reasonable period.

7) For maximum whitening, effects, it must reemit light of wavelength of 450nm.

8.8.2 Classification of fluorescent brighteners

A number of molecules have the ability to fluoresce but relatively few have achieved commercial

Importance, based on their chemical structures, fluorescent brighteners can be classified as follows:

(1) Stilbene derivatives: Over 80% of the fluorescent brightening agents in use are Stilbene derivatives. Bblankophor-R, one of the early brightening agents was made by condensing phenyl isocynate, with 4, 4'-diaminostilbene-2, 2'-disulohuric acid.

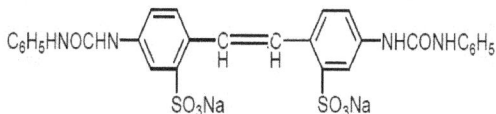

Most of the modern fluorescent brightening agents contain substituted triazinylring system and have the following general structures:

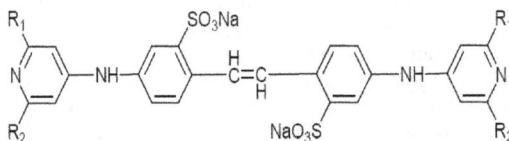

Tinopal BV   (R_1= R_2= -NH_2)

In the above formula if R_1=R_2= -NH_2 tinopal BV is obtained. It is prepared by condensing disodium salt of4, 4'-diaminostilbene-2, 2'-disulphuric acid with cyanuricchloride at 0-5°C followed by treatment with ammonia. The pH has to be maintained at 6.5-7.0 otherwise –Cl group is replaced by –OH. In other compounds R_1 may be aniline, methoxyphenyl amino or sulphophenylamino groups and R_2 methyl amino, ethyl amino or hydroxyethylamino groups. The effect of substantivity of the product, almost all these compounds are unstable to hypochlorite bleach.

(2) Stilbene triazoles: Several other products have been claimed to be brightening agents. One of these is bistriazole which obtained by oxidizing the diazodye, 4,4'-diaminostilbene -2, 2'-disulphuric acid → naphthylaminesulphuricacid (2moles) with hypochlorite. Bistrizole has a greenish fluorescence. It is useful in shading products. It is very stable to hypochlorite. Blue fluorescence and hypochlorite stability are displayed in the unsymmetrical product described in B.P.717889 (Gy) in which ring may contain A substituent like –CH_3, –CH_3O, -Cl, or –SO_3H.

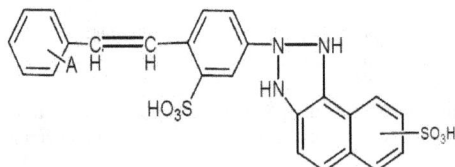

(3) Coumarin derivatives: Compounds of the type of the structure given below are suitable for nylon and wool. These are used from aqueous solution either as acid salts or where one R is alkyl and the other H, as the aldehydebisulphite (methane-w-sulphonate) compound.

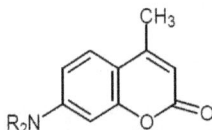

(Where R = alkyl group)

(4) Benxthiazoles: benzoxazoles and benziminazoles. CIBA claimed that the compounds having the structure given below have good fastness to chlorine, e.g. uvitex RS.

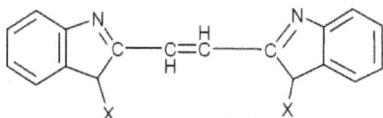

CIBA also investigated compounds of general formula (shown below) where X may be S, O, NH or N (e.g., R= -CH$_2$-CH$_2$OH). Such compounds have been found to possess good fastness to hypochlorite, substantivity for nylon. Quaternary derivatives of this series have been found to possess improved substantivity for nylon, polyester and cellulose acetate.

CIBA also reported that the compounds having the structure shown below are mass coloration of polyester and nylons.

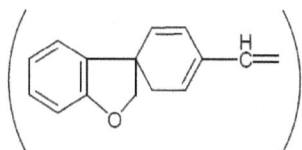

CIBA also synthesized the compound of the structure given below. This is stated to be suitable for plastics and hydrophobic fibers.

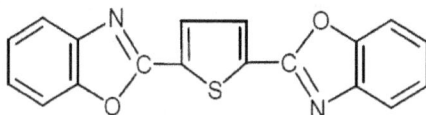

(5) Diarylpyrazolines: These compounds having the structure shown below are effective brighteners. These can be applied with detergents in the form of a dispersion or in solution when sulphonated.

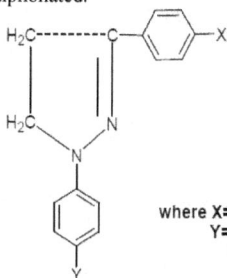

where X= H,Cl
Y=H,COOCH$_3$
SO$_2$NH$_2$ or SO$_3$H

(6) Naphthalimide derivetives: BASF claim in B.P.741798 that compounds of the type having the structure given below are suitable fluorescent brightening agents for synthetic materials such as polyesters and polyamides. They may also added to the detergents.

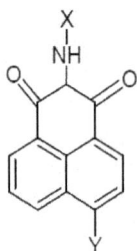

where X= alkyl or aryl
Y=alkoxyl,
NHCOCH$_3$,NHClNH$_2$
or NHCONHC$_6$H$_5$

(7) Pyrene derivatives: Triazyl derivatives of pyrene having the structure given below have been found to be highly effective brighteners for synthetic fibers. These may be applied either from aqueous dispersion of incorporated during fiber manufacture. These may also be added to the detergents.

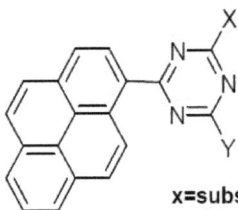

x=substituted amino, alkoxy,substituted alkoxy,
alkylthio or substituted alkylthio
Y= X or Cl

(8) Miscellaneous chemical classes: One of the early brighteners was blankophore WT which was obtained by the condensation of benzoin with urea followed by the disulphonation of the product.

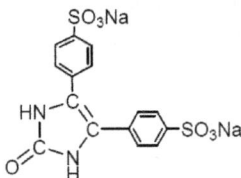

Blankophor WT

It is mainly applied to wool. There are many pyrazine derivatives which are suitable as brighteners for wool and synthetic fibers of various types.

8.9 Hair Dyes

A great majority of the dyes used in hair colouring are known as oxidation hair dyes (Corbett, 1985, 2000). A much smaller number of the commercial hair dyes are synthetic dyes that have affinity for protein substrates such as wool. Oxidation dyes, the more permanent of the two groups, are produced directly on the hair by oxidizing aromatic diamines such as *para*-phenylenediamine or 2, 5-diaminotoluene with an oxidizing agent. Suitable diamines have been referred to as "primary intermediates" and the oxidizing agents (e.g. hydrogen peroxide) as "developers." Other useful primary intermediates are aminodiphenylamines, minomethylphenols, and *para*-aminophenol.

When used alone, the primary intermediates provide a very limited shade-range following their oxidation on hair. To enhance the range of available hair colours, the primary intermediates are oxidized in the presence of suitable "couplers." While most couplers do not produce colours when exposed to developers alone, they give a wide range of shades on hair when applied in combination with primary intermediates. Appropriate couplers include 3-aminophenol, resorcinol, and α-naphthol.

The chemistry associated with the oxidation of primary intermediates is now reasonably well known. For *para*-phenylenediamine and *para*-aminophenol (*cf.* **1**), oxidation-induced self-coupling proceeds via the process outlined in Fig. 1, where it can be seen that permanent hair-dye formation involves oxidation followed by coupling to give type-**2** structures. Fig. 2 provides chemistry representative of combinations arising from joining an α-naphthol-based coupler (**3**) and a sulfonated *N*-phenyl-*p*-phenylenediamineprimary intermediate (**4**) to produce experimental dyes (**5**) and (**6**).This chemistry also illustrates the fact that oxidation dyes are often mixtures rather than single products.

An analytical method has been developed for characterizing the reactants and reaction products of oxidative hair-dye formulations. The results indicated that a significant amount, i.e. ≈20% or more of the initial concentrations of precursor(s) and coupler(s), is always present in the formulation that is not diffused into hairs.

Fig. 1 Oxidation hair-dye formation from primary intermediates (X = O, NH)

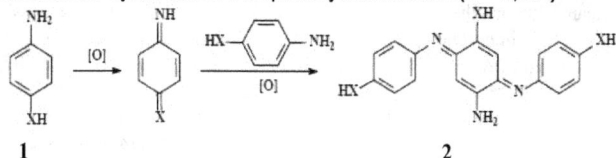

Fig. 2 Oxidation hair-dye formation from a primary intermediate and coupler

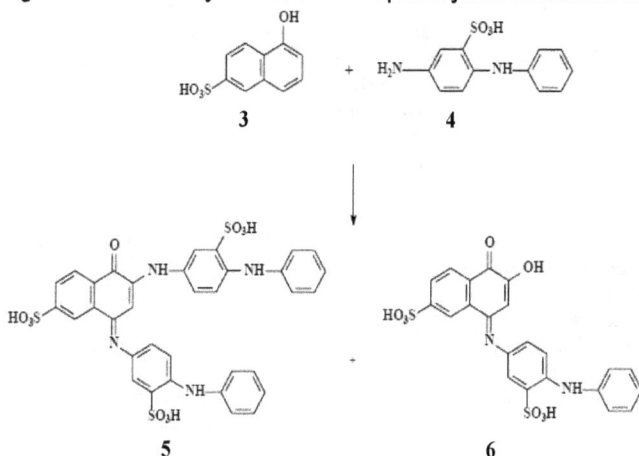

C I Basic dyes such as Yellow 57, Red 76, Blue 99, Brown 16, and Brown 17 have been used in colour refreshener shampoos and conditioners. Similarly, C I Acid dyes such as Yellow 3, Orange 7, Red 33, Violet 43, and Blue 9 have been used in shampoos, in this case to deliver highlighting effects (Corbett, 2000). Example structures of nonpermanent hair dyes are provided in Fig. 3, where it can be seen that these dyes are drawn from those known to have affinity for protein-based textile fibers.

Fig. 3 Examples of non-permanent hair dyes

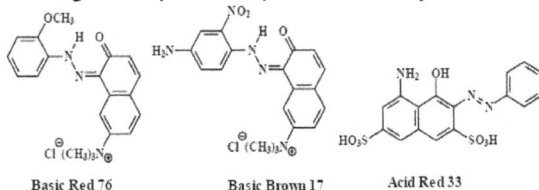

Basic Red 76   Basic Brown 17   Acid Red 33

8.10 Toxicity of Dyes

A variety of synthetic dyes and their intermediates are released by the textile industry which pose a threat to environment safety. This creates an immediate environmental concern with regard to water quality which directly influences the aquatic flora and fauna. Simultaneously, the impact of azo dyes on human health has been a concern and occupational exposure to workers in dye manufacturing and dye utilizing industries have received considerable attention. It has been found that cytotoxicity of a typical dye molecule may be low but, the toxicity related to aromatic amines is significant due to their carcinogenicity and mutagen city (Libra et al. 2004). Therefore, it is important to investigate the toxic impacts of aromatic amines or dye degradation end products for risk assessment in further operations. The toxicity of dyes and its degraded end products have been investigated by several researchers using various assays. Wang et al. (2003) performed bioluminescence of Lumistox bacteria (Microtox R) assays to reveal the toxicity of Ramazole black 5 in baffled reactor. Moran et al. (2008) used *Daphniamagna* to assess the toxicity of dye molecules in aerobic treatment of textile water. Dye containing industrial effluents alters the chemical and biological status of the soil and water which affect the growth and productivity of plants. The impact of toxicity of dyes has been investigated on plants and plant growth parameters namely percent seed germination, shoot and root length, chlorophyll content, etc. The influence of textile effluent on peanut seed germination and shoot length were reported by Saravanamoorthy and Ranjitha Kumari, (2007). Sharma et al. (2006) also examined the toxicity of both original and decolorized methyl red samples using Hydrilla and Lemna (duckweed) bioassay. Pandey et al. (2008) reported phototoxic effect on seed germination and early growth of maize and rice. Algal growth was tested against 56 commercial dyestuffs where their growth was not much affected at dye concentrations below 1 mg l-1 but basic dyes were found to be acutely toxic for algal growth (Greene and Baughman 1996).Cytotoxicity of original methyl red and decolorized dye samples was carried out usingCOS-70, an African green monkey kidney fibroblast-like cell line by Anderson et al.(2004). The toxic effect of methyl red dye was observed on COS-70 cells while decolorized dye sample was non-toxic to the cell lines. The toxic effect of azo dyes and their metabolites after bacterial treatment has been also studied by agar diffusion plate assay. Mane et al. (2008) investigated microbial toxicity study of original and biological treated reactive navy blue RX by agar diffusion plate assay. Genotoxicity based on the effect of DNA-damaging agents on a dark mutant of *Photobacterium leiognathi* and genetically engineered *Pseudomonas* containing lux operon that codes for bioluminescence was used for toxicity assay of wastewater

(Wiles et al. 2003).Earthworms are considered to be useful biomarkers to determine the risk accessibility of dyes as its skin is a significant route for the uptake of contaminants. Ramaswami and Subbram, (1992) used earthworms (*Polypheretima elongata*) to assess the toxicity of Navy blue and Direct Brown.

8.11 Pigments

These are insoluble powders of very fine particles size, i.e., as small as 0.01micron, which are used in paints, plastics, rubbers, textile, inks and other materials to impart colour, opaqueness and other desirable properties to the product. Pigments are both natural and synthetic in origin; and organic and inorganic in composition. The oxides of iron, chromium, lead and other materials give a limited range of colour with good light fastness. However, many of these changes colour with sulphur compounds found in urban atmosphere today, making them unsuitable. With the growth of dyestuff industry, a range of pigments giving bright colours of good fastness properties pigment was introduced in 1935. This was followed by introduction of several pigments. The difference between dyes and pigments is their relative solubility; dyes are soluble while the pigments are essentially insoluble in the liquid media in which they are dispersed. *Organic pigments in general have lower hiding power but tinting strength than inorganic pigments.*

Toner pigments and lakes: In the manufacture of organic pigments certain colouring materials become insoluble in the pure form whereas others require a metal or inorganic base to precipitate them. The colouring materials which are insoluble in the pure form are known as toner pigments and those which require a base are referred to as lakes.

Uses of synthetic organic pigments: Pigments find application in aqueous and non-aqueous paints, printing inks paper coating leather finishing, plastic products and other similar processes. The pigment maybe used alone or incorporated with a white pigment such as zinc oxide, titanium dioxide or white lead as a means of controlling the opacity and the depth of shade required.

Most printing inks contain pigments and are used for the printing of metal-foil, tin-plate, card-board wrapping materials and so on. Pigments are extensively, used in printing and textile in combination with resin binder. Pigments are incorporated in cellulose pulp to obtain colour paper. Similarly, mass colouration of synthetic fibers, plastics and rubber is carried out. Pigments are also used in cosmetics, soap, wax, chalks, crayons, artist's colours and so on. In all applications, the physical form, shape and size of the pigment particles are of the highest importance. Great care is exercised in standardization of manufacturing process of pigments to obtain correct crystalline structure and particle size of pigments.

8.11.1 Requirements of organic pigments

These are as follows:

(i) *Fastness to light:-* The most important criterion by which pigments are evaluated is fastness to light which depend on the medium in which the pigments are dispersed. In general pigments are less fast to light in pale tints than in dark tints.

(ii) *Fastness to heat:-* A pigment should be stable to relatively high temperatures. This applies especially in surface coating cured or polymerized by heart and in thermosetting colouration. With interior pigments, chemical decomposition may result from such heat treatment, of physical

change from one modification to another may occur causing deterioration in hue and other properties.

(iii) *Insolubility in solvents:-* An ideal pigment should be insoluble in all media. However, this condition is not completely fulfilled. Solubility in vehicle or solvent many bring about crystallization of the pigment causing a change in the colour properties of the paint.

(iv) *Fastness to acid:* Pigments must be acid-fast if they are employed in acid media or they are to be exposed to acid vapours.

(v) *Fastness to alkali:* Pigments must be alkali-fast if they are to be used in the manufacture of

distempers or the colourations of plaster surfaces.

(vi) *Insolubility in water:* Complete insolubility of pigment in water is rated as excellent. Here pigments have a slight degree of solubility in water or other liquid media they are said to show bleeding. One particular pigment cannot fulfill all the above mentioned requirements. However, to satisfy all these large numbers of organic pigments have been developed. These organic pigments have been developed. These organic pigments belong have been to most different classes of dyestuff.

8.11.2 Types of pigments: Lakes of acid or anionic dyes. These pigments are precipitated from solution of anionic dyes of the type DX+ (where X is generally sodium atom) by double decomposition with the soluble salts of heavy metals such as calcium or barium, e.g. pigment red 57.

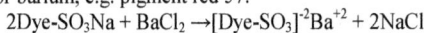

$$2Dye\text{-}SO_3Na + BaCl_2 \rightarrow [Dye\text{-}SO_3]^{-2}Ba^{+2} + 2NaCl$$

Such lackes are resistant to solvents but they are very sensitive to acids and alkalis.

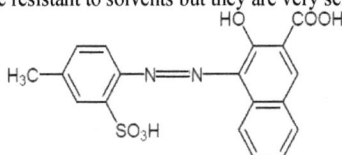

Pigment red 57

(a) Lakes of basic or cationic dyes: These pigments are precipitated from solution of cationic dyes of the type D+X- (where D is chromophoric system embodying one or more basic groups and X is a chlorine atom or similar salt forming group) by double decomposition with tannic acid or with certain inorganic poly acids. The best examples are the 'Fanal' or permanent pigments produced by BASF by precipitating a cationic dye with phosphotungstomolybadic acid (PTMA), e.g. pigment violet 1. These lakes are generally superior in fastness properties to the parent dyes but they do not attain the all-round high standards of the modern pigments, i.e., phthalocymines.

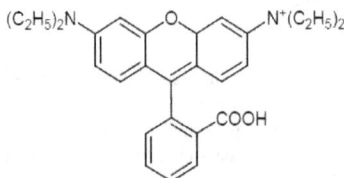

Pigment violet 1

(b) Metal complexes: These pigments have excellent light fastness property but are inferior in other fastness properties. These are co-ordination or chelate compounds and require for their formation dye molecules which contain oxygen or nitrogen atom to donate electrons to the metal atom. One example is pigment green B (C.I. pigment

green 8, 10006) which is prepared from 1-nitroso-2- naphthol and ferric salt. Another example is C.I pigment brown 2, 12071 which is the copper complex derived from the monoazo dye p-nitro aniline to 2-naphthol.

For many years, metal complexes of alizarin have been used as pigments. For example, the calcium/aluminium lake is bluish-red, chromium dull bluish-red and iron dull purple. These are all used either as such or formed on the fiber as in dyed-style printing.

(c) Neutral, metal-free compounds: These are the most widely used group of pigments. This includes mainly monoazo and bisazo dyes and also a few representatives from azine, indigo and anthraquinone classes. These are mostly dye molecules not containing solublizing groups such as -SO3H or –COOH. These provide full range hues. These pigments possess very good fastness property against acids and alkalis but poor fastness to solvents and plasticize. These are quite popular in India.

(i) *Monoazo pigments:* An example is C.I. pigment yellow 1, 11680,Hansa yellow G(4-amino-3-

nitro toluene acetanilide).Another example is C.I pigment yellow 7,12780 2-nitroaniline-2,4,-dihydroxy quinoline. It is a bright reddish yellow pigment.

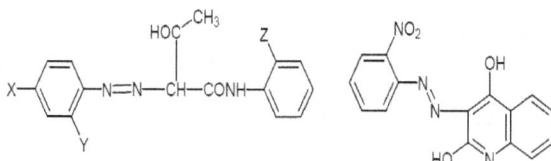

$(X = CH_3, Y = NO_2, Z= H)$ C.I pigment yellow 7

An example of the large class of pyrazolone pigment is C.I. pigment orange 6, 12730, 4-amino-3-nitrotoluene →3-methyl-1-phenyl-5-pyrazolone.

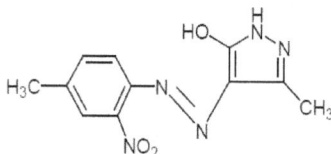

C I pigment orange 6

Another pigment derived from 2-naphthol is C I, pigment orange 5, 12075, 2, 4-dinitroaniline-2-naphthol.It is widely used for its brightness and strength. However, its other fastness, properties, particularly solvent and vehicle 'bleed' are inferior.

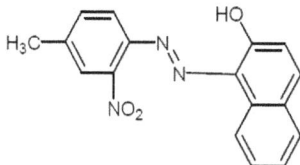

C I, pigment orange 5

The pigments derived from components of the naphthol AS type show marked improvement it solvent and vehicle fastness as in C I pigment red 2, 2,5-dichloroaniline →3hydroxy-2-naphth-anilide.

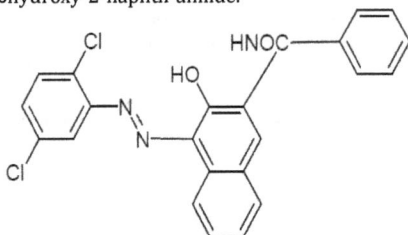

C I pigment red 2

(iii) **Diazo pigments:** These are widely used in the manufacture of printing inks and in the mass colouration of rubber. Typical examples are benzidine yellow which are well known for brilliance and high-pictorial strength. An example is C I pigment yellow 12 210090, 3, 3'-dichlorobenzidine→acetoacetanilide (2moles)

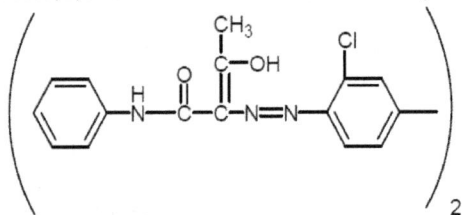

C I pigment yellow 12

Another example is C I pigment orange 13, 21100, permanent orange G. it uses a pyrazolone as coupling component.

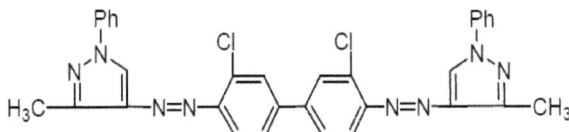

C I pigment orange 13

One more example is C I pigment blue 26, 21185, dianisidine blue. It is obtained from tetraazo anisidine and 3-hydroxynaphtho-0-anisidine.

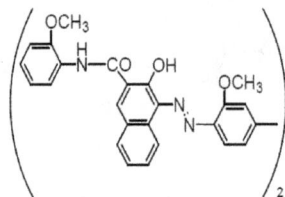

C I pigment blue 26

There are some other pigments which belong to another class. C I pigment yellow 11 is obtained from 4-chloro-2-nitroaniline by condensation with formaldehyde and it is a nitro pigment. Azine pigment is obtained by oxidation of aniline in the presence of catalyst such as copper or vanadium, e.g. aniline black.

(d) Modern high-grade pigments:
In 1935, monatral blue was marked by ICI. This was the first phthalocyamine pigment with high all round fastness properties. The structure determination was carried out by Linstead and coworkers who found a tetrabenzoprophyrazino nucleus, a structure related to that of chlorophyll and also haemin, the red colouring matter of blood corpcules.

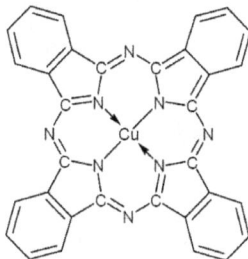

Copper phthalcyamine

Copper phthalocyamine is prepared by several methods. One method involves the heating of phthalonitrile or a related compound such as o-cynobenzamide or phthalamide with a cuprous salt. Another method called urea process for preparation of copper phthalocyamine consists in together urea, phthalic anhydride. The yield of the method is increased by catalysts like boric acid, ammonium molybdate and ammonium phosphate.

Metal-free phthaloylamine may be prepared from sodium pththalocyamine by demethylization with a strong acid. It may also be obtained directly from phthalonitrile by heating, in an inert atmosphere, under pressure, it is a bright greenish-blue pigment.

Polychloro-copper phthalocyamine is a true green pigment. It is prepared by passing chlorine gas through a melt of copper phthalocyamine, aluminium chloride and sodium chloride at 200 ^{0}C. a number of soluble dyes have been prepared from these pigments by introducing solublizing groups in the pigment. Intensive research to obtain yellow, red and violet pigments having fastness properties similar to phthalocyamines has result in the production of modern high grade pigment.

Modern high grade pigments belong to the various classes which are given in table below.

Class	Colour change
Azo coupling	yellow
Azo condensation	Yellow, orange, red
Derivatives of 4,5,6,7-tetrachloro isoindolin-1-one	Greenish-yellow, orange-red, brown
Anthraquinone	Yellow, orange, violet
Perinone, Perylene	Orange, red, violet

Quinacridone	Marron, scarlet, red-magneta, violet
Dioxazine	Violet
Phthalocyamine	Blue green

(e) Azomethine pigments: These are obtained by using more complex diazo coupling components and in some cases by forming metal complexes. These show excellent fastness properties. Nickel azo yellow (DUP) is the 1:2 nickel-dye complex which is obtained from the dye p-chloroaniline to2,4-dihydroxy-quinoline.

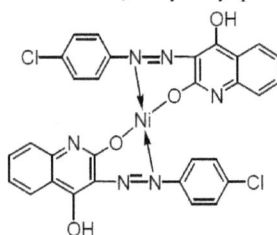

Nickel azo yellow

Benzidine yellowish Green (Hoechst) is prepared from 2, 5, 2', 5'-tetrachlorobenzidine to acetoacetate –*m*-xylidine (2moles).

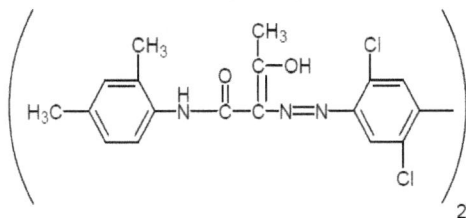

Benzidine yellowish Green

Copper complexes of 2, 2-dihydroxy azomethine dyes posses excellent light and heat fastness and are fast to cross lacquering and migration. The structure of these pigments is as follows:

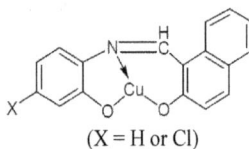

(X = H or Cl)

These pigments are obtained by condensing hydroxyl-4-nitro-aminobenzene with 2-hydroxyl-1-
naphthaldehyde in DMF followed by conversion into the copper complex. In an analogous manner, BASF prepared a pigment by condensing 1, 2, 4, 5-tetraaminobenzene 2-hydroxy-1-naphthaldehyde.This pigments has high light fastness and high light fastness.

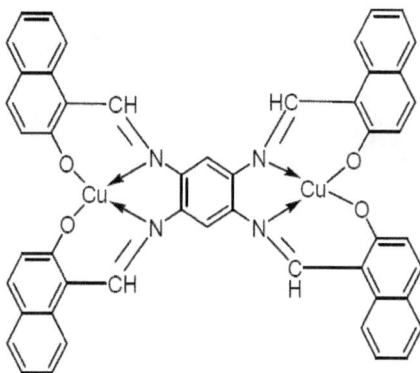

BASF BP 1195 766

(f) Azo condensations pigments: CIBA prepared these pigments by condensing 2-monoazo pigments. The following example illustrates the method.

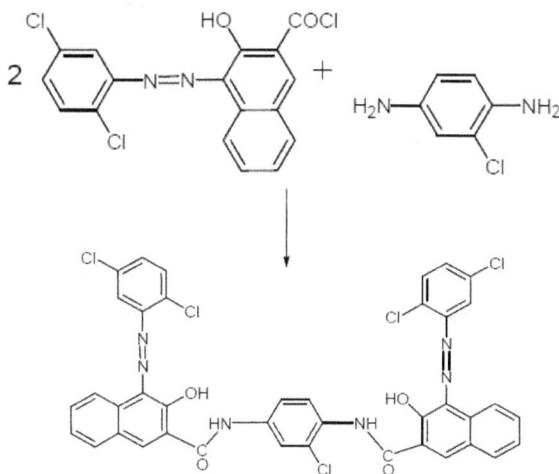

Pigment red 344

Due to insolubility of the pigment, it is very difficult to get a diazo pigment by the normal method of diazotization and coupling. The constitution of many of these pigments is not known. However, a range of yellow, oranges, red and browns of all round fastness properties are available. These pigments are used in mass colouration of polymers and high grade lacquers for automobiles.

(g) Tetrachloro-iso-indolin-1-one pigments: In 1965, these pigments were introduced by Geigy. These pigments have a new chromophoric system. These are prepared by condensing a phthalimide derivatives with *p*-phenylenediamine.

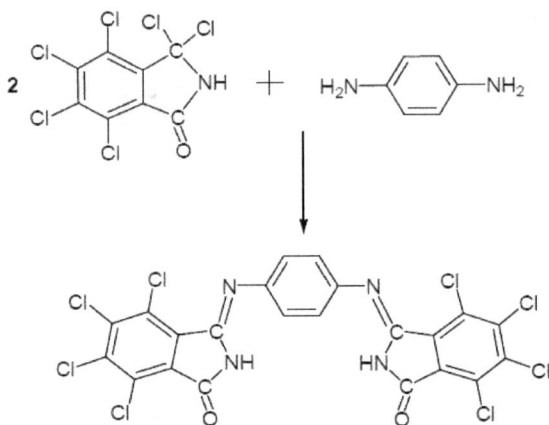

It is be noted that the presence of eight chlorine atoms in the pigment molecule is essential because their absence results in products unsatisfactory as pigments. By using other diamines, a full range of shades from greenish yellow-to orange, red, brown are obtained. Their fastness properties are very good.

Perinone and perylene pigments: Perinone pigments are derived from naphthalene-1,4,5,8-tetracarboxylic and o-phenylenediamine.

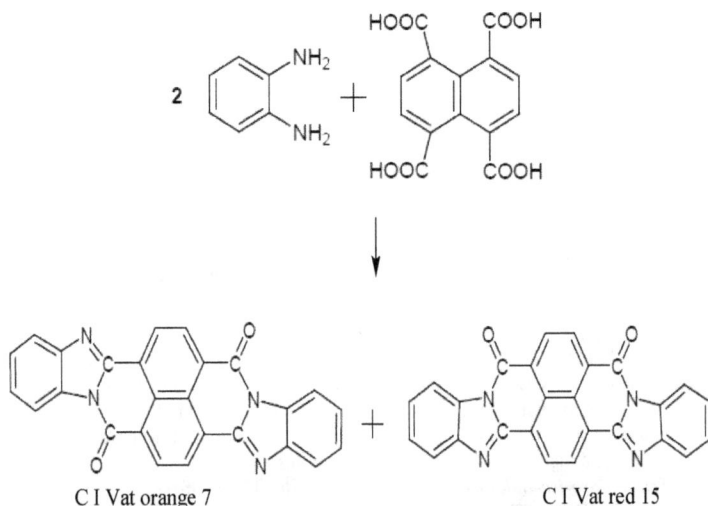

C I Vat orange 7   C I Vat red 15

The orange has superior fastness to light, heat and solvents while the red is inferior by reason of solvents bleed and migration. Hoechst developed the perylene pigments which are diimides of perylene-3, 4, 9, 10-tetracarboxylic acid. Some complexes of perylene pigments are:

Perylene red	R = CH$_2$O
Perylene marron	R = CH$_2$
Pigment red 123	R = p-substituted phenyl
Vat red 29	R = Phenyl

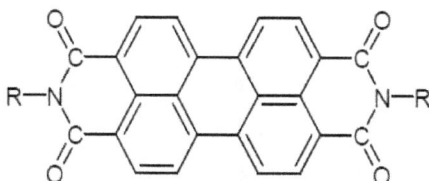

Perylene pigment

(h) Quinacridone pigments: These were introduced by DuPont in 1958. The colour of the pigment depends on the different crystal modifications. γ-modification is red while β-modifications is violet. The excellent fastness property may be due to hydrogen boning between neighboring molecules. Quinacridone pigments may be synthesized from 2,5-dibromo-terephthalic acid aniline followed by sing closure which may be achieved by heating in oleum or alluminium chloride with or without solvent.

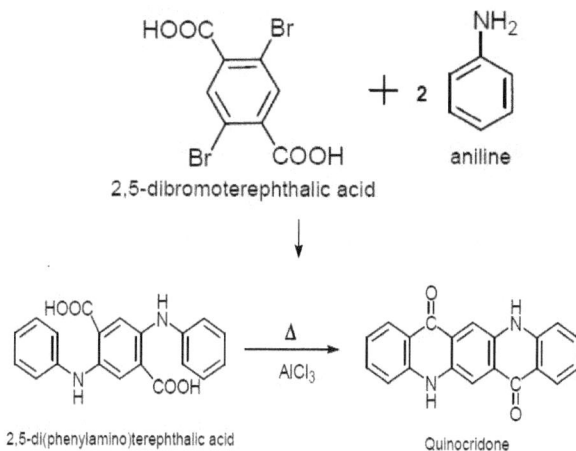

2,5-dibromoterephthalic acid   aniline

2,5-di(phenylamino)terephthalic acid   Quinocridone

(i) Thioindigo pigments: The best example of these pigments is the thioindigo Bordeaux. It is reddish-violet and has light fastness comparable, even in light tins, with phthalocyamine pigments.

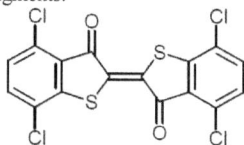

Thioindigo

If the positions of any the four chlorine atoms are altered or one of the chlorine atoms is removed or one more chlorine atom is introduced, there occurs solvent bleeding in the products.

High-grade anthraquinone pigments: Several vat dyes are used as high grade pigments e.g. inanthone, flavanthrone, halogenodibenzenthrone sulphonamide, etc. Some recent anthraquinone pigments are:

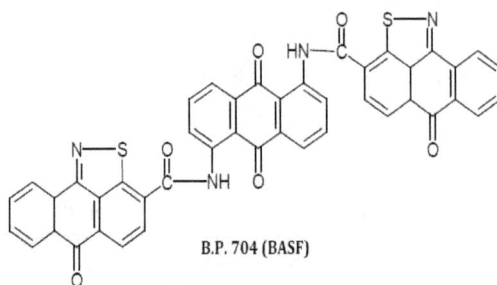

B.P. 704 (BASF)

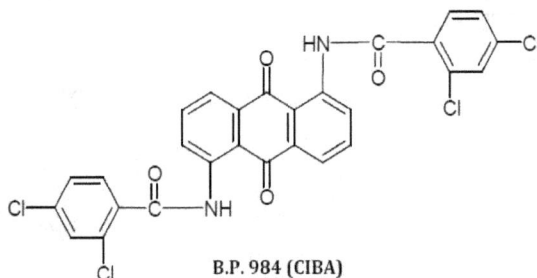

B.P. 984 (CIBA)

Dioxazine pigments: In 1952, Hoechst introduced carbazole dioazine violet.

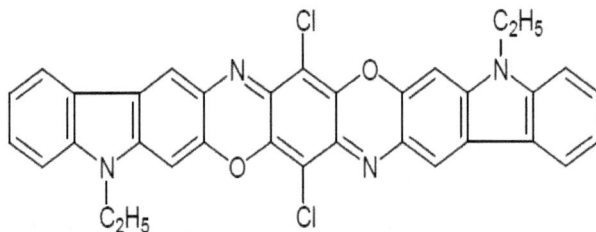

Carbazole dioxazine violet

It has high tinctorial strength and good light fastness. However, its fastness to solvents and plasticizers is inferior to other pigments.

9. Green chemistry

9.1 The Development of Organic Synthesis

The well - being of modern society is unimaginable without the myriad products of industrial organic synthesis. Our quality of life is strongly dependent on, *inter alia*, the products of the pharmaceutical industry, such as antibiotics for combating disease and analgesics or anti - infl ammatory drugs for relieving pain. The origins of this industry date back to 1935, when Domagk discovered the antibacterial properties of the red dye, prontosil, the prototype of a range of sulfa drugs that quickly found their way into medical practice.

The history of organic synthesis is generally traced back to Wohler's synthesis of the natural product urea from ammonium isocyanate in 1828. This lay to rest the *viz. vitalis* (vital force) theory, which maintained that a substance produced by a living organism could not be produced synthetically. The discovery had monumental significance, because it showed that, in principle, all organic compounds are amenable to synthesis in the laboratory.

The next landmark in the development of organic synthesis was the preparation of the first synthetic dye, mauveine (aniline purple) by Perkin in 1856, generally regarded as the first industrial organic synthesis. It is also remarkable example of serendipity. Perkin was trying to synthesize the anti - malarial drug quinine by oxidation of N - allyltoluidine with potassium dichromate. This noble but naive attempt, bearing in mind that only the molecular formula of quinine ($C_{20}H_{24}N_2O_2$) was known at the time, was doomed to fail. In subsequent experiments with aniline, fortuitously contaminated with toluidines, Perkin obtained a low yield of a purple - colored product. Apparently, the young Perkin was not only a good chemist but also a good businessman, and he quickly recognized the commercial potential of his finding. The rapid development of the product, and the process to make it, culminated in the commercialization of mauveine, which replaced the natural dye, Tyrian purple. At the time of Perkin ' s discovery Tyrian purple, which was extracted from a species of Mediterranean snail, cost more per kg than gold.

This serendipitous discovery marked the advent of the synthetic dyestuffs industry based on coal tar, a waste product from steel manufacture. The development of mauveine was followed by the industrial synthesis of the natural dyes alizarin and indigo by Graebe and Liebermann in 1868 and Adolf Baeyer in 1870, respectively. The commercialization of these dyes marked the demise of their agricultural production and the birth of a science - based, predominantly German, chemical industry.

By the turn of the 20th century the germ theory of disease had been developed by Pasteur and Koch, and for chemists seeking new uses for coal tar derivatives which were unsuitable as dyes, the burgeoning field of pharmaceuticals was unobvious one for exploitation. A leading light in this field was Paul Ehrlich, who coined the term chemotherapy. He envisaged that certain chemicals could act as 'magic bullets' by being extremely toxic to an infecting microbe but harmless to the host. This led him to test dyes as chemotherapeutic agents and to the discovery of an effective treatment for syphilis. Because Ehrlich had studied dye molecules as ' magic bullets ' it became routine to test all dyes as chemotherapeutic agents, and this practice led to the above - mentioned discovery of prontosil as an antibacterial agent. Thus, the modern pharmaceutical industry was born as a spin - off of the manufacture of synthetic dyestuffs from coal tar. The introduction of the sulfa drugs was followed by the development of the penicillin antibiotics. Fleming ' s chance observation of the anti - bacterial action of the penicillin mold in 1928 and the subsequent isolation and identification of its active constituent by Florey and Chain in 1940 marked the beginning of the antibiotics era that still continues today. At roughly the same time, the steroid hormones found their way into medical practice. Cortisone was introduced by the pharmaceutical industry in 1944 as a drug for the treatment of arthritis and rheumatic fever. This was followed by the development of steroid hormones as the active constituents of the contraceptive pill.

The penicillin and the related cephalosporins, and the steroid hormones represented considerably more complicated synthetic targets than the earlier mentioned sulfa drugs. Indeed, as the target molecules shifted from readily available natural compounds and relatively simple synthetic molecules to complex semi -synthetic structures, a key factor in their successful introduction into medical practice became the availability of a cost-effective synthesis. For example, the discovery of the regiospecific and enantiospecific microbial hydroxylation of progesterone to 11 α - hydroxyprogesterone (Fig:1) by Peterson and Murray at the Upjohn Company led to a commercially viable synthesis of cortisone that replaced a 31-step chemical synthesis from a bile acid and paved the way for the subsequent commercial success of the steroid hormones. According to Peterson, when he proposed the microbial hydroxylation, many outstanding organic chemists were of the opinion that it couldn't be done. Peterson's response was that the microbes didn't know that. Although this chemistry was invented four decades before the term Green Chemistry was officially coined, it remains one of the outstanding applications of Green Chemistry within the pharmaceutical industry.

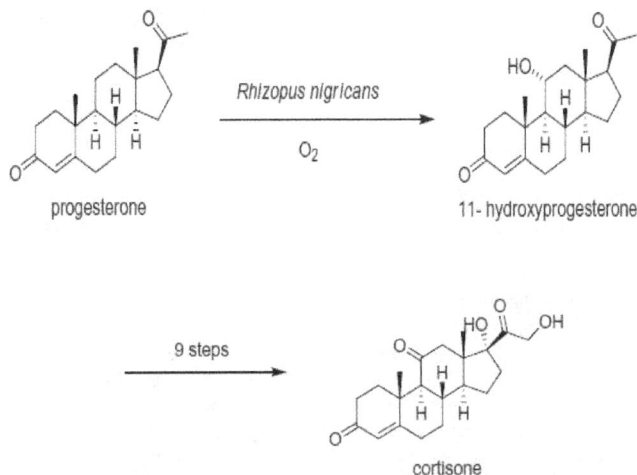

progesterone   *Rhizopus nigricans*   11- hydroxyprogesterone
                              O_2

9 steps

cortisone

Fig: 1 Cortisone synthesis

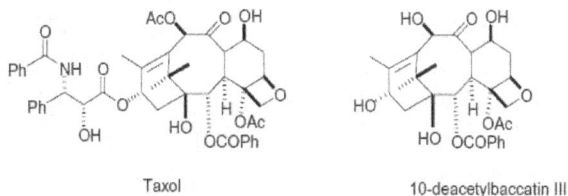

Taxol   10-deacetylbaccatin III

Fig: 2 Structure of the anticancer drug Taxol and 10 - deacetylbaccatin III

This monumental discovery marked the beginning of the development, over the following decades, of drugs of ever - increasing molecular complexity. In order to meet this challenge, synthetic organic chemists aspired to increasing levels of sophistication. A case in point is the anticancer drug, Taxol, derived from the bark of the Pacific yew tree, *Taxus brevifolia,* and introduced into medical practice in the 1990s (see Fig: 2). The breakthrough was made possible by Holton's invention of a commercially viable and sustainable semi-synthesis from 10 - deacetylbaccatin III, a constituent of the needles of the English yew, *Taxusbaccata.* The Bristol - Myers Squibb Company subsequently developed and commercialized a fermentation process that avoids the semi-synthetic process.

In short, the success of the modern pharmaceutical industry is firmly built on the remarkable achievements of organic synthesis over the last century. However, the down side is that many of these times - honored and trusted synthetic methodologies solvents were not known and the issues of waste minimization and sustainability were largely unheard of.

9.2 The Environmental Factor

In the last two decades it has become increasingly clear that the chemical and allied industries, such as pharmaceuticals, are faced with serious environmental problems. Many of the classical synthetic methodologies have broad scope but generate copious amounts of waste, and the chemical industry has been subjected to increasing pressure to minimize or, preferably, eliminate this waste. An illustrative example is provided by the manufacture of phloroglucinol, a reprographic chemical and pharmaceutical intermediate. Up until the mid-1980s it was produced mainly from 2, 4, 6 - trinitrotoluene (TNT) by the process shown in (Fig: 3) a perfect example of vintage nineteenth - century organic chemistry.

For every kg of phloroglucinol produced ca. 40 kg of solid waste, containing $Cr_2(SO_4)_3$, NH_4Cl, $FeCl_2$, and $KHSO_4$, were generated. This process was eventually discontinued as the costs associated with the disposal of this chromium-containing waste approached or exceeded the selling price of the product. That such an enormous amount of waste is formed is easily understood by examining the stoichiometric equation (see Fig: 3) of the overall process, something very rarely done by organic chemists. This predicts the formation of ca. 20 kg of waste per kg of phloroglucinol, assuming 100% chemical yield and exactly stoichiometric quantities of the various reagents. In practice, an excess of the oxidant and reductant and a large excess of sulfuric acid, this subsequently has to be neutralized with base,

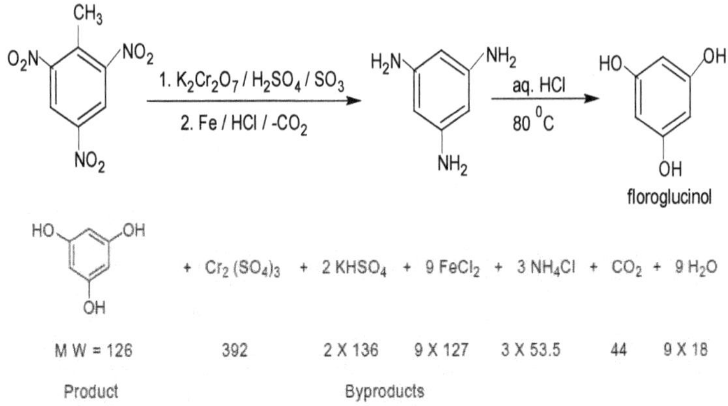

CH_3
O_2N
NO_2
NO_2
1. $K_2Cr_2O_7$ / H_2SO_4 / SO_3
2. Fe / HCl / -CO_2
H_2N
NH_2
NH_2
aq. HCl
80 ^{0}C
HO
OH
OH
floroglucinol
+ $Cr_2(SO_4)_3$ + 2 $KHSO_4$ + 9 $FeCl_2$ + 3 NH_4Cl + CO_2 + 9 H_2O
M W = 126
392
2 X 136
9 X 127
3 X 53.5
44
9 X 18
Product
Byproducts

Atom efficiency + 126 / 2282 = ca. 5%

E factor = ca. 40

Fig: 3 Manufacture of floroglucinol from TNT

Table: 1 E factor in chemical industry

Industry segment	Volume (ty^{-1})[a]	E factor (kg waste/kg product)
Bulk chemicals	10^4–10^6	<1–5
Fine chemicals industry	10^2–10^4	5–> 50
Pharmaceutical industry	10–10^3	25–> 100

(a) Annual production of the product worldwide or at a single site is used, and the isolated yield of phloroglucinol is less than 100%. This explains the observed 40 kg of waste per kg of desired product. Indeed, an analysis of the amount of waste formed in processes for the manufacture of a range of fine chemicals and pharmaceuticals intermediates has revealed that the generation of tens of kilograms of waste per kilogram of desired product was not exceptional in the fine chemical industry. This led to the introduction of the E (environmental) factor (kilograms of waste per kilogram of product) as a measure of the environmental footprint of manufacturing processes in various segments of the chemical industry (Table:1).

The E factor represents the *actual amount* of waste produced in the process, defined as everything but the desired product. It takes the chemical yield into account and includes reagents, solvent losses, process aids, and, in principle, even fuel. Water was generally excluded from the E factor as the inclusion of all process water could lead to exceptionally high E factors in many cases and make meaningful comparisons of processes difficult. A higher E factor means more waste and consequently, a larger environmental footprint. The ideal E factor is zero. Put quite simply, it is the total mass of raw materials minus the total mass of product, all divided by the total mass of product. It can be easily calculated from knowledge of the number of tons of raw materials purchased and the number of tons of product sold, the calculation being for a particular product or a production site or even a whole company.

It is clear from Table:1 that the E factor increases substantially ongoing from bulk chemicals to fine chemicals and then to pharmaceuticals. This is partly are reflection of the increasing complexity of the products, necessitating multistep syntheses, but is also a result of the widespread use of stoichiometric reagents (see below). A reduction in the number of steps of a synthesis will in most cases lead to a reduction in the amounts of reagents and solvents used and hence a reduction in the amount of waste generated. This led Wender to introduce the concepts of step economy and function oriented synthesis (FOS) of pharmaceuticals. The central tenet of FOS is that the structure of an active lead compound, which may be a natural product, can be reduced to simpler structures designed for ease of synthesis while retaining or enhancing the biological activity. This approach can provide practical access to new (designed) structures with novel activities while at the same time allowing for a relatively straightforward synthesis.

As noted above, knowledge of the stoichiometric equation allows one to predict the theoretical minimum amount of waste that can be expected. This led to the concept of *atom economy* or *atom utilization* to quickly assess the environmental acceptability of alternatives to a particular product before any experiment is performed. It is a theoretical number, that is, it assumes a chemical yield of100% and exactly stoichiometric amounts and disregards substances which do not appear in the stoichiometric equation.

In short, the key to minimizing waste is precision or *selectivity* in organic synthesis which is a measure of how efficiently a synthesis is performed. The standard definition of selectivity is the yield of product divided by the amount of substrate converted, expressed as a percentage. Organic chemists distinguish between different categories of selectivity:

- Chemo selectivity (competition between different functional groups)
- Regioselectivity (selective formation of one regioisomer, for example *ortho* vs *para* substitution in aromatic rings)
- Diastereoselectivity (the selective formation of one diastereomer)
- Enantioselectivity (the selective formation of one of a pair of enantiomers)

However, one category of selectivity was, traditionally, largely ignored by organic chemists: the *atom selectivity* or *atom utilization* or *atom economy*. The virtually complete disregard of this important parameter is the root cause of the waste problem in chemicals manufacture. As Lord Kelvin remarked, 'To measure is to know'. Quantification of the waste generated in chemicals manufacturing, by way of E factors, served to bring the message home and focus the attention of fine chemical and pharmaceutical companies on the need for a paradigm shift from a concept of process efficiency, which was exclusively based on chemical yield, to one that is motivated by elimination of waste and maximization of raw materials utilization. Indeed, the E factor has been widely adopted by the chemical industry and the pharmaceutical industry in particular. To quote from a recent article: 'Another aspect of process development mentioned by all pharmaceutical process chemists who spoke with Chemical and Engineering News is the need for determining an E factor. '

The Green Chemistry Institute (GCI) Pharmaceutical Roundtable has used the Process Mass Intensity (PMI), defined as the total mass used in a process divided by the mass of product (i.e. PMI = E factor + 1) to benchmark the environmental acceptability of processes used by its members (see the GCI website).The latter include several leading pharmaceutical companies (Eli Lilly, GlaxoSmithKline, Pfizer, Merck, AstraZeneca, Schering Plow, and Johnson and Johnson). The aim was to use this data to drive the greening of the pharmaceutical industry. We believe, however, that the E factor is to be preferred over the PMI since the ideal E factor of 0 is a better reflection of the goal of zero waste. The E factor, and derived metrics, takes only the mass of waste generated into account. However, the environmental impact of waste is determined not only by its amount but also by its nature. Hence, we introduced the term 'environmental quotient', EQ, obtained by multiplying the E factor by an arbitrarily assigned unfriendliness quotient, Q. For example, one could arbitrarily assign a Q value of 1 to NaCl and, say, 100 – 1000 to a heavy metal salt, such as chromium, depending on factors like its toxicity or ease of recycling. Although the magnitude of Q is debatable and difficult to quantify, 'quantitative assessment' of the environmental impact of waste is, in principle, possible. Q is dependent on, *inter alia*, the ease of disposal or recycling of waste and, generally speaking, organic waste is easier to dispose of or recycle than inorganic waste.

9.3 The Role of Catalysis
The main source of waste is inorganic salts such as sodium chloride, sodium sulfate, and ammonium sulfate that are formed in the reaction or in downstream processing. One of the reasons that the E factor increases dramatically on going from bulk to fine chemicals and pharmaceuticals is that the latter are more complicated molecules that involve multi-step syntheses. However, the larger E factors in the fine chemical and pharmaceutical industries are also a consequence of the widespread use of classical stoichiometric reagents rather than catalysts. Examples which readily come to mind are metal (Na, Mg, Zn, Fe) and metal hydride ($LiAlH_4$, $NaBH_4$) reducing agents and oxidants such as permanganate, manganese dioxide, and chromium (VI) reagents. For example, the phloroglucinol process (see above) combines an oxidation by stoichiometric chromium (VI) with a reduction with Fe/HCl. Similarly, a plethora of organic reactions, such as sulfonations, nitrations, halogenations, diazotizations, and Friedel - Crafts acylations, employ stoichiometric amounts of mineral acids (H_2SO_4, HF, H_3PO_4) or Lewis acids ($AlCl_3$, $ZnCl_2$, BF_3)and are major sources of inorganic waste.

Once the major cause of the waste has been recognized, the solution to the waste problem is evident: the general replacement of classical syntheses that use stoichiometric amounts of inorganic (or organic) reagents by cleaner, catalytic alternatives. If the solution is so simple, why are catalytic processes not as widely used in fi ne and specialty chemicals manufacture as they are in bulk chemicals? One reason is that the volumes involved are much smaller, and thus the need to minimize waste is less acute than in bulk chemicals manufacture. Secondly, the economics of bulk chemicals manufacture dictate the use of the least expensive reagent, which was generally the most atom economical, for example O_2 for oxidation H_2 for reduction, and CO for C – C bond formation.

A third reason is the pressure of time. In pharmaceutical manufacture 'time to market' is crucial, and an advantage of many time - honored classical technologies is that they are well tried and broadly applicable and, hence, can be implemented rather quickly. In contrast, the development of a cleaner, catalytic alternative could be more time consuming. Consequently, environmentally (and economically) inferior technologies are often used to meet stringent market deadlines, and subsequent process changes can be prohibitive owing to problems associated with FDA approval.

J. J. Berzelius 1779-1848

Organic Chemistry (1807)   Catalysis (1835)

Urea synthesis   1828   ca. 1900   Catalysis definition
(Wöhler)   (Ostwald)
                                               Catalytic Hydrogenation
First synthetic dye   1856   (Sabatier)
Aniline purple   ca. 1920   Petrochemicals
(Perkin)

Dyestuffs Industry   1936   Catalytic cracking
(based on coal-tar)   1949   Catalytic reforming
                                  1955   Ziegler-Natta catalysis

Fine Chemicals   Bulk Chemicals & Polymers

Catalysis in Organic Synthesis

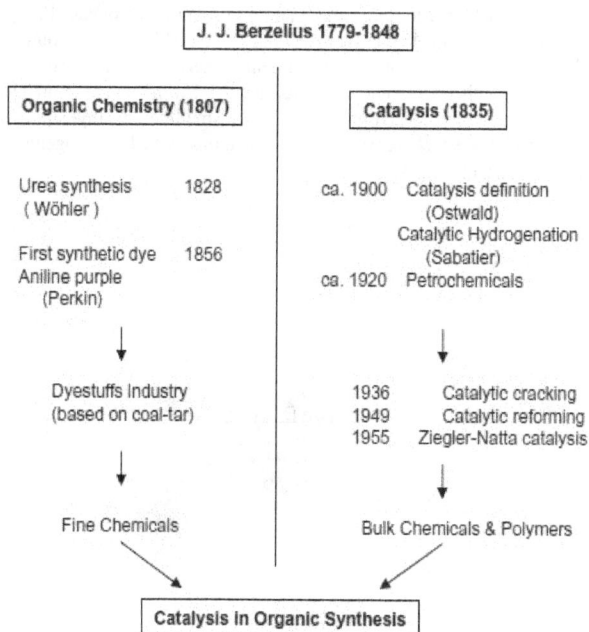

Fig: 4 the development of organic synthesis and catalysis

Another reason, however, is the more or less separate paths of development of organic synthesis and catalysis (Fig: 4) since the time of Berzelius, who coined the terms 'organic chemistry' and 'catalysis' in 1807 and 1835, respectively. Subsequently, catalysis developed largely as a sub discipline of physical chemistry. With the advent of petrochemicals in the 1930s, catalysis was widely applied in oil refining and bulk chemicals manufacture. However, the scientists responsible for these developments were, generally speaking, not organic chemists but were chemical engineers and surface scientists.

Industrial organic synthesis, in contrast, followed a largely 'stoichiometric' line of evolution that can be traced back to Perkin ' s synthesis of mauveine, the subsequent development of the dyestuffs industry based on coal tar, and the fine chemicals and pharmaceuticals industries, which can be regarded as spin offs from the dyestuffs industry. Consequently, fine chemicals and pharmaceuticals manufacture, which is largely the domain of synthetic organic chemists, is rampant with classical 'stoichiometric' processes. Until fairly recently, catalytic methodologies were only sporadically applied, with the exception of catalytic hydrogenation which, incidentally, was invented by an organic chemist, Sabatier, in 1905.

The desperate need for more catalytic methodologies in industrial organic synthesis is nowhere more apparent than in oxidation chemistry. For example, as any organic chemistry textbook will tell you, the reagent of choice for the oxidation of secondary alcohols to the corresponding ketones, a pivotal reaction in organic synthesis, is the Jones reagent. The latter consists of chromium trioxide and sulfuric acid and is reminiscent of the phloroglucinol process referred to earlier. The introduction of the storage-stable pyridiniumchlorochromate (PCC) and pyridinium dichromate (PDC) in the 1970s, represented a practical improvement, but the stoichiometric amounts of carcinogenic chromium (VI) remain a serious problem. Other stoichiometric oxidants that are popular with synthetic organic chemists are the Swern reagent and Dess Martin Periodinane (DMP). The former produces the evil smelling dimethyl sulfide as the byproduct, the latter is shock sensitive, and oxidations with both reagents are abominably atom in efficient (Fig: 5).

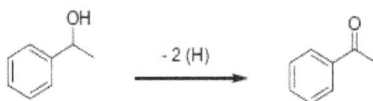

Oxidant	Atom Efficiency
CrO_3 / H_2SO_4	44%
(DMP structure)	22%
$(CH_3)_2 SO / (COCl)_2$	37%
NaOCl	48%
O_2	87%

Fig: 5 Atom efficiencies of alcohol oxidations

Obviously there is a definite need in the fine chemical and pharmaceutical industry for catalytic systems that are green and scalable and have broad utility. More recently, oxidations with the inexpensive household bleach (NaOCl) catalyzed by stable nitroxyl radicals, such as TEMPO and PIPO, have emerged as more environmentally friendly methods. It is worth noting at this juncture that 'greenness' is a relative description and there are many shades of green. Although the use of NaOCl as the terminal oxidant affords NaCl as the by - product and may lead to the formation of chlorinated impurities, it constitutes a dramatic improvement compared to the use of chromium (VI) and other reagents referred to above. Moreover, we note that, in the case of pharmaceutical intermediates, the volumes of NaCl produced as a byproduct on an industrial scale are not likely to present a problem. Nonetheless, catalytic methodologies employing the green oxidants, molecular oxygen (air) and hydrogen peroxide, as the terminal oxidant would represent a further improvement in this respect. However, as Dunn and coworkers have pointed out, the use of molecular oxygen presents significant safety issues associated with the flammability of mixtures of oxygen with volatile organic solvents in the gas phase. Even when these reduced by using 10% oxygen diluted with nitrogen, these methods are on the edge of acceptability. An improved safety profile and more acceptable scalability are obtained by performing the oxidation in water as an inert solvent. For fine chemicals or large volume pharmaceuticals the environmental and cost benefits of using simple air or oxygen as the oxidant would justify the capital investment in the more specialized equipment required to use these oxidants on a large scale.

9.4 Green Chemistry: Benign by Design

In the mid - 1990s Anastas and coworkers at the United States Environmental Protection Agency (EPA) were developing the concept of *benign by design*, that is designing environmentally benign products and processes to address the environmental issues of both chemical products and the processes by which they are produced. This incorporated the concepts of atom economy and E factors and eventually became a guiding principle of *Green Chemistry* as embodied in the 12 Principles of Green Chemistry, the essence of which can be reduced to the useful working definition:

Green chemistry efficiently utilizes (preferably renewable) raw materials, eliminates waste, and avoids the use of toxic and/or hazardous reagents and solvents in the manufacture and application of chemical products.

Raw materials include, in principle, the source of energy, as this also leads to waste generation in the form of carbon dioxide. Green Chemistry is primary pollution prevention rather than waste remediation (end of pipe solutions). More recently, the twelve Principles of Green Engineering were proposed, which contain the same underlying features – conservation of energy and other raw materials and elimination of waste and hazardous materials – but from an engineering standpoint. Poliakoff and coworkers proposed a mnemonic, *productively*, which captures the spirit of the twelve Principles of Green Chemistry in a single slide. Another concept which has become the focus of attention, both in industry and society at large, in the last decade or more is that of sustainable development, first introduced in the Brundtl and report in the late 1980s and defined as:

Meeting the needs of the present generation without compromising the ability of future generations to meet their own needs.

Sustainable development and Green Chemistry have now become a strategic industrial and societal focus, the former is our ultimate goal and the latter is a means to achieve it.

9.5 Ibuprofen Manufacture

An elegant example of a process with high atom efficiency is provided by the manufacture of the over the counter, non-steroidal anti-inflammatory drug, ibuprofen. Two routes for the production of ibuprofen via the common intermediate, p - isobutylacetophenone, are compared in (Fig: 6). The classical route, developed by the Boots Pure Drug Company (the discoverers of ibuprofen), entails 6 steps with stoichiometric reagents, relatively low atom efficiency, and substantial inorganic salt formation. In contrast, the elegant alternative, developed by the Boots Hoechst Celanese (BHC) Company, involves only three catalytic steps. The first step involves the use of anhydrous hydrogen fluoride as both catalyst and solvent in a Friedel-Crafts acylation. The hydrogen fluoride is recyclable and waste is essentially eliminated. This is followed by two catalytic steps (hydrogenation and carbonylation), both of which are 100% atom efficient.

The BHC ibuprofen process was commercialized in 1992 in a ca. 4000 tons per annum facility in Texas. The process was awarded the Kirkpatrick Achievement Award for outstanding advances in chemical engineering technology in 1993 and a Presidential Green Chemistry Challenge Award in 1996. It represents a bench mark in environmental excellence in chemical processing technology that revolutionized bulk pharmaceutical manufacturing. It provides an innovative and

excellent solution to the prevalent problem of the large volumes of waste associated with the traditional stoichiometric use of auxiliary chemicals. The anhydrous hydrogen fluoride is recovered and recycled with greater than 99.9% efficiency, no other solvent is used in the process, simplifying product recovery and minimizing fugitive emissions. This combined with the almost complete atom utilization of this streamlined process truly makes it a waste - minimizing, environmentally friendly technology and a source of inspiration for other pharmaceutical manufacturers.

9. 6 The Question of Solvents: Alternative Reaction Media

Another important issue in green chemistry is the use of organic solvents. The use of many traditional organic solvents, such as chlorinated hydrocarbons, has been severely curtailed. Indeed, so many of the solvents that are favored by organic chemists have been blacklisted that the whole question of solvent use requires rethinking and has become a primary focus, especially in the manufacture of pharmaceuticals. In our original studies of E factors of various processes,

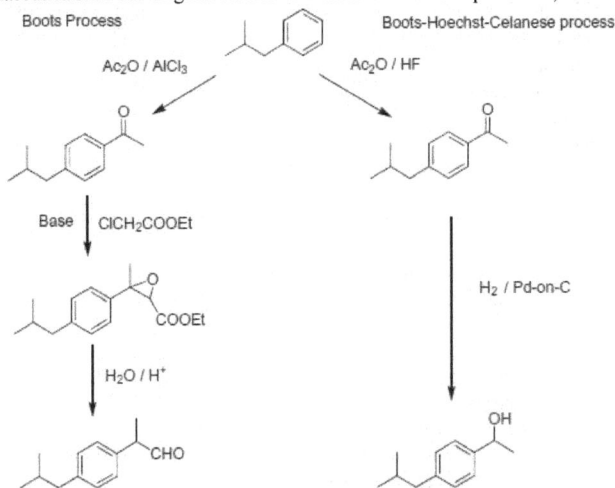

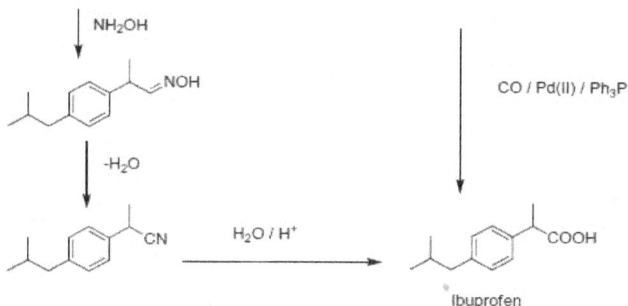

Fig: 6 Two processes for ibuprofen

We assumed, if details were not known, that solvents would be recycled by distillation and that this would involve a 10% loss. However, the organic chemist's penchant for using different solvents for the various steps in multistep synthesis makes recycling difficult owing to cross contamination. A benchmarking exercise performed by the GCI Pharmaceutical Roundtable (see above) revealed that solvents were a major contributor to the E factors of pharmaceutical manufacturing processes. Indeed, it has been estimated by GlaxoSmithKline workers that ca.

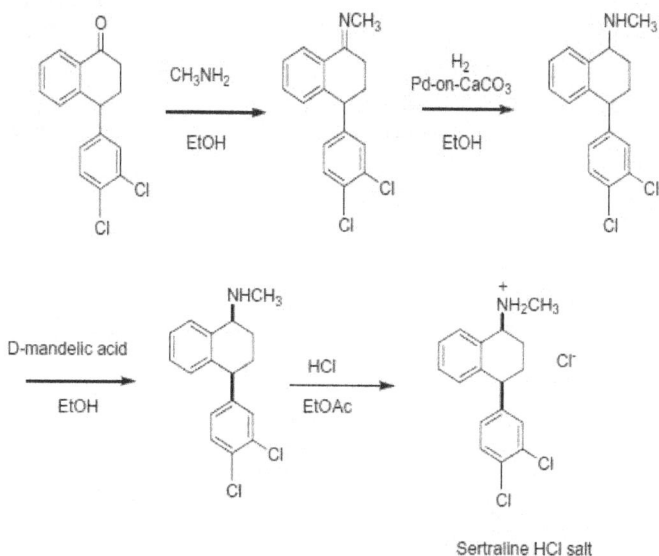

Fig: 7 the new sertraline process

85% of the total mass of chemicals involved in pharmaceutical manufacture comprises solvents. Consequently, pharmaceutical companies are focusing their effort on minimizing solvent use and in replacement of many traditional organic solvents, such as chlorinated and aromatic hydrocarbons, by more environmentally friendly alternatives.

An illustrative example is the redesign of the sertraline manufacturing process, for which Pfizer received a Presidential Green Chemistry Challenge Award in 2002. Among other waste -minimizing improvements, a three step sequence was streamlined by employing ethanol as the sole solvent (Fig: 7). This eliminated the need to use, distill, and recover four solvents (methylene chloride, tetrahydrofuran, toluene, and hexane) and resulted in a reduction in solvent usage from 250 to 25 liters per kilogram of sertraline.

Similarly, Pfizer workers also reported impressive improvements in solvent usage in the process for sildenafil (Viagra) manufacture, reducing the solvent usage from 1700 liters per kilogram of product used in the medicinal chemistry route to 7 L kg 1 in the current commercial process with a target for the future of4 L kg $-$ 1 . The E factor for the current process is 8, placing it more in the lower end of fine chemicals rather than with typical pharmaceutical manufacturing processes.

These issues surrounding a wide range of volatile and nonvolatile, polar aprotic solvents have stimulated the fine chemicals and pharmaceutical industries to seek more benign alternatives. There is a marked trend away from hydrocarbons and chlorinated hydrocarbons toward lower alcohols, esters, and, in some cases, ethers. Inexpensive natural products such as ethanol have the added advantage of being readily biodegradable, and ethyl lactate, produced by combining two innocuous natural products, is currently being promoted as an environmentally attractive solvent for chemical reactions. The problem with solvents is not so much in their use but in the seemingly inherent inefficiencies associated with their containment, recovery, and reuse.

The best solvent is no solvent at all, but if a solvent is needed there should be provisions for its efficient removal from the product and reuse. The subject of alternative reaction media also touches on another issue that is important from both an environmental and an economic viewpoint: recovery and reuse of the catalyst. An insoluble solid that is heterogeneous, catalyst is easily separated by centrifugation or filtration. A homogeneous catalyst, in contrast, presents more of a problem, the serious shortcoming of homogeneous catalysis being the cumbersome separation of the catalyst from the reaction products audits quantitative recovery in an active form. In pharmaceutical manufacture, another important issue is contamination of the product. Attempts to heterogeneous, homogeneous catalysts by attachment to organic or inorganic supports have, generally speaking, not resulted in commercially viable processes for a number of reasons, such as leaching of the metal, poor catalyst productivity, irreproducible activity and selectivity, and degradation of the support. There is a definite need, therefore, for systems that combine the advantages of high activity and selectivity of homogeneous catalysts with the facile recovery and recycling characteristic of their heterogeneous counterparts. This can be achieved by employing a different type of heterogeneous system, namely liquid – liquid biphasic catalysis, whereby the catalyst is dissolved in one liquid phase and the reactants and product(s) are in a second liquid phase. The catalyst is recovered and recycled by simple phase separation. Preferably, the catalyst solution remains in the reactor and is reused with a fresh batch of reactants without further treatment or, ideally, it is adapted to continuous operation.

Various nonconventional reaction media have been intensively studied in recent years, including *water, supercritical CO_2, fluorous biphasic, and ionic liquids* alone or in liquid - liquid biphasic combinations. The use of water and supercritical carbon dioxide as reaction media fits with the current trend toward the use of renewable, biomass - based raw materials, which are ultimately derived from carbon dioxide and water.

Water has many benefits: it is nontoxic, nonflammable, abundantly available, and inexpensive. Furthermore, performing the reaction in an aqueous biphasic system, whereby the catalyst resides in the water phase and the product is dissolved in the organic phase, allows for recovery and recycling of the catalyst by simple phase separation. A case in point is the BHC process for ibuprofen manufacture (see above).The key carbonylation step involves a homogeneous palladium catalyst, and contamination of the product (the active pharmaceutical ingredient) with unacceptably high amounts of palladium necessitates an expensive purification. Replacing the organic soluble palladium (0) triphenylphosphine complex with an analogous complex of the water - soluble trisulfonates, triphenylphosphine, TPPTS, affords a catalytic system for the aqueous biphasic carbonylation of alcohols. For example, when the above - mentioned ibuprofen synthesis was performed with TPPTS in an aqueous biphasic system, product contamination by the catalyst was essentially eliminated.

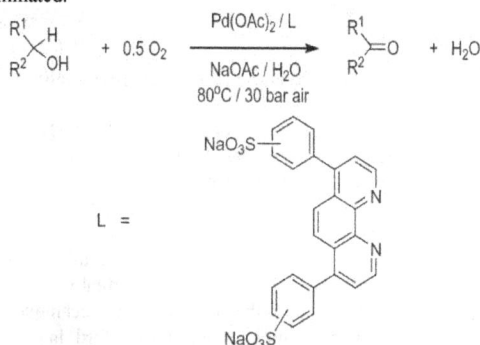

Fig: 8 aqueous biphasic aerobic oxidation of alcohols

Similarly, a water - soluble palladium complex of a sulfonated phenanthroline ligand catalyzed the highly selective aerobic oxidation of primary and secondary alcohols in an aqueous biphasic system in the absence of any organic solvent (Fig: 8). The liquid product could be recovered by simple phase separation, and the aqueous phase, containing the catalyst, used with a fresh batch of alcohol substrate, affording a truly green method for the oxidation of alcohols.

9.7 Biocatalysis: Green Chemistry Meets White Biotechnology

Biocatalysis has many attractive features in the context of green chemistry: reactions are generally performed in water under mild conditions of temperature and pressure using an environmentally compatible, biodegradable catalyst (an enzyme) derived from renewable raw materials. High activities and chemo, regio, and stereoselectivities are obtained in reactions of multifunctional molecules without the need for the functional group activation and protection often required in traditional organic syntheses. This affords more environmentally attractive and cost - effective processes with fewer steps and, hence, less waste. Illustrative examples are provided by the substitution of classical chemical processes with enzymatic counterparts in the synthesis of semi synthetic penicillins and cephalosporins.

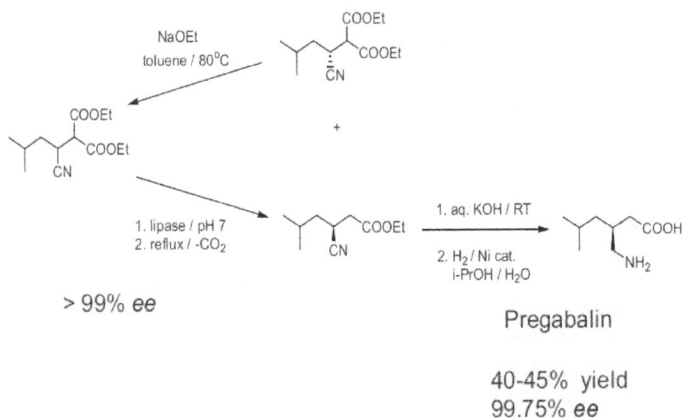

Fig: 9 Chemoenzymatic process for pregabalin

If biocatalysis is so attractive, why was it not widely used in the past? The answer is that only recent advances in biotechnology have made it possible. First, the availability of numerous whole - genome sequences has dramatically increased the number of potentially available enzymes. Second, *in vitro* evolution has enabled the manipulation of enzymes such that they exhibit the desired properties: substrate specificity, activity, stability, and pH profile. Third, recombinant DNA techniques have made it, in principle, possible to produce virtually any enzyme for a commercially acceptable price. Fourth, the cost - effective techniques that have now been developed for the immobilization of enzymes afford improved operational stability and enable their facile recovery and recycling.

An illustrative example of the replacement of a traditional organic synthesis by a more economically and environmentally attractive chemoenzymatic process is provided by the manufacture of pregabalin. The key step is an enzymatic kinetic resolution of an ester (Fig: 9) using the readily available lipase from *Thermomyces lanuginosus* (Lipolase). The stereochemistry at C_2 is not important as it is lost in the subsequent thermal decarboxylation step. The unreacted substrate was racemized by heating with a catalytic amount of sodiumethoxide in toluene at 80 ° C and was then recycled to the resolution step. Subsequent hydrolysis and hydrogenation affords pregabalin in 40 – 45% overall yield.

The chemoenzymatic route afforded a dramatic improvement in process efficiency compared to the first - generation process. This was reflected in the E factor which decreased 7 - fold, from 86 to 12, and the substantial reduction in organic solvent usage resulting from a largely aqueous reaction medium. The enzymes found in Nature are the result of aeons of cumulative natural selection, but they were not evolved to perform biotransformations of non -natural, pharmaceutical target molecules. In order to make them suited to these tasks they generally need to be re - evolved, but we don't have millions of years to do it. Fortunately, modern advances in biotechnology have made it possible to accomplish this in weeks in the laboratory using *in vitro* techniques such as gene shuffling.

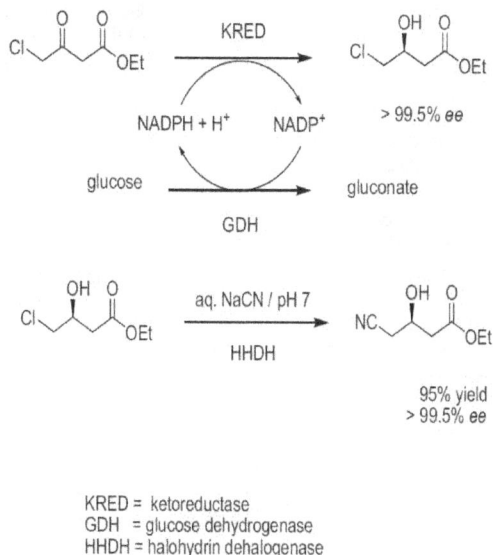

KRED = ketoreductase
GDH = glucose dehydrogenase
HHDH = halohydrin dehalogenase

Fig: 10 Codexis process for atorvastatin intermediate

An illustrative example is provided by the Codexis process for the production of an intermediate for Pfizer's blockbuster drug Atorvastatin (Lipitor). The two – step process (Fig: 10), for which Codexis received a 2006 Presidential Green Chemistry Challenge Award, involves three enzymes (one for cofactor regeneration). The low activities of the wild - type enzymes formed a serious obstacle to commercialization, but *in vitro* evolution of the individual enzymes, using gene shuffling, afforded economically viable productivities.

The highly selective biocatalytic reactions afford a substantial reduction in waste. The overall isolated yield is greater than 90%, and the product is more than 98%chemically pure with an enantiomeric excess of > 99.9%. All three evolved enzymes are highly active and are used at such low loadings that counter - current extraction can be used to minimize solvent volumes. Moreover, the butyl acetate solvent is recycled with an efficiency of 85%.The E factor (kgs waste per kg product) for the overall process is 5.8 if process water is excluded (2.3 for the reduction and 3.5 for the cyanation). If process water is included, the E factor for the whole process is 18 (6.6 for the reduction and 11.4 for the cyanation). The main contributors to the E factor are solvent losses which accounted for 51% of the waste, sodium gluconate (25%), NaCl and Na 2 SO 4 (combined circa. 22%). The three enzymes and the NADP cofactor account for < 1% of the waste. The main waste streams are aqueous and directly biodegradable.

9. 8 Conclusions and Prospects

Over the last fifteen years the manufacture of pharmaceuticals has undergone revolutionary changes. Target molecules have become increasingly complex, and legislative pressure, starting in the late 1980s, has stimulated the marketing of chiral molecules as pure enantiomers. This, in turn, stimulated the development of cost -

effective methods for the manufacture of enantiomerically pure compounds. On top of this, there has been a paradigm shift from the traditional concept of process efficiency to one that assigns economic value to conserving energy and raw materials, eliminating waste, and avoiding the use of toxic and/or hazardous chemicals. Indeed, the concepts of E factors, atom economy, and step economy have gradually become incorporated into mainstream organic synthesis in both industry and academia.

The pharmaceutical industry has risen to the occasion and is making substantial progress in replacing traditional processes with greener, more sustainable alternatives, though there is much still to do. It has adopted the E factor, or its direct equivalent, as its measuring staff, and recent publications have identified key areas where improvement is most needed. In short, we conclude that the challenge of sustainability and Green Chemistry is leading to fundamental, game -changing innovations in organic synthesis that will ultimately lead to economic, environmental, and societal benefits in the pharmaceutical industry and in the chemical and allied industries at large.

www.ingramcontent.com/pod-product-compliance
Lightning Source LLC
Chambersburg PA
CBHW060324200326
41519CB00011BA/1833